DATE DUE

DEMCO 38-296

The Language of
BIOTECHNOLOGY

SECOND EDITION

The Language of
BIOTECHNOLOGY
A DICTIONARY OF TERMS
SECOND EDITION

John M. Walker
University of Hertfordshire

Michael Cox
University of Hertfordshire

Allan Whitaker, Contributor
University of Hertfordshire

Stephen Hall, Contributor
University of Hertfordshire

ACS Professional Reference Book

American Chemical Society, Washington, DC 1995

Library of Congress Cataloging-in-Publication Data

Walker, John M., 1948–
     The language of biotechnology: a dictionary of terms/John M. Walker; Michael Cox, editors; Allan Whitaker, Stephen Hall, contributors.
        p.   cm.
     Previously published: 1988.
     ISBN (invalid) 084122957
     ISBN 0–8412–2982–1
1. Biotechnology—Dictionaries.
     I. Cox, Michael, 1933–. II. Whitaker, Allan. III. American Chemical Society. IV. Title.
TP248.16.W35   1995   94–23812
660′.6′03—dc20   CIP

The paper used in this publication meets the minimum requirements of American National Standard for Information Sciences—Permanence of Paper for Printed Library Materials, ANSI Z39.48–1984. ⊗

About the Authors

JOHN M. WALKER joined the Division of Biosciences at the University of Hertfordshire (formerly Hatfield Polytechnic) in 1980 and is currently their Head of Biochemistry and Director of the Science Training Center in the School of Natural Sciences. He was awarded a readership in protein chemistry in 1986 and was awarded the title of Professor in 1991.

Before that he was a postdoctoral research scientist at the Institute of Cancer Research, London, for eight years. Walker received a First Class Honors degree in chemistry and a Ph.D. in biological chemistry from University College, London.

For many years he was involved in studies on nonhistone chromosomal proteins. At the University of Hertfordshire he has been increasingly interested in algal biotechnology, particularly the use of algae as a source of novel pharmaceutical compounds, and in the studies of the factors responsible for protein stability.

Since 1981 he has been the organizer of the highly successful Techniques in Protein Chemistry laboratory workshops. These workshops are held annually at Hatfield and have also been run in Yugoslavia, Malaysia, Indonesia, Spain, Turkey, Egypt, Italy, Pakistan, and India. He has served as editor of a wide range of books on molecular biology methodology and is currently editor-in-chief of the journal *Molecular Biotechnology*.

MICHAEL COX has been Professor of Applied Chemistry at the University of Hertfordshire since 1987, having obtained M.Sc. and Ph.D. degrees in inorganic chemistry at University College, London. His research interest in separation processes followed a sabbatical at Warren Spring Laboratory, Department of Industry, in 1969. In collaboration with industry, he has developed novel processes for both liquid–liquid extraction and ion exchange and is the author of over 50 research papers and review articles. He is also a contributing author to a number of texts on liquid–liquid extraction. In 1992 he was awarded the Memorial Medal of the Kurnakov Institute of General and Inorganic Chemistry of the Russian Academy of Sciences for his work in this field. He has been a regular contributor to the International Solvent Extraction Conferences (ISEC) and was Secretary to the Conference Committee of ISEC '93. Currently he is Secretary to the International Committee for Solvent Extraction and a member of the editorial boards of the journals *Hydrometallurgy* and *Solvent Extraction and Ion Exchange*.

Preface

TO DEFINE THE TERM "BIOTECHNOLOGY" or to describe the role of a biotechnologist is never easy. Such broad terms often mean different things to different people. As a starting point, biotechnology can be defined as the practical application of biological systems to the manufacturing and service industries and to the management of the environment.

Yet such a brief description ignores the wide spectrum of related disciplines that must come together to commercialize a biological process. At one end of the spectrum, biotechnology involves work at the laboratory bench level, often aimed at identifying or constructing microorganisms or cells capable of providing products of economic value. This aim can be achieved by the molecular biologist who applies new methodologies that arise from developments in the biological sciences, especially in genetic engineering, to microorganisms, plant cells, or animal cells. Alternatively, microbial screening programs (which can involve a multidisciplinary team of microbiologists, pharmacologists, biochemists, and chemists) can result in the identification of microorganisms producing potentially useful metabolites.

Once the organism or cell of interest is obtained, it is necessary to develop the growth of the microorganism, animal cell, or plant cell on a large scale in an economically viable process. However, in plant biotechnology, growth of appropriately modified cells to give a plant with novel characteristics can sometimes be the ultimate goal.

Scale-up involves both the microbiologist's knowledge of microbial physiology–biochemistry and fermentation technology and the biochemical or chemical engineer's knowledge of reactor design, heat and mass transfer, and process control. Recovery of the desired product involves downstream processing, which requires expertise in areas such as chemistry, biochemistry, and chemical engineering to develop the various large-scale methodologies for concentrating, extracting, and purifying the required product as economically as possible.

The diverse applications of biotechnological processes range from waste treatment and odor control to the production of reagents for the diagnosis and treatment of disease. Enzymes, for example, are used in a variety of processes, including enzyme reactors, washing powders, biosensors, and diagnostic kits. The use of genetic engineering, via site-directed mutagenesis, to design more stable enzymes for these biotechnological processes takes us back to the beginning of our biotechnology spectrum.

The typical biotechnologist will probably be an expert in one of these areas but will need to integrate specialized work with studies being carried out by collaborators in related fields. Each worker will have to understand the language (the "jargon") of all the collaborators if their individual areas of research are to be integrated successfully.

In this dictionary we have therefore attempted to define routinely used specialized language in the various areas of biotechnology. Involved scientists should be able to read and appreciate texts relating to a particular area of biotechnology, even when this area is unrelated to their own disciplines. We will no doubt be accused of the sin of omission in our choice of terms to include in this dictionary. Because of the diverse nature of biotechnology, it is often difficult to know where to draw the line. Given the size constraints inherent in any book, this had to be a matter of personal choice.

We thank Allan Whitaker for his role as contributor in producing this book, as well as Stephen Boffey (University of Hertfordshire), John Melling (Warren Spring Laboratory), and Michael Verrall (Beecham Pharmaceuticals) for their useful comments and advice.

The first edition of this book was published in 1988, and since then, the promise of biotechnology as a commercial success has continued to be realized, particularly in the areas of health care and agricultural biotechnology. Many of these developments have gone hand in hand with either new applications or the development of new techniques in molecular biology. This new edition has therefore been extensively updated in all these areas; for example with the introduction of new technology such as the polymerase chain reaction (PCR), antisense technology, and the biolistic process. Many new therapeutic agents, such as growth factors, derivatives of monoclonal antibodies, and new developments in vaccine production are included, as well as gene therapy and new approaches to disease diagnosis. The expansion of areas such as transgenic plants and animals and glycobiology has also led to new nomenclature, and the Human Genome Project has of course been introduced since the last edition. Many of the original definitions have been extended to expand on new applications and uses of existing techniques and processes. Finally we have welcomed observations from colleagues and reviewers alike on individual entries that were deemed absent from the first edition. Much has changed in the field of biotechnology over the past 6 years, and we hope we have encapsulated these changes in this second edition. We welcome Stephen Hall as a new contributor to this volume.

JOHN M. WALKER
MICHAEL COX
School of Natural Sciences
University of Hertfordshire
Hatfield, Herts AL10 9AB
England
March 29, 1995

Pronunciation Symbols

For more information see the Guide to Pronunciation.

ə .... banana, collide, abut

ˈə, ˌə .... humdrum, abut

ə .... immediately preceding \l\, \n\, \m\, \ŋ\, as in battle, mitten, eaten, and sometimes open \ˈō-pᵊm\, lock and key \-ᵊŋ-\; immediately following \l\, \m\, \r\, as often in French table, prisme, titre

ər .... further, merger, bird

ˈər-
ˈə-r } .... as in two different pronunciations of hurry \ˈhər-ē, ˈhə-rē\

a .... mat, map, mad, gag, snap, patch

ā .... day, fade, date, aorta, drape, cape

ä .... bother, cot, and, with most American speakers, father, cart

à .... father as pronounced by speakers who do not rhyme it with bother; French patte

aů .... now, loud, out

b .... baby, rib

ch .... chin, nature \ˈnā-chər\

d .... did, adder

e .... bet, bed, peck

ˈē, ˌē .... beat, nosebleed, evenly, easy

ē .... easy, mealy

f .... fifty, cuff

g .... go, big, gift

h .... hat, ahead

hw .... whale as pronounced by those who do not have the same pronunciation for both whale and wail

i .... tip, banish, active

ī .... site, side, buy, tripe

j .... job, gem, edge, join, judge

k .... kin, cook, ache

k̲ .... German ich, Buch; one pronunciation of loch

l .... lily, pool

m .... murmur, dim, nymph

n .... no, own

ⁿ .... indicates that a preceding vowel or diphthong is pronounced with the nasal passages open, as in French un bon vin blanc \œ̃ⁿ -bōⁿ -vaⁿ -bläⁿ\

ŋ .... sing \ˈsiŋ\, singer \ˈsiŋ-ər\, finger \ˈfiŋ-gər\, ink \ˈiŋk\

ō .... bone, know, beau

ȯ .... saw, all, gnaw, caught

œ .... French boeuf, German Hölle

œ̄ .... French feu, German Höhle

ȯi .... coin, destroy

p .... pepper, lip

r .... red, car, rarity

s .... source, less

sh .... as in shy, mission, machine, special (actually, this is a single sound, not two); with a hyphen between, two sounds as in grasshopper \ˈgras-ˌhä-pər\

t .... tie, attack, late, later, latter

th .... as in thin, ether (actually, this is a single sound, not two); with a hyphen between, two sounds as in knighthood \ˈnīt-ˌhůd\

th̲ .... then, either, this (actually, this is a single sound, not two)

ü .... rule, youth, union \ˈyün-yən\, few \ˈfyü\

ů .... pull, wood, book, curable \ˈkyůr-ə-bəl\, fury \ˈfyůr-ē\

ue .... German füllen, hübsch

ūe .... French rue, German fühlen

v .... vivid, give

w .... we, away

y .... yard, young, cue \ˈkyü\, mute \ˈmyüt\, union \ˈyün-yən\

ʸ .... indicates that during the articulation of the sound represented by the preceding character the front of the tongue has substantially the position it has for the articulation of the first sound of yard, as in French digne \dēnʸ\

z .... zone, raise

zh .... as in vision, azure \ˈa-zhər\ (actually, this is a single sound, not two); with a hyphen between, two sounds as in hogshead \ˈhȯgz-ˌhed, ˈhägz-\

\ .... slant line used in pairs to mark the beginning and end of a transcription: \ˈpen\

ˈ .... mark preceding a syllable with primary (strongest) stress: \ˈpen-mən-ˌship\

ˌ .... mark preceding a syllable with secondary (medium) stress: \ˈpen-mən-ˌship\

- .... mark of syllable division

( ) .... indicate that what is symbolized between is present in some utterances but not in others: factory \ˈfak-t(ə-)rē\

÷ .... indicates that many regard as unacceptable the pronunciation variant immediately following: cupola \ˈkyü-pə-lə, ÷-ˌlō\

The system of indicating pronunciation is used by permission. From Merriam-Webster's Collegiate® Dictionary, Tenth Edition ©1994 by Merriam-Webster Inc.

$\overset{\circ}{A}$ \'aŋ-strəm\ *See* angstrom.

A-factor \'ā 'fak-tər\ A bioregulator (3S-isocapryloyl-4S-hydroxymethyl-γ-butyrol-actone) produced by *Streptomyces griseus* and other *Streptomyces* species. Low concentrations of this compound induce morphological differentiation and strep-tomycin production in certain mutant strains lacking the A-factor.

abomasum \ab-ō-'mā-səm\ The fourth stomach of the calf. The source of rennet, the commercial enzyme preparation used in cheese making.

abrasion \ə-'brā-zhən\ The rubbing or wearing of barley during the malting process, a treatment resulting in mild damage to the husk and underlying layers. This treatment is carried out before the barley is sprayed with or steeped in gibberellic acid. Abrasion allows the gibberellic acid to pass directly to the aleurone layer and thus accelerates the malting process.

abrin \'ab-rin\ A protein, consisting of two nonidentical subunits, purified from the seeds of the plant *Abrus precatorius.* Abrin is toxic to mammalian cells because of its ability to inhibit protein synthesis. It is one of a number of proteins being studied as the toxic component in antibody-toxin conjugates for use in chemotherapy. *See also* bispecific antibodies, chimeric antibody.

abscisic acid (ABA) \ab-'sis-ik 'as-əd ('ā 'bē 'ā)\ An endogenous substance (hormone) that inhibits plant growth. ABA, which is usually associated with phenomena such as fruit drop, leaf senescence, and bud and seed dormancy, has found use in plant tissue culture media to reduce growth rates, thus increasing subculture intervals. Growth limitation, induced by hormonal inhibitors at low temperatures and under oxygen limitation, is also a reliable and cost-effective way of maintaining stable germplasm.

absolute filter \'ab-sə-ˌlüt 'fil-tər\ A filter used to remove microorganisms from an airstream; the filter pore sizes are smaller than the particles to be removed (*compare with* air filter). The steam-sterilizable cartridge filter may be made of polytetrafluoroethylene or a similar material, with a uniform pore size of 0.2 μm. Pressure drops are greater than those with an ordinary air filter.

absorbance \ab-'sȯr-bən(t)s\ A measure of the extent to which a beam of radiation, usually ultraviolet, visible, or infrared, is attenuated by transmission through an absorbing medium, which can be a solid, liquid or solution, or a gas. The absorbance (A) is defined as:

$$A = \log I_o / I$$

where I_o is the intensity of the incident radiation, and I is the intensity of the transmitted radiation. Absorbance has now largely replaced the earlier term, *optical density*. The variation of absorbance with concentration of an absorbing species often follows the Beer–Lambert law. This variation allows for determination of the concentration of an unknown solution. *See also* absorption coefficient.

absorbent \ab-'sȯr-bənt\ A material that can be used to absorb another substance.

absorptiometer \ab-ˌsȯrp-tē-'ä-məd-ər\ An optical instrument that measures the attenuation of visible radiation passing through a liquid sample. It can be used to measure turbidity as the incident radiation is scattered by the suspended particles. Thus, the amount of attenuation can be related to the concentration of suspended solids. It is an instrument commonly used to monitor waste-treatment processes.

absorption coefficient \ab-'zȯrp-shən kō-ə-'fish-ənt\ The proportionality coefficient of the Beer–Lambert law, the units of which depend on the units of path length (l) and solution concentration (c) employed. Thus

$$\text{absorption coefficient} = \frac{A}{cl}$$

where A is absorbance, c is in moles per cubic meter, and l is in meters, the absorption coefficient becomes the molar absorption coefficient ϵ, with the units of square meters per mole (SI system). It is very important to specify units when quoting absorption coefficients because misleading information can be implied with regard to the relative intensity of spectral bands. The absorption coefficient used to be called the extinction coefficient.

absorption spectrum \ab-'zȯrp-shən 'spek-trəm\ A measure of the variation of absorbance with wavelength, frequency, or energy used to characterize atoms or molecules.

acceleration phase \ak-ˌsel-ə-'rā-shən 'fāz\ The period of gradually increasing growth prior to the log phase in a microbial culture. After the introduction of a microorganism into a nutrient medium, there is an initial lag phase of no growth. This phase is followed by the acceleration phase, an interval during which cell growth rate gradually increases until the cells reach a constant maximum rate. At this point the cells are said to be in log phase.

acclimatization \ə-ˌklī-mət-ə-'zā-shən\ The biological process whereby an organism adapts to a new environment. This development includes the processes whereby a microbial culture adapts to growing on a new energy source by the induction of the appropriate degradative enzymes. Also known as adaptation. *See also* enzyme induction.

ACE \\'a 'sē 'ē\\ *See* angiotensin-converting enzyme.

acellular \\ā-'sel-yə-lər\\ Not composed of cells. The term describes any tissue or organism that consists of a mass of protoplasm that is not divided into cells, for example, the plasmodium of the slime mold *Physarum* and the hyphae of some fungi.

acesulfames A class of sweeteners used in the food industry, derived from oxathiazinone. Acesulfame-K, the potassium salt of the 6-methyl derivative, is about 130 times as sweet as sucrose.

acetic acid (ethanoic acid) \\ə-'sēt-ik 'as-əd (ˌeth-ə-'nō-ik 'as-əd)\\ CH_3COOH, industrially the most important organic acid. It is used in the manufacture of a range of chemical products, such as plastics and insecticides. Acetic acid was originally produced by the microbial oxidation of ethanol, but this approach is no longer economically viable (except for vinegar manufacture).

***Acetobacter* species** \\ə-'sēt-ō-ˌbak-tər 'spē-shēz\\ Gram-negative ellipsoidal or rod-shaped bacteria that oxidize a variety of compounds to organic acids (e.g., acetic acid). Abundant in fermenting plant materials and important in vinegar manufacture.

acetoclastic bacteria \\ə-'sēt-ō-ˌklas-tik bak-'tir-ē-ə\\ Organisms that use only acetic acid and produce methane during anaerobic digestion.

acetogenic bacteria \\ə-'sēt-ō-ˌjen-ik bak-'tir-ē-ə\\ Bacteria capable of reducing carbon dioxide to acetic acid or converting sugars quantitatively into acetate.

acetogenin \\a-'sēt-ō-jen-ən \\ *See* polyketides.

acetone–butanol fermentation \\'as-ə-tōn-ˌbyüt-əⁿn-òl ˌfər-men-'tā-shən\\ The production of a mixture of acetone and butanol by anaerobic fermentation of glucose by *Clostridium acetobutylicum*. However, few industrial plants still produce these solvents microbiologically. Most acetone and butanol is currently produced by synthetic processes that use petrochemical raw materials.

acetyl coenzyme A (acetyl CoA) \\ə-'sēt-əl kō-'en-ˌzīm 'ā (ə-'sēt-əl ˌkō-'ā)\\ *See* coenzyme A.

acetyl reduction assay \\ə-'sēt-əl ri-'dək-shən 'as-ā\\ A method for directly measuring nitrogen fixation activity in plants. The reduction of acetylene to ethylene by the enzyme nitrogenase is determined by gas chromatography with a flame ionization detector.

acid proteases \\'as-əd 'prōt-ē-ās-əs\\ Protein-hydrolyzing enzymes that have maximum activity at low pH (usually in the pH range 2–6). The most extensively studied acid protease is pepsin, obtained from the stomach, although commercially used acid proteases are prepared from fungal sources. Acid proteases from *Mucor pusillus* and *Mucor miehei* are used as rennet substitutes in milk coagulation. *M. pusillus* is grown on semisolid medium, whereas *M. miehei* is grown in submerged culture. Acid protease is also produced commercially by the growth of *Aspergillus oryzae* on semisolid media. It is used to hydrolyze soybean protein in soy sauce manufacture, for improving the baking properties of flour, and as a digestive aid. *See also* rennet.

acidogenic fermentation \a-'sid-ō-ˌjen-ik ˌfər-men-'tā-shən\ Any fermentation process whereby one or more weak acids (e.g., acetic, propionic, or butyric) are produced.

acidophile \a-'sid-ə-ˌfil\ A classification used to describe any organism that grows optimally under acid conditions. Acidophiles have an optimum growth pH below 6 and typically grow poorly or not at all above pH 7. Obligate acidophiles that cannot grow at neutral pH include members of the bacterial genera *Sulfolobus*, *Thermoplasma*, and *Thiobacillus*.

acidulants \ə-'sij-ə-lənts\ Chemicals added to food as flavoring agents to impart a "sharp" taste to the food. The most commonly used acidulant is citric acid, produced by the fermentation of molasses or glucose hydrolysate by *Aspergillus niger*. Fumaric, itaconic, and malic acids are also used.

acoustic conditioning \ə-'kü-stik kən-'dish-(ə-)niŋ\ A technique used to increase the size of suspended particles by the aggregation of small particles as a result of the application of low-frequency (50–60 Hz) vibrations. The technique is used in broth conditioning to improve the filterability of the system.

acridine orange \'ak-rə-dēn 'är-inj\ 3,6-bis(dimethylamino)acridinium chloride. A basic dye and fluorochrome used to stain DNA and RNA. When intercalated into double-stranded DNA it fluoresces green; when bound to the phosphate groups of single-stranded DNA or RNA it fluoresces orange-red. The compound is also mutagenic.

acrylamide \ə-'kril-ə-ˌmīd\ *See* polyacrylamide gel.

acrylamide gel \ə-'kril-ə-ˌmīd 'jel\ *See* polyacrylamide gel.

actinomycetes \ak-tin-ō-'mī-sēts\ Gram-positive mycelial or rod-shaped bacteria with a tendency to branch. Some actinomycetes form spores in small groups, in chains of indefinite numbers, or in sporangia. The order includes the genera *Actinoplanes*, *Micromonospora*, *Nocardia*, *Streptomyces*, and *Streptosporangium*. Many species produce antibiotics.

Activase \'ak-tə-ˌvās\ Genentech trade name for tissue plasminogen activator. *See* tissue plasminogen activator.

activated carbon \'ak-tə-ˌvā-təd 'kär-bən\ A form of finely divided carbon capable of absorbing substances. The carbon is sometimes specified by its source (e.g., coconut, bone) and is produced by high-temperature treatment in the presence of steam, air, or carbon monoxide. Activated carbon can be used for decoloring solutions and for removing organic compounds from potable waters by its incorporation in percolating filters.

activated sludge process \'ak-tə-ˌvā-təd 'sləj 'präs-es\ A process used in sewage and wastewater treatment to remove organic matter, whereby actively growing aerobic microorganisms are brought into contact with the effluent in the presence of dissolved oxygen. The microorganisms break down the organic matter into carbon dioxide, water, and simple salts. A sludge recycling system is used to maintain high biomass levels in the reactor and thereby increase the rate of treatment. New biomass is ideally produced at a rate commensurate with the loss of solids from the effluent. This ideal is not always the case because the rate depends on the plant's mode of operation. The process can be either a batch or a continuous system that is open to the air and agitated to increase air entrapment, thus optimizing the oxidative breakdown of organic matter. A certain degree of recycling is necessary to increase the contact time between the settled solids from the effluent, the active biomass, and dissolved oxygen. Many variations on the basic design exist, including contact stabilization to reduce reactor volume; stepped feed aeration to balance supply and demand for oxygen; pure oxygen systems to increase oxygen mass transfer; and the deep-shaft process to increase oxygen mass transfer. *See*, for example, deep-shaft airlift fermentor.

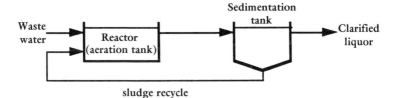

sludge recycle

activation energy \ˌak-tə-'vā-shən 'en-ər-jē\ The energy needed to raise reactants, or an enzyme–substrate complex, to the transition state where it has an equal likelihood of conversion to product or reversion to reactants.

active biomass \'ak-tiv 'bī-ō-ˌmas\ The proportion of a cell culture that has microbiological activity.

active site \'ak-tiv 'sīt\ In enzymes, the region where the substrate binds during catalytic conversion to the product.

active transport \'ak-tiv 'tran(t)s-pō(ə)rt\ The movement of a solute across a biological membrane against its concentration gradient so that the concentration on the product side is greater than that on the feed side. This process requires the expenditure of some form of energy because it cannot proceed spontaneously.

activity of an enzyme \ak-'tiv-ət-ē 'əv 'an 'en-ˌzīm\ The amount of enzyme present in a particular preparation is usually expressed in terms of units of activity based on the rate of the reaction the enzyme catalyzes. The international unit of activity, U, is defined as the amount of enzyme that will convert 1 μmol of substrate to product in 1 min under defined conditions (usually 25 °C and optimum pH).

activity of immobilized enzymes \ak-'tiv-ət-ē 'əv im-'ō-bə-ˌlīzd 'en-ˌzīmz\ *See* effectiveness factor.

activity, thermodynamic \ak-'tiv-ət-ē ˌthər-mō-dī-'nam-ik\ *See* thermodynamic activity.

activity coefficient \ak-'tiv-ət-ē kō-ə-'fish-ənt\ *See* thermodynamic activity.

activity yield of immobilized enzymes \ak-'tiv-ət-ē 'yē(ə)ld 'əv im-'ō-bə-ˌlīzd 'en-ˌzīmz\ *See* effectiveness factor.

adaptamer \ə-'dap-tə-mər\ *See* aptamer.

adaptation \ˌad-ˌap-'tā-shən\ *See* acclimatization.

adaptive enzyme \ə-'dap-tiv 'en-ˌzīm\ *See* enzyme induction.

adaptors \ə-'dap-tərz\ Short synthetic oligonucleotides with one blunt end and one sticky end. The blunt ends are ligated to the blunt ends of a DNA fragment and thus produce a DNA fragment with sticky ends. This fragment can be joined to an appropriately cleaved vector by base-pairing of sticky ends followed by ligation.

adjuvant \'aj-ə-vənt\ A material that is added to a vaccine to increase the immune response. Ideally, the use of an adjuvant should produce a greater amount of antibody and require fewer doses and less antigen than would be needed if an adjuvant was not used. Adjuvants appear to act by stimulating the division of lymphocytes and causing a slow release of antigen into the immune system by prolonging retention of the antigen at the site of action. In humans, the most widely used adjuvants are aluminum phosphate and aluminum hydroxide gel. Freund's complete adjuvant is used exclusively in laboratory animals; it cannot be used in humans because the oil component is not biodegradable and may cause autoimmune and allergic responses. Freund's complete adjuvant is a mixture of a white mineral oil, an emulsifier, and heat-killed *Mycobacterium tuberculosis* cells, which act as an immunostimulant. Freund's incomplete adjuvant lacks the *Mycobacterium* cells.

adsorption, in chromatography \ad-'zȯrp-shən 'in ˌkrō-mə-'täg-rə-fē\ The binding of a ligate to the surface of an affinity adsorbent.

adsorption, in microbiology \ad-'zȯrp-shən 'in ˌmī-krō-bī-'äl-ə-jē\ The process whereby bacteriophages attach to specific receptors on the host cell prior to infection by injecting their nucleic acid.

adsorption, in surface chemistry \ad-'zȯrp-shən 'in 'sər-fəs 'kem-ə-strē\ The accumulation of molecules at the surface of a liquid or solid. This accumulation occurs because the atoms at the surface have different properties from those in the bulk phase. The excess free energy of these surface atoms tends to attract foreign atoms or molecules. When they attach to the surface, these atoms lower the free energy of the surface. Adsorption may involve either chemical or physical forces. *See also* chemisorption.

adsorption fermentation \ad-'zȯrp-shən ˌfər-men-'tā-shən\ A process by which products are removed from the culture broth by adsorption onto polymers, carbon, or other materials. Two configurations have been used, suspension of the adsorbent in the fermentor and circulation of a stream from the fermentor through an adsorbent bed. Stream circulation is generally preferred because adsorbents may be toxic to microorganisms. A modification of the system uses ion-exchange resins to remove charged products. Also known as extractive fermentation.

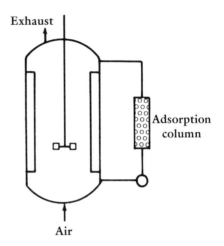

adsorption isotherm \ad-'zȯrp-shən 'ī-sə-ˌthərm\ An equation that represents the adsorption of a substance onto a surface, which may be a solid or liquid, at constant temperature. Various isotherms have been devised to represent different forms of adsorption (e.g., Langmuir, Freundlich, and BET). The adsorption isotherm is a factor in determining the shape of the breakthrough profile of an adsorption column. *See also* BET isotherm, breakthrough profile, Freundlich isotherm, Langmuir adsorption isotherm.

adventitious \ad-ven-'tish-əs\ produced in an uncharacteristic position on a plant, for example, shoots from roots.

aeration \e(-ə)r-'ā-shən\ The introduction of air or oxygen into a liquid, fermentor, or activated sludge plant. *See also* sparger.

aeration number (N_a) \e(-ə)r-'ā-shən 'nəm-bər ('en 'ā)\ A number used in aeration–agitation studies in fermentors to relate gas-flow rates to impeller speed and diameter.

$$N_a = \frac{G}{ND_i^3}$$

where G is the gas-flow rate m^3/s, N is the impeller speed (s^{-1}), and D_i is the impeller diameter (m). The aeration number is used to relate power consumption in gassed and ungassed liquids

$$\frac{P_g}{P} = f(N_a)$$

where P and P_g are the ungassed and gassed power consumptions, and f is a proportionality constant.

aerobe \'e(-ə)r-ˌōb\ An organism that requires the presence of molecular oxygen for growth.

aerobic reactor \e(-ə)r-'ō-bik rē-'ak-tər\ A reactor that is operated under aerobic conditions, with air usually used as the source of dissolved oxygen.

aerobic waste treatment \e(-ə)r-'ō-bik 'wāst 'trēt-mənt\ The degradation of waste material by the use of aerobic microorganisms. There are several designs for waste-treatment plants. *See* activated sludge process, deep-shaft airlift fermentor, oxidation pond, percolating filter.

aerosol \'ar-ȯ-ˌsäl\ A dispersion of finely divided particles of solid or liquid in a gas. The particles are often of colloidal size, as in smoke.

affinity adsorbent \ə-'fin-ət-ē ad-'sȯr-bənt\ A support matrix coated with an affinity ligand that is capable of selectively adsorbing substances such as proteins, enzymes, and antibodies. *See* affinity chromatography.

affinity chromatography \ə-'fin-ət-ē ˌkrō-mə-'täg-rə-fē\ A chromatographic technique that depends on the specific interaction of one molecule with another. Affinity chromatography uses the covalent attachment to an insoluble inert suppoⁱᵗ of a specific ligand. The ligand must have a unique affinity for the molecule to be purified (ligate). Because of this specific interaction the technique is capable of extreme selectivity, but it is also likely to be expensive. An example of the process is the separation of an enzyme by binding an analogue of its substrate to an inert matrix, thereby providing an opportunity for the enzyme to be retained on the support. A crude sample containing the enzyme (ligate) is run through the column under conditions that encourage binding of enzyme to ligand. Unretarded contaminants pass through the matrix and are completely removed by washing. The enzyme is then eluted by changing the environment to favor desorption (e.g., change in ionic strength of the buffer, change in pH of the buffer, or addition of a compound that competes for the binding to the ligand). *See also* boronates, dye–ligand chromatography.

affinity partitioning \ə-'fin-ət-ē pär-'tish-ən-iŋ\ A liquid–liquid separation process in which one phase has been chemically modified to include a ligand that has a specific affinity for the molecule to be separated. Thus, this compound tends to be concentrated in one phase to the exclusion of others in the feed mixture. After separating the two liquid phases, the separated molecule may be released from the affinity ligand by altering the solution conditions, thus breaking any chemical interaction between the ligand and the desired molecule. *See also* boronates, two-phase aqueous partitioning.

affinity precipitation \ə-'fin-ət-ē pri-ˌsip-ə-'tā-shən\ A process that enhances the separation of biological products by precipitating compounds as a result of inter-actions between two types of molecules, e.g., antibody–antigen. *See also* affinity chromatography.

affinity tag \ə-'fin-ət-ē 'tag\ A peptide or protein sequence added to a cloned protein to facilitate purification of the protein by affinity chromatography.

This fusion protein is constructed by combining the gene that specifies the protein of interest with the DNA coding for the affinity tag polypeptide. The affinity sequence is cleaved from the fusion protein after the protein has been purified. For example, the enzyme glutathione S-transferase (GST) can be used as an affinity tag based on its specific affinity for glutathione. The fusion protein is purified using affinity chromatography on immobilized glutathione. Excess glutathione is used to elute the fusion protein. GST is cleaved from the protein of interest by treatment with the enzyme thrombin (a thrombin-specific cleavage sequence was introduced between the genes from GST and the protein of interest). After cleavage, GST is separated from the affinity tag by affinity chromatography on immobilized glutathione.

affinity tailing \ə-'fin-ət-ē 'tā-liŋ\ *See* affinity tag.

aflatoxins \af-lə-'täk-sənz\ A group of fungal toxins produced by strains of *Aspergillus flavus* and *A. parasiticus.* Aflatoxins are highly potent human liver carcinogens. The producing fungi grow easily on poorly stored grain (e.g., in damp and humid conditions), and aflatoxins present in such contaminated foodstuffs are thought to make a major contribution to the high incidence of liver cancer in many Third World countries.

agar \'äg-ər\ A polysaccharide mixture isolated from certain agar-bearing red algae, especially *Gelidium* and *Gracilaria* species. Agar comprises two fractions: a highly charged agaropectin fraction and a neutral agarose fraction. When agar is heated in aqueous solution and cooled, a gel is formed. If appropriate nutrients are included in this solution, the set gel forms a surface upon which to grow microorganisms. When the gel is set in a Petri dish, this provides the *agar plate* much used by microbiologists. Because of the charged nature of agar, it is of little use as a support for electrophoresis or immunodiffusion. Agarose, purified from agar, is used for this purpose. Agar has also been used for the immobilization of cells in the form of spherical beads, blocks, or membranes. In the food industry it is used in soups, jellies, ice cream, and meat and fish pastes. It is not digestible by human beings. Also known as agar-agar.

Agaricus bisporus \ə-'gär-ə-kəs bis-'pōr-əs\ The common edible mushroom. Commercial cultivation of this mushroom accounts for about 75% of the mushrooms produced worldwide. *See also Lentinus edodes, Pleurotus ostreatus, Volvariella volvacea.*

agarobiose \ag-ə-rō-'bī-os\ *See* agarose.

agarose \'ag-ə-rōs\ A purified linear galactan made up of the basic repeating unit agarobiose. It is purified from agar and agar-containing seaweed. When it is heated in aqueous solution and cooled, a gel is formed that is ideal as an inert support for electrophoresis, immunoelectrophoresis, and immunodiffusion. The gelling properties are attributed to both inter- and intramolecular hydrogen bonding. Agarose, a more porous medium than acrylamide, is usually used at concentrations around 1%. Substitution of the alternating sugar residues with carboxyl, methoxyl, pyruvate, and especially sulfate occurs to varying degrees. This substitution can result in electroendosmosis during electrophoresis and ionic interactions between the gel and sample in all uses, both unwanted effects. Agarose is therefore sold in different purity grades, based on the sulfate concentration; the lower the sulfate content, the higher the purity.

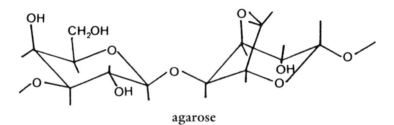

agarose

agglomeration \ə-ˌgläm-ə-ˈrā-shən\ *See* flocculation.

agglutination \ə-ˌglüt-ᵊn-ˈā-shən\ The formation of clumps of cells or microorganisms by linking cell-surface antigens with antibodies. Many immunoassay techniques involve the agglutination of synthetic particles such as plastic beads coated with enzymes or antibodies. Agglutination reactions are commonly used to identify blood groups and bacteria.

aggregation \ag-ri-ˈgā-shən\ The formation of large assemblies of cells, molecules, or particles. With particles, the process consists of two steps, flocculation and coagulation.

agitator (impeller) \ˈaj-ə-ˌtāt-ər (im-ˈpel-ər)\ A device used to introduce turbulence in a mixture by shaking or stirring. Different commercially available agitator–impeller designs include flat blade, marine screw, turbine, and variable pitch screw. Selection of the appropriate device depends on a number of factors, including processing requirements, flow properties of the fluids, and materials of construction. The final choice often involves experience with performance of related systems.

aglucone \a-ˈglü-ˌkōn \ *See* aglycon.

aglycon (aglucone) \a-ˈglī-ˌkōn (a-ˈglü-ˌkōn)\ The noncarbohydrate portion of a glycoside.

agonist \ˈag-ə-nəst\ Any molecule, such as a drug or a hormone, that enhances the activity of another molecule. For example, a hormone acts as an agonist when it binds to its receptor, triggering the usual response. *See also* antagonism, synergism.

Agrobacterium tumefaciens   \ag-rō-bak-'tir-ē-əm ˌtü-mə-'fā-shē-ənz\   A Gram-negative soil microorganism that causes crown gall disease (transformation of plant cells to proliferating tumor cells at the crown, the junction of the root and stem) in many species of dicotyledonous plants. Crown gall disease is caused by the Ti plasmid (tumor-inducing) within the bacterium. After infection, part of the 200-kilobase pair plasmid, also known as the T-DNA (15–30 kbp), is integrated into the plant chromosome and maintained in a stable form in daughter cells. The T-DNA genes expressed in the plant cells are responsible for the malignant transformation of the plant cells. Transformation of plant cells with *A. tumefaciens* containing engineered Ti plasmid vectors (e.g., "disarmed" plasmids that have had some or all of the T-DNA genes deleted, thus eliminating cancerous growth of cells) has potential for plant genetic engineering. However, at present the method is almost entirely limited to the transformation of dicotyledonous plants, whereas most important crops (e.g., barley, wheat, and rice) are monocotyledonous.

agrochemicals   \ag-rō-'kem-i-kəlz\   Chemicals used for crop protection. They include pesticides, herbicides, and fungicides.

agronomy   \ə-'grän-ə-mē\   The branch of agriculture dealing with crop production and soil management.

air filter   \'a(ə)r 'fil-tər\   A filter used to remove microorganisms from an airstream connected to a fermentor, a clean room, or a sterile cabinet. The pore sizes in the filter are larger than the diameter of the particles to be removed (*compare with* absolute filter), and therefore this system relies on the depth of the filter bed to remove particles completely by adsorption onto the fibrous structure. The filter bed is made of a fibrous material such as cotton, glass, slag, or steel wool. The gaps between the fibers are usually in the range 0.5–15 μm. Fibrous filters are preferred to absolute filters in the fermentation industry because they are more robust and cheaper and produce lower pressure drops. The filters may be steam-sterilized before use.

airlift fermentors   \'a(ə)r-ˌlift fər-'ment-ərz\   Bioreactors designed for aerobic fermentation where agitation is provided solely by the sparge air. Three basic designs exist. In the **concentric draft tube (internal loop) design**, the bioreactor contains a concentric draft tube, and air is sparged into the base of the draft tube leading to density differences through the fermentor, which causes circulation of the media as shown in Figure a. Alternatively, air can be introduced at the base of the annulus between the draft tube and column wall to regulate the direction of circulation, as shown in Figure b. The **tubular loop (external loop) fermentor** (Figure c) consists of two columns in parallel, connected top and bottom. Several variations of these basic designs exist. Because shear forces are small, this method is ideal for the growth of animal and plant cells, which are very fragile. *See* deep-shaft airlift fermentor, gas lift.

airlift fermentor

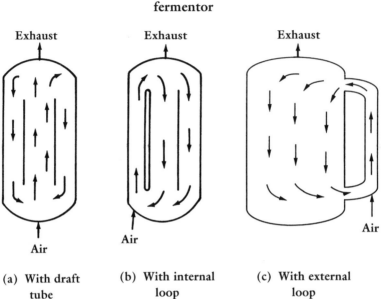

| (a) With draft tube | (b) With internal loop | (c) With external loop |

albumin \al-'byü-mən\ One of the four major categories of seed storage proteins. *See also* seed storage proteins.

Alfa Laval extractor \'al-fə lə-'väl 'kän-tak-tər\ A centrifugal device for liquid–liquid extraction, consisting of a vertically mounted bowl rotating about a central shaft. The bowl consists of concentric channels with spirally wound baffles to control the flow paths of the two fluids. The baffles also provide regions of intense mixing as the heavier fluid moves toward the bowl periphery and the lighter fluid moves toward the bowl axis. This device is now obsolete. *See also* Podbielniak extractor.

algae \'al-jē\ A large group of unicellular and multicellular eukaryotic aquatic plants, usually photosynthetic and pigmented. Some classification systems also include the prokaryotic *Cyanophyta* under the name blue-green algae, although ideally they are bacteria or cyanobacteria. Algae exist in both freshwater and marine environments. They are used commercially for the production of a range of compounds including agar, agarose, alginate, and carrageenan. They are also used as a source of pigments (e.g., β-carotene) and have been used to provide single-cell protein, particularly in tropical and subtropical areas (e.g., *Spirulina* is produced in Israel and Mexico, and *Chlorella* species have been used as food in Japan). The fouling of water systems by algae can be a problem in industrial processes, swimming pools, and reservoirs.

algal oxidation pond \'al-gəl ˌäk-sə-'dā-shən 'pänd\ A large, shallow lagoon used for wastewater treatment. Wastewater is treated by the combined action of bacteria and photosynthetic algae. The algae produce oxygen and therefore maintain aerobic conditions. *See also* oxidation pond.

algicides \'al-jə-ˌsīdz\ Chemical agents that selectively kill algae.

alginate \'al-jə-ˌnāt\ A polysaccharide comprising D-mannuronic and L-guluronic acids. Traditionally it has been produced from marine algae (e.g., *Laminaria* species and *Macrocystis pyrifera*), but the source is subject to considerable variation. Alginate is also produced commercially from *Azobacter vinelandii*, but this microbial source differs from the algal source in having *O*-acetyl groups associated with D-mannuronic acid residues. Alginates are used as thickening and gelling agents in dairy products and in textile printing, and alginate gels have been used to immobilize cells and microorganisms by entrapment methods.

alkaline phosphatase (E.C. 3.1.3.1) \'al-kə-līn 'fäs-fə-ˌtās\ An enzyme (esterase) usually isolated from *Escherichia coli* or calf intestinal tissue, which hydrolyzes phosphate esters and has an optimum pH in the range 9–11. It is used to remove phosphate groups from 5′ termini of linear DNA molecules (e.g., a restricted plasmid molecule). This removal prevents recircularization of the restricted plasmid molecule by ligase during gene cloning experiments, thus ensuring that the intact circular molecules generated by the ligase contain an inserted gene. The enzyme is also commonly used in enzyme–antibody conjugates in techniques such as ELISA and immunohistochemistry.

alkaline proteases \'al-kə-līn 'prōt-ē-ˌās-əz\ Proteases with an optimum pH in the range 8–11. Industrially important alkaline proteases are produced from *Bacillus* species, especially *Bacillus licheniformis*, which produces the enzyme *subtilisin Carlsberg* (E.C. 3.4.21.14), one of the major constituents of enzyme detergents. Alkaline proteases are used in washing powders and for dehairing hides.

alkaloids \'al-kə-ˌlȯidz\ Naturally occurring organic compounds that possess marked pharmacological activity (not necessarily beneficial) in humans. Alkaloids are generally heterocyclic compounds of complex structure. Many alkaloids are found in plants, and a number have found use as drugs (e.g., atropine, colchicine, morphine, quinine, and scopolamine). Other alkaloids include cocaine, caffeine, nicotine, opium, and strychnine.

allele \ə-'lē(ə)l\ Any of one or more alternative forms of a given gene. They occur by mutation, where deletions, substitutions, or insertions have altered the original specific sequence of nucleotides. In a diploid cell or organism, the two alleles of a given gene occur in corresponding positions (loci) on a pair of homologous chromosomes. If these alleles are genetically identical, the cell or organism is said to be *homozygous*. If they are genetically different, it is *heterozygous* with respect to that gene.

allogenic \ˌal-ō-jə-'nē-ik\ Genetically dissimilar.

allograft (allogenic graft) \'al-ə-graft (ˌal-ō-jə-'nē-ik 'graft)\ A graft between genetically dissimilar individuals of the same species.

allosteric enzymes \al-ō-'ster-ik 'en-zı̄mz\ Enzymes that contain regions, in addition to substrate binding sites, to which small molecules (effectors) may bind, resulting in the enhancement or inhibition of the catalytic activity of the enzyme. *See also* allostery, cooperativity, effector.

allostery \'al-ō-ster-ē\ The phenomenon whereby a low-molecular-weight ligand binds to a receptor site on a protein thus affecting the specificity of a second site on the protein. This second site could be a receptor site or the active site of an enzyme. Binding of the allosteric effector to the first site causes a change in shape of the protein, thus modifying the shape and affinity of the second binding site. *See also* feedback inhibition.

alpha \'al-fə\ Compounds beginning with α- are listed under their roman names.

alpha viruses \'al-fə 'vı̄-rəs-əs\ A class of RNA viruses. Alpha viruses are the major genus of the Togavirus family and include Sindbis virus, Semliki Forest virus, and the human pathogens eastern and western equine encephalitis viruses. The first two viruses are currently being developed as vectors for the expression of heterologous genes.

Alu PCR \'ā 'el 'yü 'pē 'sē 'är\ *See* polymerase chain reaction.

Alu sequence \'ā 'el 'yü 'sē-kwen(t)s\ A member of a family of interspersed repetitive dimeric DNA sequences about 300 base pairs long in the human genome. Alu sequences form the major family of human short interspersed repeat sequences and are dispersed throughout the genome, with an average of 4 kilobase pairs between copies.

ambident anions \'am-bə-dənt 'an-ı̄-ənz\ Groups of atoms carrying an overall negative charge that are capable of donating electrons to a metal atom or other electron acceptors; for example, the thiocyanate ion: $S\!\!-\!\!C\!\!\equiv\!\!N$ (S-bonded); $S\!\!=\!\!C\!\!=\!\!N$ (N-bonded)

ambident ligands \'am-bə-dənt 'lig-əndz\ Ligands that are capable of electron donation through more than one site; for example, NH_2CH_2COOH (glycine) can coordinate either via O or N.

amensalism \ā-'men-sə-ḷiz-əm\ A system in a culture containing two microbial species in which one species suffers as a result of its interaction with the other species. The opposite of commensalism. *See also* commensalism, mutualism, neutralism.

amino acid analysis \ə-'mē-nō 'as-əd ə-'nal-ə-səs\ The analytical determination of the relative amounts and types of amino acids present in a protein or peptide. Traditionally, proteins were hydrolyzed to their constituent amino acids, and the amino acids were separated by ion-exchange chromatography. The amino acids were quantified by reaction with ninhydrin as they eluted from the ion-exchange column, and the blue color that developed was quantified by colorimetric analysis. Quantitation of each amino acid was achieved by comparison with the color yields achieved by a standard mixture of amino acids of known concentration. However, the use of ninhydrin for postcolumn derivatization has now been generally superseded by more sensitive reagents such as *o*-phthalaldehyde and fluorescamine, both of which produce fluorescent derivatives with amino acids.

These fluorescent methods can detect less than 50 pmol of an amino acid. An alternative strategy is to use precolumn derivatization, in which the amino acids in the protein hydrolysate are first coupled with a hydrophobic molecule and then separated by HPLC on a reversed-phase column. Commonly used precolumm derivatizing agents include *o*-phthalaldehyde (OPA), phenylisothiocyanate (PITC), and 9-fluorenylmethyl chloroformate (FMOC). OPA derivates are detected by their fluorescence; PITC and FMOC derivatives are detected by their UV absorbance.

amino sugar \ə-'mē-nō 'shůg-ər\ A monosaccharide in which one or more hydroxyl groups have been replaced by an amino group.

amino terminal \ə-'mē-nō 'tərm-ən-ᵊl\ The end of a polypeptide chain (peptide or protein) that has a free amino group. Often referred to as the N-terminus. *Compare with* carboxy terminal.

aminoglycoside antibiotics \ə-ˌmē-nō-'glī-kə-ˌsīd ˌant-i-bī-'ät-iks\ A group of microbially produced antibiotics that contain an amino sugar, an amino- or guanido-substituted inositol ring, and one or more residues of other sugars. They are all bacteriocidal. Aminoglycoside antibiotics are broad-spectrum antibiotics that function by binding to bacterial ribosomes, thus inhibiting protein synthesis, but their toxicity limits their use in the clinic. Compounds include gentamicin, kanamycin, neomycin, and streptomycin.

6-aminopenicillanic acid (6-APA) \ˌsiks ə-ˌmē-nō-ˌpen-ə-sil-'an-ik 'as-əd (ˌsiks 'ā 'pē 'ā)\ The basic penicillin nucleus and, therefore, an important substrate for the generation of a range of clinically important penicillin analogues. It is produced industrially by the hydrolysis of penicillin G by using organisms that produce penicillin acylase or by using immobilized penicillin acylase.

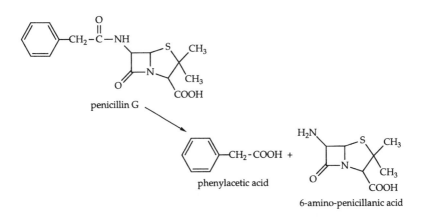

ammonium sulfate precipitation \ə-'mō-nē-əm 'səl-fāt pri-sip-ə-'tā-shən\ The use of ammonium sulfate to salt out proteins from solution. Precipitated proteins are collected by centrifugation. The method is often used as an initial fractionation step in the purification of proteins. Other salts, or polymers such as polyethylene glycol, can also be used. *See also* salting out.

amniocentesis \am-nē-ō-ˌsen-'tē-səs\ The aspiration of amniotic fluid (liquid surrounding the fetus in the uterus) via a needle inserted through the abdominal wall. Fibroblasts found in this fluid are fetal in origin and may be separated and cultured for biochemical or molecular analysis. Predominantly used as a method for obtaining a fetal DNA sample for prenatal diagnosis, the procedure is carried out in the second trimester. With the time required for fibroblast culture and subsequent analysis, results of prenatal diagnosis are not obtained until 20 weeks of pregnancy or later, disturbingly late if termination is required. *Compare with* chorionic villus sampling, fetoscopy.

amperometric sensor \ˌam-pə-rə-'me-trik 'sen(t)-sər\ An electrochemical sensor in which the output from the device is in the form of an electric current that is proportional to the concentration of the analyte.

amphipathic \am-fə-'path-ik\ Possessing both a hydrophobic and a polar moiety. The term is used to describe compounds such as fatty acids. In aqueous solutions, such compounds aggregate to form micelles. *See also* micelle.

ampholyte \'am-fə-līt\ An amphoteric electrolyte; a molecule that carries a net charge that varies with the pH of the surrounding environment until at a particular value, the isoelectric point, the net charge is zero. A series of these compounds may be used to establish a pH gradient in a supporting gel for isoelectric focusing techniques. *See also* isoelectric focusing.

ampicillin \ˌam-pə-'sil-ən\ D-(—)-α-aminobenzylpenicillin. A clinically important antibiotic produced from penicillin G in a two-step process. Penicillin G is hydrolyzed to 6-aminopenicillanic acid by a mutant of *Kluyvera citrophila*, after which a mutant of *Pseudomonas melanogenum* is added to the fermentation, together with the methyl ester of DL-phenylglycine, resulting in the formation of ampicillin. *See also* 6-aminopenicillanic acid, penicillin G.

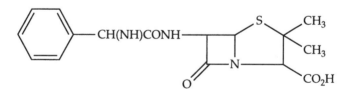

amplification of DNA \ˌam-plə-fə-'kā-shən 'ev 'dē-'en-'ā\ Replication of a DNA sample. Any specific DNA sequence can now be amplified many millionfold using PCR technology. *See also* polymerase chain reaction.

amplification of plasmids \ˌam-plə-fə-'kā-shən 'ev 'plaz-mədz\ Increase in the copy number of a plasmid in a bacterial culture. Some multicopy plasmids (i.e., plasmids with high copy numbers, 20 or more) are able to replicate in the absence of protein synthesis, whereas the replication of genomic DNA relies on protein synthesis. After sufficient bacterial cell density has been achieved, an inhibitor of protein synthesis (e.g., chloramphenicol) is added to the culture, and incubation is continued for a further 12 hours. During this time, plasmid molecules continue to replicate in the absence of chromosomal replication and cell division, resulting in plasmid copy numbers often of many thousands. This method is used to increase the level of multicopy plasmids before plasmid isolation.

α-amylase (E.C. 3.2.1.1)   \\'al-fə 'am-ə-lās\\ An enzyme (endoamylase) that cleaves internal α-1,4-glucosidic bonds of starch (amylose and amylopectin) to yield oligosaccharides of varying chain length. Thermostable α-amylases exist, e.g., from *Bacillus amyloliquefaciens* and *Bacillus licheniformis*. The latter is used in the high-temperature liquefaction of starch because it can operate at temperatures above 100 °C. Other sources of the enzyme include bacteria or fungi such as *Bacillus subtilis* or *Aspergillus oryzae*, both of which produce the enzyme extracellularly. The enzyme has a wide range of uses, including production of corn syrup (starch hydrolysis), low-carbohydrate beer, wallpaper removers, and cold-soluble laundry starch; starch degradation in baking; saccharification in brewing (mashing); desizing of fabrics; and as an aid to digestion. *See also* starch.

β-amylase (E.C. 3.2.1.2)   \\'bāt-ə 'am-ə-lās\\ An enzyme with *exo-α*-1,4-glucanase activity. It acts on linear α-1,4-linked glucans, cleaving alternate bonds from the nonreducing end of the chain to form maltose. When amylopectin or glycogen is the substrate, the action of the enzyme is halted at α-1,6-branch points, thus producing maltose and limit dextrins. The enzyme, common in plants but rare in microorganisms, is prepared by germinating barley.

amyloglucosidase   \\am-ə-lō-glü-'kō-si-ˌdās\\ *See* glucoamylases, starch.

amylopectin   \\am-ə-lō-'pek-tən\\ *See* starch.

amyloplasts   \\'am-ə-(ˌ)lō-ˌplasts\\ *See* plastid.

amylose   \\'am-ə-lōs\\ *See* starch.

anabolic pathway   \\an-ə-'bäl-ik 'path-ˌwā\\ Any metabolic pathway concerned with the synthesis of organic molecules. *Compare with* catabolic pathway.

anabolism   \\ə-'nab-ə-ˌliz-əm\\ The metabolic processes that result in the formation of cellular compounds from low-molecular-weight precursors.

anaerobe   \\'an-ə-ˌrōb\\ A microorganism that grows in the absence of molecular oxygen. *See also* facultative anaerobe, obligate anaerobe.

anaerobic   \\an-ə-'rō-bik\\ In the absence of oxygen, used to describe a process, e.g., corrosion, or an organism that can grow in the absence of molecular oxygen.

anaerobic bioreactor   \\an-ə-'rō-bik ˌbī-ō-rē-'ak-tər\\ *See* anaerobic digester.

anaerobic digester   \\an-ə-'rō-bik dī-'jes-tər\\ A bioreactor designed for the anaerobic digestion of organic wastewaters from domestic (e.g., municipal sewage) and industrial sources. The process produces a mixture of methane and carbon dioxide (biogas) and new microbial cells. As much as 90% of the chemical energy in the wastewater can be converted to methane. For industrial uses, the design is usually a continuously stirred tank reactor (CSTR) with an inflow and outflow. Anaerobic filters that use a microbial population attached to an inert support matrix can be used for wastewaters with low solids content. The wastewater is mixed with an appropriate anaerobic inoculum and held in the bioreactor for sufficient time to allow efficient wastewater treatment and high methane yield before being expelled and replaced by further wastewater. Batch processes tend to be used for farm-scale or small-community use.

anaerobic digestion \an-ə-'rō-bik dī-'jes(h)-chən\ The bacterial fermentation of organic matter to methane and carbon dioxide that occurs in the near-absence of oxygen. Five major bacterial groups must act in synchrony in the breakdown of complex organic compounds to give methanol and carbon dioxide: acetoclastic methanogens, homoacetogens, hydrogen-producing acetogens, hydrogen-using methanogens, and hydrolytic bacteria.

analyte \'an-ə-ˌlīt\ The substance that is determined in an analytical procedure.

anaplerotic sequences (pathways) \ˌan-ə-plər-'ä-tik 'sē-kwen-sez ('path-ˌwāz)\ Metabolic pathways that are used to replenish the pools of intermediates that are diminished in major metabolic reactions. The word is derived from a Greek word that means "filling up". Intermediates are often withdrawn from metabolic pathways for use in biosynthetic processes and need to be replenished. For example, oxaloacetic acid produced in the citric acid cycle, is used to form aspartate and other amino acids by transamination. In animals, one of the most important anaplerotic reactions is the action of pyruvate carboxylase on pyruvate to form further oxaloacetate to replace that lost from the tricarboxylic acid cycle.

anchorage dependence \'an-k(ə)-rij di-'pen-dən(t)s\ The requirement of some mammalian cell cultures to grow as monolayers anchored to a glass or plastic substrate. Only transformed cells, hybridomas, and hematopoietic cells can easily be propagated in suspension culture. Anchorage dependence causes problems in the scale-up of mammalian cell culture and requires the design of vessels with large surface areas. However, where possible, suspension culture is the preferred method for scaling up mammalian cell culture. *See also* bead-bed reactor, hollow-fiber reactor, microcarrier, roller bottles, suspension culture.

aneuploid \'an-yü-plôid\ Having an abnormally high or low number of chromosomes, with the individual chromosomes not in their normal proportions, used to refer to cells. *See also* ploidy.

angiotensin-converting enzyme (ACE) \an-jē-ō-ˌten(t)-sən-kən-'vərt-iŋ 'en-ˌzīm ('ā 'sē 'ē)\ A human metalloprotease that converts the decapeptide angiotensin I to the octapeptide angiotensin II. Angiotensin II is a powerful hypertensive agent causing increase in blood pressure. Inhibitors of the enzyme ACE are used clinically to treat high blood pressure.

angstrom (Å) \'aŋ-strəm ('ā)\ A unit of length formerly used for atomic dimensions and light wavelengths. 1 (Å) = 10^{-8} cm. Usually the picometer (10^{-12} m) is now used to measure these distances.

animal cell culture \'an-ə-məl 'sel 'kəl-chər\ The large-scale growth of mammalian cells for the production of protein products. Mammalian cells are generally fragile, and fermentors designed for the growth of microorganisms are often unsuitable for growth of mammalian cells. Fermentor designs for animal cell culture are of two basic types:

(1) Cells are either immobilized or trapped and then perfused with culture medium; or

(2) Cells are in free suspension, but mixed and aerated by a very gentle technique.

An airlift fermentor for the latter type has been designed, whereas hollow-fiber cartridges, roller bottles, bead-bed reactors, the encapsulation of cells in alginate spheres, and static maintenance reactors are methods used for the growth of trapped or immobilized cells. *See also* microcarrier.

anionic detergent \an-ī-'än-ik di-'tər-jent\ *See* anionic surfactant.

anionic surfactant \an-ī-'än-ik sər-'fak-tənt\ A detergent with a negatively charged functional group. They are often alkyl or alkaryl sulfonates, for example, sodium dodecylbenzenesulfonate. *See also* sodium dodecyl sulfate, surfactant.

anisotropic \an-ī-sə-'träp-ik\ Varying with the angle of observation, used to describe a structure.

anisotropic membrane \an-ī-sə-'träp-ik 'mem-ˌbrān\ A synthetic membrane that has an asymmetric pore structure through the film thickness. This asymmetry usually consists of a thin film with a tight pore structure backed by a thicker section with more open pores (e.g., a polymeric ultrafiltration membrane with a very fine pore size on one face and an open structure on the other). These membranes are used for ultrafiltration and reverse osmosis with the open structure side facing the feed stream and the tight side, the product stream. *Compare with* isotropic membrane.

anistreplase \an-i-'strep-lās\ A streptokinase derivative used as a thrombolytic agent.

annealing \ə-'nē(ə)l-iŋ\ The association of two single-stranded nucleic acid molecules by hydrogen bonding between complementary bases on the respective strands.

anomer \'an-ə-mər\ One of two isomeric forms of a carbohydrate (called α and β), which differ only in the configuration about the anomeric carbon of the ring structure. In the α-isomer the hydrogen at the anomeric carbon is above the plane of the ring; in the β-isomer it is below.

anomeric carbon \an-ə-'mer-ik 'kär-bən\ The carbon atom of the carbonyl group in a carbohydrate.

anoxic culture \a-'näk-sik 'kəl-chər\ A culture of microorganisms that show anaerobic respiration, using inorganic materials other than oxygen as the terminal electron acceptor, for example, the use of a denitrification zone in the activated sludge process.

anoxic reactor \a-'näk-sik rē-'ak-tər\ A reactor used for effluent treatment in which no dissolved oxygen is present and biochemical oxidation occurs because facultative anaerobes use nitrate ions as the oxygen source. Activated sludge processes are often modified to incorporate an anoxic zone for denitrification of the wastewater:

$$NO_3^-(aq) \longrightarrow NO_2^-(aq) \longrightarrow N_2(g)$$

ansamycins \an-sə-'mī-sinz\ Antibiotics that consist of a planar aromatic moiety. One side of this moiety is linked to the other by an aliphatic bridge. The classification is based solely on chemical structure. This group includes both substances that act on bacteria and not on higher cells and substances that inhibit eukaryotic cells. The best known member of this group is the antibiotic rifampicin, which is used to treat tuberculosis.

antagonism \an-'tag-ə-ˌniz-əm\ The interaction of substances (antagonists) such that one partially or completely inhibits the effect of the other, or even reverses the initial effect, for example, a drug that blocks a hormone receptor site. *Compare with* agonist, synergism.

antagonist \an-'tag-ə-nist\ *See* antagonism.

anther culture \'an(t)-thər 'kəl-chər\ A technique in which immature pollen is induced to divide and generate tissue, either on solid media or in liquid culture. Anthers that contain pollen are simply removed from the plant and placed on a culture medium, where some microspores survive and develop. The generated tissue can be either embryo tissue, in which case it is transferred to an appropriate medium to allow root and shoot development to take place, or callus tissue, in which case the tissue is placed in an appropriate solution of plant hormones to induce the differentiation of shoots and roots.

anthracyclines \an-thrə-'sī-klēnz \ A group of O-glycoside antibiotics. They consist of a colored aglycone (an anthracyclinone) and one or more sugar residues, one of them usually an amino sugar. In addition to antibacterial activity, daunorubicin and adriamycin also have antitumor activity and are used in the treatment of human tumors. To date, anthracyclines have been found only in actinomycetes.

anthropogenic \an(t)-thrə-pə-'jen-ik\ *See* xenobiotic.

antibiotic \ˌant-i-bī-'ät-ik\ A substance produced by a living organism that inhibits the growth of or kills another living organism. Many such compounds, particularly from *Streptomycetes* species, have found use in treatment of microbial disease in humans. *See also* ampicillin, cephalosporins, penicillin G.

antibiotic resistance \ˌant-i-bī-'ät-ik ri-'zis-tən(t)s\ Biochemical or structural resistance to antibiotics within a microbial population. Such resistance may arise because of natural resistance, by mutations occurring within a sensitive population, or by transfer of genetic resistance factors to sensitive microbial cells. The changes enable the microorganism to grow in the presence of a specific antibiotic. For example, many microorganisms are resistant to β-lactam antibiotics because they synthesize a β-lactamase enzyme.

antibody \'ant-i-ˌbäd-ē\ An immunoglobulin present in the serum of an animal and synthesized by plasma cells in response to invasion by an antigen, conferring immunity against subsequent infection by the same antigen. *See also* B-lymphocyte, immunoglobulin G.

anticipatory control \an-'tis-ə-pə-ˌtōr-ē kən-'trōl\ A system for controlling a continuous process in which control is exercised predictively rather than retroactively (feedforward rather than feedback). This type of control system requires an accurate predictive model of the process or its environment to enable precise control of the process so that corrective action can be taken and deviation from the prescribed conditions anticipated. *See also* feedback control, feedforward control.

anticodon \ant-i-'kō-dän\ A series of three bases found on a given tRNA molecule, which are complementary to, and therefore base-pair with, a codon on an mRNA molecule. This recognition occurs during protein synthesis at the ribosome and causes the amino acid carried by the tRNA molecule to be incorporated into the growing polypeptide chain. *See also* codon, transfer RNA, translation.

antifoam agents \ant-i-'fōm 'ā-jənts\ Surface-active agents used to reduce surface tension of foams formed on the surface of broths as a consequence of aeration and agitation in fermentors. The compounds most suitable for use include stearyl and octyl decanol, cottonseed oil, linseed oil, soybean oil, silicones, sulfonates, and polypropylene glycol. Their presence can sometimes cause problems with downstream processing of the resulting products because of their surface-active nature.

antifreeze protein \'ant-i-frēz 'prō-ṭēn\ An extracellular glycoprotein found in the blood of some arctic and antarctic fish species. The protein depresses the freezing point by inhibiting the formation of ice crystals. *See also* ice-nucleating bacteria.

antigen \'ant-i-jən\ A molecule that stimulates the production of neutralizing antibody proteins when injected into a vertebrate. Antigens are usually of high molecular weight and can be purified molecules (e.g., protein, carbohydrates) or surface molecules on invading organisms such as bacteria, fungi, and viruses. *See also* adjuvant.

antigenic determinant \ant-i-'jen-ik di-'tərm-(ə-)nənt\ *See* epitope.

antigenicity \ant-i-jə-'nis-ət-ē\ The potential of an antigen to stimulate an immune response in a particular host.

anti-idiotype antibodies \an-tī-'id-ē-ə-ṭīp 'ant-i-ḅäd-ēz\ Antibodies produced against the unique amino acid sequence that forms the antigen binding site in the variable region of an antibody. Anti-idiotype (anti-Id) antibodies can bear a structural resemblance to the original antigen, i.e., anti-anti-A (anti-Id) may sometimes be structurally similar to A. Anti-Id antibodies may therefore be used to generate an immune response that recognizes the original antigen, i.e., anti-idiotype antibodies may be used as vaccines.

antioxidant \ant-ē-'äk-səd-ənt\ A substance that prevents or reduces oxidation, especially of organic material. The mode of action depends on the actual oxidation mechanism involved. Thus, a wide range of chemicals can be effective either by being more easily oxidized than the product to be protected or by deactivating catalysts for the oxidation reaction, e.g., by sequestering active metals that may be present. Examples are long-chain phenols, vitamin C, and esters of hydroxyacids (propyl gallate). *See also* glucose oxidase.

antisense RNA \'ant-ē-ṣen(t)s 'är 'en 'ā\ RNA that is complementary to messenger RNA and therefore forms a stable base-paired structure with the mRNA. This reaction has the consequence of preventing the mRNA from being translated.

Many cell types contain ribonucleases that degrade double-stranded RNA, thus also preventing translation. Specific genes can be inhibited by the presence of antisense RNA. A particular antisense RNA is produced by inserting the gene's coding region in reverse orientation into the genome of the target organism. This approach may allow the control of diseases where excessive (e.g., the expression of oncogenes in cancer) or unwanted (e.g., the expression of a viral genome) gene expression leads to the disease. This approach has been used to control adverse effects in commercial processes, such as eliminating the enzymes involved in fruit softening. A considerable proportion of agricultural production is lost after harvesting due to overripening. In tomatoes, ripening is induced by ethylene production in the plant. The two enzymes involved in ethylene production are aminocyclopropane-1-carboxylate (ACC) and ethylene-forming enzyme (EFE). Transgenic plants that express antisense constructs to both ACC and EFE have been produced in which ethylene production is effectively switched off. Transgenic tomatoes from such plants last without decaying much longer than those from control plants and can be transported with less damage. *See also* aptamer, code-blocker therapeutics, transgenic plant.

antiserum \\'an-tī-ser-əm\\ A serum containing antibodies against a specific antigen. Because most antigens have a large number of epitopes, an antiserum contains many different antibodies against a given antigen, each antibody having been produced by a single clone of plasma cells. Such an antiserum is therefore referred to as a polyclonal antiserum. *Compare with* monoclonal antibody.

α-1-antitrypsin (α_1AT) \\'al-fə-wən-ˌan-tī-'trip-sən, ('al-fə-wən 'ā 'tē)\\ A single chain, 394 amino acid glycoprotein, MW 51,000, primarily synthesized in the liver, and found in human serum at levels of 2 g/L. It is the major serum antiprotease, accounting for ~90% of the protease-inhibiting capacity in serum from normal subjects. One of the main functions of this protein is to inhibit leukocyte elastase. Genetic deficiencies in circulating α_1AT are one of the most common lethal hereditary disorders to affect Caucasian males of European descent. In such patients, leukocyte elastase breaks down the lining of the lung, resulting in life-threatening emphysema. Replacement therapy using human plasma-derived α_1AT has been appraised in the United States. However, the large numbers of affected individuals (>20,000) and the large amounts needed (200 g/patient/yr) demand alternative sources. The production of α_1AT in the milk of transgenic sheep has been achieved at levels up to 35 g/L and may prove a more abundant source in the future. Oxidizing agents in cigarette smoke are also thought to inactivate α_1AT by oxidizing a functionally important methionine residue, thus predisposing cigarette smokers to emphysema. Also known as α-1-protease inhibitor (α_1PI).

6-APA \\'siks 'ā 'pē 'ā \\ *See* 6-aminopenicillanic acid.

apical dominance \\'ap-i-kəl 'däm(-ə)-nən(t)s\\ The phenomenon in which axillary bud growth is suppressed in the presence of a terminal bud on a branch.

apoenzyme \\ap-ō-'en-ˌzīm\\ *See* cofactor.

apparent viscosity \\ə-'par-ənt vis-'käs-ət-ē\\ The ratio of shear stress to shear rate in a fluid, when this ratio is dependent on the rate of shear.

aptamer   \'ap-tə-mər \ A synthetic oligonucleotide selected from a large pool of randomly generated polynucleotide sequences because of its ability to bind to a particular target molecule, e.g., a metabolite or protein. For example, 15 base pair oligomers of DNA can be randomly synthesized to produce a pool of approximately 10^{13} different 15mers. The target molecule is then used to select a population of those oligonucleotides (aptamers) that coincidentally bind to the target molecule. These oligonucleotides are amplified using the polymerase chain reaction, subjected to further selection to find the highest affinity molecules and then purified and sequenced. Aptamer screening can also be applied to populations of single-chain DNA or RNA. Aptamers are of interest because they may have uses as therapeutic agents. For example, single-chain DNA molecules that inhibit thrombin have been identified using this approach. *Aptamer* is a reduction of the word adaptamer.

apurinic DNA   \'ā-pyú(ə)r-ịn-ik 'dē 'en 'ā \ A DNA molecule that has lost its purine bases (adenine and guanine). This loss can be caused by acid treatment due to the acid-labile nature of the glycosidic bond to purines.

aqueous two-phase partitioning systems   \'ak-wē-əs 'tü-ͺfāz pər-'tish-ən-iŋ 'sis-təmz\ *See* two-phase aqueous partitioning.

archae   \'är-kē\ A class of bacteria that are neither prokaryotes nor eukaryotes, first identified as a discrete and identifiable grouping in the late 1970s. They have some characteristics of prokaryotes, some of eukaryotes, and some that are unique to *Archaebacteria*. With prokaryotes and eukaryotes, they are believed to represent a third primary kingdom. *Archaebacteria* can be divided into three broad phenotypes: extreme halophiles, methanogens, and thermophiles. Each group has unique biochemical features that can be exploited for use in biotechnological industries. These include the extreme stability of enzymes from thermophiles, the novel C_1 pathways of the methanogens, and the synthesis of organic polymers by some halophiles. *See also* crenotes, euryotes.

ARS elements   \'ā 'är 'es 'el-ə-mənts\ *See* autonomously replicating sequence.

arthrospore   \'är-thrə-ͺspō(ə)r\ A cell formed by fragmentation of hyphae in fungi.

ascitic tumor   \ə-'sit-ik 't(y)ü-mər\ A tumor growing in fluid in the peritoneal cavity of a mammal. Hybridoma cells are often grown as ascitic tumors as a means of producing monoclonal antibodies.

ascomycete   \as-kō-'mī-sēt\ A fungus belonging to the Ascomycotina, the largest fungal subdivision, with more than 15,000 known species. Classified on the basis of the production of ascospores within an ascus (a spherical or cylindrical saclike structure that contains a definite number of ascospores). It includes the following genera: *Chaetomium* (cellulases), *Claviceps* (ergot alkaloids), *Neurospora* (fungal genetics), *Sordaria* (fungal genetics), and some yeast genera.

L-ascorbic acid (vitamin C) \\'el ə-'skȯr-bik 'as-əd ('vīt-ə-mən 'sē)\\ A water-soluble vitamin essential to humans and found in fruits and vegetables. The exact metabolic role of ascorbic acid is not known, but it appears to play a role in the synthesis of cartilage and bone. A deficiency of this vitamin leads to scurvy in humans. It is a strong reducing agent. Precursors for the chemical synthesis are obtained by biological methods. Sorbitol is converted into L-sorbose by fermentation with *Acetobacter* species. The sorbose is then converted chemically via 2-oxo-L-gulonic acid into ascorbic acid.

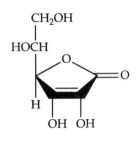

asepsis \\ā-'sep-səs\\ Freedom from contamination by undesirable microorganisms.

aseptic technique \\ā-'sep-tik tek-'nēk\\ Precautionary measures taken in microbiological work and clinical practice to prevent the contamination of cultures or sterile media and the infection of people by extraneous microorganisms.

asexual propagation \\ā-'seksh-(ə-)wəl ‚präp-ə-'gā-shən\\ *See* vegetative propagation.

asexual reproduction \\ā-'seksh-(ə-)wəl ‚rē-prə-'dək-shən\\ Reproduction, either vegetative or involving spore formation, that occurs without fusion of two nuclei (sexual).

aspartame \\'as-pər-‚tām\\ A synthetic dipeptide ester, *N*-L-α-aspartyl-L-phenylalanine methyl ester (L-asp-L-phe-OMe), used as an artificial sweetener; about 200 times as sweet as 4% sucrose in aqueous solution. Aspartame is synthesized by the condensation of *N*-benzyloxycarbonyl-L-aspartic acid with L-phenylalanine methyl ester, followed by removal of the *N*-benzyloxycarbonyl group. The condensation is catalyzed by thermolysin under conditions that encourage the equilibrium toward peptide bond synthesis rather than the hydrolysis of peptide bonds. Metalloproteases such as thermolysin are particularly useful for this type of synthesis because, unlike most other proteases, they do not have esterase activity.

L-aspartic acid \\'el ə-'spärt-ik 'as-əd\\ An amino acid widely used in medicine (e.g., for treating cramps, liver disease, and anemia), foods (as a seasoning), and cosmetics. L-Aspartic acid is produced commercially by the fermentation of fumaric acid, or from fumaric acid and ammonia, by using immobilized microbial cells with high aspartase activity. The introduction of the sweetener aspartame has increased the demand for aspartic acid.

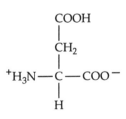

aspect ratio of fermentors \\'as-pekt 'rā-shē-ō 'ev fər-'ment-ərz\\ The height-to-diameter ratio of a tower fermentor.

***Aspergillus* species** \\as-pər-'jil-əs 'spē-sēz\\ A genus of filamentous fungi belonging to the *Deuteromycotina*. The fungi are usually easy to grow and produce a wide range of enzymes and metabolites. *Aspergillus niger*, for example, is used for the industrial production of citric acid, gluconic acid, amylase, proteases, glucose oxidase lactases, and pectic enzymes. *A. oryzae* is used as a source of α-amylase.

association \\ə-sō-sē-'ā-shən\\ Combination of molecules to produce larger, more complex species. Often used in relation to liquids. In an associated liquid the molecules are linked by weak chemical bonds (hydrogen bonds). This weak linkage produces physical properties that are anomalous when compared with nonassociated liquids. Typical examples are water and low-molecular-weight alcohols.

association constant (K_a) \\ə-sō-sē-'ā-shən 'kän(t)-stənt ('kā 'ā)\\ The reciprocal of dissociation constant. *See* dissociation constant.

ATCC \\'ā 'tē 'sē 'sē\\ Abbreviation for the Amercian Type Culture Collection. *See also* NCTC.

athymic mice \\ā-'thī-mik 'mīs\\ *See* nude mice.

ATP \\'ā 'tē 'pē\\ Adenosine triphosphate.

atrial natriuretic peptide (ANP) \\'ā-trē-əl ˌnā-trē-yủ-'ret-ik 'pep-ˌtīd ('ā 'en 'pē)\\ A peptide cardiac hormone with 28 amino acid residues that regulates salt and water balance and blood pressure. ANP is derived from a precursor protein produced by the cardiac atrium and at lower levels by other tissues. ANP has potential for the treatment of heart failure, renal failure, and fluid excess states.

autoanalysis \\ȯt-ō-ə-'nal-ə-səs \\ Analytical techniques that can be adapted to provide automatic sampling and quantitative analysis. The instruments often involve a continuous flow of liquids through small-bore tubing with manifolds for mixing of reagents, dialysis, and liquid-liquid extraction, coupled to suitable detectors (e.g., colorimetric, fluorescent, or electrolytic) equipped with flow-through cells and automatic printout of results.

autocide \\'ȯt-ō-ˌsīd\\ Any substance produced by bacteria that is toxic to the producing organism and related strains but is inactive against other bacteria.

autoclave \\'ȯt-ō-ˌklāv\\ An apparatus in which objects or materials may be sterilized by air-free saturated steam under pressure at temperatures above 100 °C. Water is heated in a closed system (with appropriate pressure-release valves) to generate steam temperatures suitable for sterilization (e.g., 121 °C). The smallest and simplest form of autoclave, commonly used in laboratories for sterilizing small samples, is the domestic pressure cooker.

autolysis \\ȯ-'täl-ə-səs\\ The process of self-digestion of cellular components effected by enzymes naturally present in tissues or microorganisms, usually following the death of the cell or tissue.

automatic cell counter \\ȯt-ə-'mad-ik 'sel 'kaủnt-ər\\ *See* cell sorter, Coulter counter.

autonomously replicating sequence (ARS) \ȯ-'tän-ə-məs-lē 'rep-lə-ˌkāt-iŋ 'sē-kwən(t)s ('ā 'är 'es)\ A chromosomal sequence that can allow autonomous replication of plasmids in yeast. Such sequences have been isolated from many regions of the yeast genome and other eukaryotic species. ARSs were thought to be chromosomal origins of replication but may only act fortuitously as such in yeast. *See also* yeast replicative plasmids.

autoradiography (radioautography) \ȯt-ō-ˌrād-ē-'äg-rə-fē (ˌrād-ē-ō-ȯ-'täg-rə-fē)\ The process whereby a photographic film is used to locate radioactively labeled substances, e.g., radiolabeled DNA bands on a gel. The specimen is overlayered with a photographic film and left for a length of time determined by the amount of radioactivity present. When the film is developed, an image is produced that corresponds to the location of the radioactive sample.

autotroph \'ȯt-ə-ˌtrōf\ An organism that uses the CO_2 of its environment as its carbon source. Autotrophs include green plants, algae, and various bacteria. *Compare with* heterotroph. *See also* chemoautotroph, photoautotroph.

auxins \'ȯk-sənz\ A group of endogenous plant-growth substances, chemically related to indoleacetic acid (IAA), which is itself the principal auxin of many plants. Auxins are characterized by their ability to promote plant growth by processes such as cell elongation and root initiation. Among the auxins commonly used in plant cell and tissue culture are 1-naphthaleneacetic acid (NAA) and 2,4-dichlorophenoxyacetic acid (2,4-D).

auxochrome \'ȯk-sə-ˌkrōm\ A group of atoms attached to a chromophore that modifies the light absorption of the chromophore.

auxostat \'ȯk-sə-ˌstat\ A continuous culture system in which the concentration of a component of the culture is predetermined and all other parameters are adjusted accordingly. *See,* for example, pH auxostat, turbidostat.

auxotroph \'ȯk-sə-ˌtrōf\ A mutant microorganism that will grow only in the presence of a nutrient not required by the wild-type form of the microorganism (e.g., histidine auxotrophs require histidine for growth).

avidin \'av-əd-ən\ A glycoprotein (MW 67,000) found in egg white that binds to the vitamin biotin with high affinity (dissociation constant (K_d) = 10^{-15} M). Both proteins and nucleic acids can be linked to biotin (biotinylated). The high affinity between avidin and biotin is used to amplify the identification of interactions such as antigen-antibody reactions or the hybridization of DNA strands. For example, a biotinylated DNA probe will hybridize to complementary DNA on a filter. The presence of this biotinylated DNA is detected by treatment of the filter with avidin linked to a marker enzyme such as horseradish peroxidase (HRP), followed by the addition of a substrate for HRP that is converted to a colored product by the enzyme. The production of color indicates the position of the biotinylated DNA. Because protein and DNA can have many biotin molecules bound to one molecule, many avidin molecules can bind to one biotinylated molecule, thus considerably amplifying the signal.

avidity \ə-'vid-ət-ē\ The net combining power of an antibody molecule with its antigen.

axenic culture \ā-'zen-ik 'kəl-chər\ A microbial culture that contains only one species of organism.

axial mixing \'ak-sē-əl 'miks-iŋ\ In fluid flow, the phenomenon by which elements of the fluid are either retarded (back mixing) or accelerated (forward mixing) relative to the average retention time. Axial mixing tends to affect both reactor efficiency and throughput; it is generally regarded as a disadvantage to processing. *See also* agitator, back mixing, forward mixing, plug flow.

azeotropic distillation \'ā-zē-ə-ˌtrōp-ik ˌdis-tə-'lā-shən\ A method for separating the components of an azeotropic mixture. When an azeotropic mixture is formed, it is impossible to separate the components of this mixture by fractional distillation. However, if a third component is added such that the added liquid forms azeotropic mixtures with the original two components, two new azeotropic mixtures are formed that can now be separated from each other by distillation. Changes in the vapor–liquid equilibrium composition with pressure may also be used advantageously for the complete separation of some mixtures by fractional distillation.

azeotropic drying \'ā-zē-ə-ˌtrōp-ik 'drī-iŋ\ A process whereby water can be removed from a liquid by the addition of a second liquid that forms an azeotropic mixture with the water, thereby allowing the removal of water at a lower temperature than 100 °C.

azeotropic mixture \'ā-zē-ə-ˌtrōp-ik 'miks-chər\ A mixture of liquids in which the composition of the vapor is the same as that of the liquid phase. Thus, the mixture distills without any change in composition. Also known as constant boiling mixture.

azide \'ā-zīd\ *See* sodium azide.

azoprotein \ˌā-zō-'prō-tēn\ A protein modified by coupling an aromatic diazo compound to the protein. For example, azocasein is used as a substrate to detect proteolytic activity. Cleavage of the azoprotein produces small, acid-soluble, colored peptides that can be measured spectrophotometrically.

***Azotobacter* species** \a-'zō-tə-ˌbak-tər 'spē-shēz\ Large, oval, pleomorphic, Gram-negative bacteria that are obligate aerobes and able to fix atmospheric nitrogen in the presence of carbohydrates or other energy sources. They are cultured on a large scale and used as soil inocula or seed dressings. *See also Rhizobium.*

B

B-cells \\'bē 'selz\\ *See* B-lymphocytes.

B-lymphocytes \\'bē 'lim(p)-fə-ˌsītz\\ Cells circulating in the bloodstream whose main role is to synthesize and secrete antibodies. They are derived from the bone (hence "B") marrow, where they mature and then migrate directly to the spleen and lymph nodes, where they are ready to react with foreign substances (antigens). When an antigen binds to a B-lymphocyte (by a specific interaction between the antigen and surface immunoglobulin that recognizes the antigen), cell proliferation is triggered, giving rise to a clone of cells, all of which secrete the same antibody. These specialized antibody-secreting cells are known as plasma cells. Because a single antigen (e.g., protein) often has many epitopes, it will often stimulate many different B-cells, resulting in the production of a number of specific antibodies against the antigen. The serum sample containing these antibodies is referred to as a polyclonal antiserum. Also known as B-cells. *Compare with* T-lymphocytes.

Bacillus \\bə-'sil-əs\\ A genus of Gram-positive, rod-shaped bacteria that produce endospores. Some species are used for the large-scale production of amylases, proteases, and microbial insecticides. Because of the heat resistance of the spores, *Bacillus* strains are a common source of contamination.

Bacillus amyloliquifaciens \\bə-'sil-əs ˌam-ə-lō-ˌli-kwi-'fā-shē-ənz\\ A Gram-positive bacterium used for the industrial production of α-amylase and serine protease.

Bacillus licheniformis \\bə-'sil-əs ˌlī-ken-i-'fórm-is\\ A Gram-positive bacterium that is the source of the protease subtilisin Carlsberg (E.C. 3.4.21.14), one of the major enzymes used in biological washing powders and liquids.

Bacillus popilliae \\bə-'sil-əs pä-'pil-i-ē\\ A bacterium that attacks only the Japanese beetle (*Popillia japonica*). It has been used for controlling Japanese beetles in the United States. The infected beetle larvae become milky white because of bacterial spore production in the hemolymph.

Bacillus subtilis \\bə-'sil-əs səb-'til-əs\\ A Gram-positive, spore-forming, rod-shaped bacterium that is nonpathogenic in humans and animals. Because of the considerable knowledge of the genetics of this organism and its ability to secrete considerable amounts of protein, it is often used as a vehicle for genetic engineering. Both phages and plasmids will replicate in *B. subtilis.*

Bacillus thuringiensis **(BT)** \bə-'sil-əs th(y)ůr-ịn-jē-'en-sis ('bē 'tē)\ A Gram-positive bacterium with potential for improving disease resistance in plants. Resting spores of the bacterium contain crystalline protein inclusion bodies (δ-endotoxins) that are toxic against a variety of insects. Various strains of *B. thuringiensis* differ in their spectra of insecticidal activity. Most are active against *Lepidoptera*, but some strains specific to *Diptera* and *Coleoptera* have been identified. Both preparations of the endotoxins and spores from *B. thuringiensis* have been produced as pest-control agents. Genes for the toxic proteins have also been cloned, and the transfer of these genes to plants has been carried out in a number of species. For example, transgenic tobacco plants have been developed that express one of these toxin genes and synthesize insecticidal protein. These plants have been shown to be protected from feeding damage by larvae of the tobacco hornworm.

bacitracins \ˌbas-ə-'trās-ənz\ Branched, cyclic antibiotic peptides active against Gram-positive bacteria and produced by various strains of *Bacillus licheniformis.*

back mixing \'bak 'mik-siŋ\ In fluid transport, the phenomenon in which some elements of the fluid are retarded relative to the average element and thus have a higher residence time in the reactor or pipe. This residence time leads to a reduction in throughput of the system. *See also* axial mixing, forward mixing.

back mutation \'bak myü-'tā-shən\ A mutation that causes a mutant gene to regain its wild-type function.

backflush \'bak-ˌfləsh\ The removal of highly retained analytes from a chromatographic column by reversing the flow of the mobile phase and flushing the compounds to waste or to another column for further analysis.

backflushing \'bak-fləsh-iŋ\ The reversal of a process against the normal service flow to dislodge particulate material. This reversal allows the cleaning of a filter, membrane, or column of adsorbent or ion-exchange material. Also known as backwashing.

backwashing \'bak-ˌwäsh-iŋ\ *See* backflushing.

bactericide \bak-'tir-ē-ˌsīd\ A compound capable of killing bacteria (e.g., bactericidal and germicidal agents).

bacteriophage \bak-'tir-ē-ə-ˌfāj\ *See* phage.

bacteriophage M13 \bak-'tir-ē-ə-ˌfāj 'em-thər(t)-ˌtēn\ *See* M13.

bacteriostatic \bak-ˌtir-ē-ō-'stat-ik\ Capable of inhibiting bacterial growth but not of killing the cells.

baculoviruses \ˌbak-yü-lō-'vī-rəs-əz\ A group of viruses that infect only arthropods (e.g., spiders, crabs, insects, millipedes, and centipedes). This selectivity, and in particular the fact that they do not infect vertebrates, has led to their investigation as pest-control agents. Baculovirus vectors are also used to express foreign genes in eukaryotic cells.

baffles \\'baf-əlz\\ Strips, often metal, that are attached radially to the walls of an agitated vessel to prevent vortex formation and thereby improve mixing and aeration.

bagasse \\bə-'gas\\ The fibrous part of the sugar cane stem, essentially lignocellulose, that remains when the juice has been extracted. It is generally burned to supply energy for crushing more canes or to distill alcohol produced by fermentation of the sugar juice, but it is also used as cattle feed and in the preparation of paper and fiberboard.

baker's yeast \\'bā-kərz 'yēst\\ The pressed yeast (*Saccharomyces cerevisiae*) that is grown specifically for use in baking. The culture is grown under stirred aerobic conditions with controlled addition of nutrients in stages to produce maximum biomass and minimum alcohol. *See also* molasses.

Bal 31 \\'bē 'ā 'el 'thərt-ē-ˌwən\\ An exonuclease, isolated from *Brevibacterium albidum*, that cleaves both the 5′ and 3′ strands from both ends of a DNA duplex. It forms a shorter DNA molecule, the size of which is determined by the time of digestion. The enzyme is used to make deletions in cloned DNA molecules.

ball mill \\'bȯl 'mil \\ Device used for the crushing and grinding of materials. It consists of a drum rotating about a horizontal axis. The drum contains a number of hardened balls that crush the material by attrition, either between themselves or between the side of the drum and the balls.

BAP \\'bē 'ā 'pē\\ Abbreviation for bacterial alkaline phosphatase. *See* alkaline phosphatase. Also an abbreviation for benzylaminopurine.

base-pairing \\'bās 'pa(ə)r-iŋ\\ The formation of hydrogen bonds between adenine (A) and thymine (T) or between guanine (G) and cytosine (C) bases in a DNA molecule. There are two hydrogen bonds in an A–T base pair (bp) and three in a G–C base pair. Such base-pairing is responsible for maintaining the double-stranded form of DNA. The size of a DNA molecule is normally referred to in terms of the number of base pairs (or thousands of base pairs, kbp), rather than the absolute molecular weight, which is normally a large and unwieldy figure. Most workers abbreviate kbp to kb when referring to the size of DNA. If base-pairing is disrupted, the DNA is said to be denatured or melted. *See also* complementary base-pairing.

basidiocarp \\bə-'sid-ē-ə-ˌkärp\\ A fungal fruiting body of a basidiomycete, which bears or contains basidia. *See also* basidium.

basidiomycete \\bə-ˌsid-ē-ō-'mī-ˌsēt\\ A fungus that belongs to *Basidiomycotina*, a group of approximately 3000 species. This group includes many of the large toadstools; edible fungi such as *Agaricus bisporus, Lentinus edodes, Pleurotus ostreatus,* and *Volvariella volvacea;* and microfungi and plant pathogens. These fungi produce basidia and basidiospores.

basidiospore \\bə-'sid-ē-ə-ˌspō(ə)r\\ The sexual spore of a basidiomycete (fungus) borne on the outside of a basidium as a result of karyogamy and meiosis.

basidium \\bə-'sid-ē-əm\\ A short, clublike cell produced by basidiomycetes (fungi). It bears four tiny projections (*sterigmata*), on each of which a spore is borne. Two nuclei fuse in the basidium; the fusion gives rise to four daughter nuclei that go into each developing spore.

basket centrifuge \\'bas-kət 'sen-trə-fyüj\\ A centrifuge design in which the bowl or basket is perforated to allow passage of the filtrate and is lined with some form of filter cloth to retain the solids. Thus the centrifuge operates essentially as a centrifugal filter and is used to collect solid matter. It operates at low rotational speeds, about 1000 rpm. The disadvantages of the basket centrifuge include the need for batch operation and a low capacity for solids, typically about 10 kg.

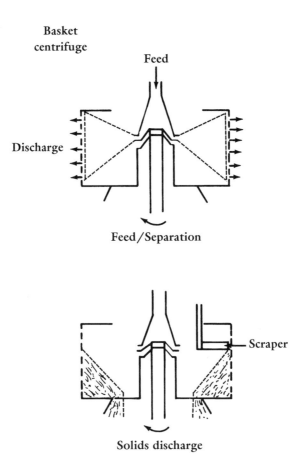

Basket
centrifuge

Feed

Discharge

Feed/Separation

Scraper

Solids discharge

batch culture \\'bach 'kəl-chər\\ A fermentation process in which all the necessary materials, with the exception of oxygen for aerobic processes, are placed in the reactor at the start of the operation. The fermentation is allowed to proceed until completion, at which point the product is harvested. *Compare with* continuous culture.

batch process \\'bach 'präs-es\\ A discontinuous process whereby a single loading of the reactor is allowed to proceed to completion and is then discharged before any further addition of reactants. *Compare with* continuous process.

bathochromatic shift \\bath-ō-krō-'mad-ik 'shift\\ The movement of a characteristic spectral band to longer wavelength, lower frequency, or lower energy following the insertion of a substituent into the molecular structure. The converse of hypsochromic shift. Also known as red shift.

BCG \\'bē 'sē 'jē\\ (**Bacillus Calmette–Guerin**) An attenuated strain of the bacterium *Mycobacterium tuberculosis*, originally produced and used by Calmette and Guerin to immunize against tuberculosis. *See also* adjuvant.

bead-bed reactor \\'bēd-'bed rē-'ak-tər\\ A reactor design for the large-scale growth of mammalian cells. Cells are grown on the surface of glass beads (3–5-mm in diameter) packed in a glass column, through which medium is continuously pumped. When growth is confluent, cells can be recovered by treatment with trypsin.

bead mill \\'bēd 'mil\\ A device used to rupture cells by using a ball-type mill loaded with glass or ceramic beads. These mills are available for both laboratory- and large-scale processing. Laboratory-scale processors include the Mickle tissue disintegrator and the Braun homogenizer. A major disadvantage of these mills is the increase of sample temperature during disruption. Devices are, however, available that incorporate cooling jackets. The large-scale units incorporate both glass beads and rotating disks in the grinding chamber. The suspension of cells is disrupted by collisions between shear force layers generated by the high-speed rotation of the disks and by grinding of the glass beads. Many designs are available with varying geometries of the agitator disks and orientations of the grinding chamber. One problem with this type of disintegrator is the possibility of denaturation, apparently caused by the shear forces and local heating effects.

bed volume \\'bed 'väl-yüm\\ The geometric volume of the packing in a packed column, as in ion exchange or chromatography; also a unit of volume for fluids used in processing such columns. Also known as column volume. *See also* void volume.

Beer–Lambert law \\'bi(ə)r-'lam-bərt 'lȯ\\ A law relating the absorption of radiation passing through an absorbing fluid with the thickness (path length) of the fluid and the concentration of the absorbing substance. This law allows the determination of the concentration of substances in solution and applies to many spectrophotometric systems, in particular to ultraviolet and visible radiation. The law is represented by the equation:

$$A = \epsilon c l$$

where A is absorbance, c is concentration (mol/m^3), l is path length (m), and ϵ is a constant (m^2/mol), the molar absorption coefficient. *See also* absorbance, absorption coefficient, isosbestic point.

beet molasses \\'bēt mə-'las-əz\\ The concentrated liquor remaining after crystallization of the sucrose from beet sugar solutions. It contains approximately 48% wt/vol sucrose, 1% wt/vol raffinose, and 1% wt/vol invert sugar. The remainder is noncarbohydrate.

belt filter \\'belt 'fil-tər\ *See* belt press filter.

belt press filter \\'belt 'pres-,fil-tər\ A mechanical device for filtering and dewatering slurries, broths, and other liquids. The feed is passed between two belts that rotate continuously under tension, thus squeezing out the contained liquid. An alternative design squeezes the slurry between a belt and a perforated drum. A *belt filter* differs from a belt press filter in that it is a vacuum device.

benzylaminopurine (BAP) \\ben-zəl-,am-ēn-ō-'pyu̇(ə)r-ēn ('bē 'ā 'pē)\ *See* cytokinins.

benzylpenicillin \\ben-zəl-pen-ə-'sil-ən\ *See* penicillin G.

berberine \\'bər-bə-,rēn\ A yellow isoquinoline alkaloid found in several plants, including *Coptis japonica*, *Phellodendron amurense*, and *Thalictrum rugosum*. It is used as an intestinal antiseptic in Japan. The production of berberine by cell culture is currently being investigated.

BET isotherm \\'bē 'ē 'tē 'ī-sə-,thərm\ An adsorption isotherm based on a theory by S. Brunauer, P. H. Emmett, and E. Teller concerning multilayer adsorption onto a surface. The authors postulated a number of simultaneous Langmuir-type adsorptions between each two successive molecular layers. The expression for the BET isotherm for gas adsorption has the form

$$V = \frac{V_m C p}{(p_o - p)[1 + (C-1)(p/p_o)]}$$

where V is volume of gas adsorbed at pressure p, V_m is monolayer capacity of the substrate, p_o is saturated vapor pressure of gas, and C is constant for any given gas. Similar equations can be deduced for other adsorbed fluids.

beta- \\'bāt-ə\ Compounds beginning with β- are listed under their roman names.

BHK cells \\'bē 'āch 'kā 'selz\ Baby hamster kidney cells. BHK cells were one of the first cell lines to be grown in suspension culture on a large scale. They are currently grown in fermenters up to 10,000 dm^3 for purposes such as the production of foot-and-mouth disease vaccines.

Bingham equation \\'biŋ-əm i-'kwā-zhən\ A linear equation relating to plastic fluid flow of liquids such as those found in fermentation broths. These fluids follow the equation:

$$\tau = k\gamma + \tau_o$$

where τ is the applied stress, γ is the shear rate, τ_o is the yield stress, and k is the plastic viscosity. A fluid that follows this equation is termed a Bingham plastic or Bingham fluid. An example of this behavior is found with

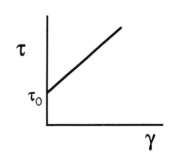

kanamycin fermentation using *Streptomyces kanamyceticus*, and also with penicillin G fermentation using *Penicillium chrysogenum*.

Bingham fluid \\'biŋ-əm 'flü-əd\\ *See* Bingham equation.

Bingham plastic \\'biŋ-əm 'plas-tik\\ *See* Bingham equation.

binodal curve \\bī-'nōd-ᵊl 'kərv\\ A curve denoting the limiting values of concentrations at which a single phase is transformed into two phases. For instance, with fluids that are partially miscible, there is a range of concentrations over which two phases are formed.

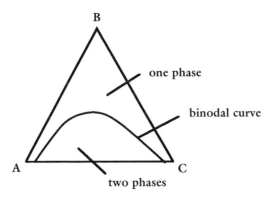

bioaffinity chromatography \\bī-ō-ə-'fin-ət-ē ‚krō-mə-'täg-rə-fē\\ *See* affinity chromatography.

bioaffinity sensor \\bī-ō-ə-'fin-ət-ē 'sen(t)-sər\\ A biosensor in which molecular recognition is used to generate the biochemical signal. Thus, immobilized hormone receptors, drug receptors, or antibodies can be used to detect hormones, drugs or drug metabolites, or antigens, respectively.

bioassay \\bī-ō-'as-ā\\ Any method for determining the concentration or activity of a substance by measuring its effect on a living organism.

biocatalyst \\bī-ō-'kat-ᵊl-əst\\ A catalyst that is or is derived from a living organism, tissue, or cell culture. *See also* cell immobilization, enzyme immobilization, gas flooding.

biochemical fuel cells \\bī-ō-'kem-i-kəl 'fyü(-ə)l 'selz\\ Any system that uses biological reactions to effect the conversion of fuel or biomass (chemical energy) to electricity (electrical energy). The ultimate aim is to be able to generate electricity from any form of industrial waste or sewage. Commercially available cells exist, but they have yet to be evaluated for routine practical uses. Also known as biofuel cells, electrochemical reactors.

biochemical oxygen demand (BOD) \\bī-ō-'kem-i-kəl 'äk-si-jən di-'mand ('bē 'ō 'dē)\\ A measure of the quantity of oxygen required for microbial oxidation of organic matter in water. Samples are incubated in the dark for 5 days at 20 °C, and the oxygen consumed is measured in milligrams per cubic decimeter of sample. BOD is often used as a measure of the extent of organic pollution of water. *See also* chemical oxygen demand.

bioconversion \bī-ō-kən-'vər-zhən\ *See* biotransformation.

biodegradable \bī-ō-di-'grād-ə-bəl\ Capable of being broken down by living organisms; refers to compounds.

biodeterioration \bī-ō-di-ṭir-ē-ə-'rā-shən\ Any undesirable changes in the properties of a material brought about by the activities of organisms. This broad term covers a wide range of situations, including microbial corrosion of stone and metal, mold and bacterial growth on leather, microbial degradation of rubber and plastics, fungal growth in aviation fuels and oils, rotting of wood, the breakdown of lubricants by microorganisms, crop damage by microorganisms, blocking of pipes by microbial growth, staining of materials by microbial contamination, and the growth of microorganisms in foodstuffs.

biofilm \bī-ō-'film\ An actively growing pure or mixed microbial population irreversibly adhering to a living or nonliving surface. Adhesion is mediated by microbial production of polysaccharides and glycoproteins. Biofilms cause biofouling in industrial water systems. This fouling leads to corrosion, increased power consumption, and structural damage to cooling towers. Aerosols produced in air-conditioning plants may contain *Legionella* species. Biofilms are also important in sewage filter beds and in some cell-immobilization techniques. *See also* percolating filter, rotary-disk biological contactor.

biofilter \bī-ō-'fil-tər\ A type of filter consisting of a tank filled with a solid material on which an active biomass is supported. The biomass reduces the organic content of the water, as well as removing particulate matter. Biofilters are often used for wastewater treatment. *See also* percolating filter, rotary-disk biological contactor.

biofouling \bī-ō-'fau(ə)l-iŋ\ The blocking or surface coating of a system or process caused by unwanted microbial growth.

biofuel \bī-ō-'fyü(ə)l\ Any fuel obtained from biological raw materials or by biological processing. The major biofuels are biogas produced by anaerobic digestion and ethanol produced by the fermentation of sugar, molasses, or hydrolyzed starch.

biofuel cells \bī-ō-'fyü(ə)l 'selz\ *See* biochemical fuel cells.

biogas \'bī-ō-gas\ A mixture of methane and other gases (primarily carbon dioxide) produced by anaerobic fermentation of organic matter such as sewage or agricultural byproducts. The methane concentration is usually between 50 and 70%. The gas can be used either directly for cooking or lighting, or as a raw material in the production of other compounds. The methane concentration can be enriched by removing the remaining gases and then used to power engines. In many rural areas in underdeveloped countries, biogas provides a major source of energy. *See also* anaerobic digestion.

biogenic \bī-ō-'jen-ik\ *See* xenobiotic.

biolistic process \bī-ō-'lis-tik 'präs-es\ The process used to shoot biological materials into living targets. The term *biolistic* is derived from "biological ballistics". High-velocity microprojectiles (e.g., 4-μm tungsten particles or gold particles) coated with DNA are shot into cells, resulting in cell transformation. The method is also known as the particle gun method, the gene gun method, the microprojectile method, and the bioblaster method; it should not be confused with shotgun cloning.

biological control \bī-ə-'läj-i-kəl kən-'trōl\ The deliberate use of one species of organism to control or eliminate another. Biological control takes advantage of the fact that certain microorganisms can inhibit the growth of others by either antagonistic or competitive mechanisms. One microorganism can therefore be used to control a specific microbial population. For example, inoculation of the fungus *Trichoderma lignorum* into wet soil suppresses damping-off of seedlings. This fungus produces a toxin that kills other fungi involved in the damping-off process. *See also Bacillus thuringiensis*, nuclear polyhedrosis virus, *Verticillium lecanii*.

biological oxygen demand \bī-ə-'läj-i-kəl 'äk-si-jən di-'mand\ *See* biochemical oxygen demand.

bioluminescence \bī-ō-lü-mə-'nes-ᵊn(t)s\ The emission of visible light by living organisms. It is achieved by the oxidation of an ATP–luciferin complex by the enzyme luciferase. This process is used as the basis of an assay sensitive for ATP, using the luciferase–luciferin complex of the firefly. The sensitivity of the assay allows it to be used for the identification and quantitation of populations of microorganisms. *See also* chemiluminescence, reporter gene.

biomass \'bī-ō-ˌmas\ Plants or biological wastes, such as those produced from agriculture or food processing, that are grown to be available for conversion by biotechnological processes to high-grade fuels and specialty chemicals. More specifically, in microbiology, the cellular component of a biotechnological process (e.g., the bacteria in a fermentation process). Biomass may be either an unwanted waste product, the required product (e.g., single-cell protein), or a source of the required product (e.g., an intracellular product such as an enzyme).

biophotolysis \bī-ō-fō-'täl-ə-səs\ The use of the photosystems of photosynthetic organisms to cleave water into hydrogen and oxygen, which can then be used as an energy source.

biopolymer \bī-ō-'päl-ə-mər\ A large-molecular-weight polymeric compound (e.g., polysaccharide or nucleic acid) that is produced from biological sources. In terms of commercial usage, the most important biopolymers are polysaccharides. *See also* alginate, dextran, polytran, xanthan gum, zanflo.

bioprobe \'bī-ō-ˌprōb\ *See* biosensor.

biopsy \'bī-äp-sē\ Removal of a fragment of tissue from a living patient for analysis.

bioreactor \bī-ō-rē-'ak-tər\ A containment vessel for biological reactions, used in particular for fermentation processes and enzyme reactions. Bioreactors are usually constructed of stainless steel or glass and have such features as provision for inoculation and sampling, variable aeration (for aerobic processes), variable agitation, and temperature control. For fermentation processes it must be possible to sterilize the system and operate under sterile conditions, as well as to fill and empty the system aseptically. Also, the design should allow the operator to monitor and optimize growth and product formation. Various designs are available, including activated sludge processes, anaerobic digesters, continuously stirred tank reactors, loop (recycle) reactors, and tower bioreactors.

bioremediation \bī-ō-ri-ˌmēd-ē-'ā-shən\ The biological degradation and detoxification of hazardous waste such as organic contaminants, heavy metals, and nitrogenous wastes. For example, in groundwater the anaerobic reductive dehalogenation of chlorinated products and the complete aerobic oxidation of chlorinated solvents and aliphatic hydrocarbon contaminants.

bioscrubbing \bī-ō-'skrəb-iŋ\ The process of removing toxic or odorous wastes by using biological systems, particularly microorganisms. Many industrial processes produce toxic or odorous gases (particularly reduced sulfur compounds such as hydrogen sulfide, dimethyl sulfide, and mercaptans) and toxic wastes such as cyanide, many of which can be smelled by humans at very low concentrations (ppm or ppb levels). The malodors of animal wastes are due to reduced sulfur compounds.

The scrubber compartment is often a spray or packed column in which finely distributed water droplets flow countercurrently to the waste gas. This flow results in a continuous mass transfer of pollutants and oxygen from the waste gas to the liquid phase. Oxidative reactions using microorganisms then take place in the aqueous phase. Many potential uses of microorganisms for scrubbing exist, such as anaerobic desulfurization using *Thiobacillus* species and conversion of cyanide to thiocyanate by rhodanase, an enzyme found in *Bacillus stearothermophilus*. However, these systems have not yet been fully developed for widespread application.

bioselective chromatography \bī-ō-sə-'lek-tiv ˌkrō-mə-'täg-rə-fē\ *See* affinity chromatography.

biosensor \bī-ō-'sen(t)-sər\ An analytical tool or system consisting of an immobilized biological material (e.g., enzyme, antibody, whole cell, or organelle) in intimate contact with a suitable transducer device that converts the biochemical signal into a quantifiable electrical signal. Several biosensor devices have been described, based on many transducer systems, including conductimetric, redox-mediated, field-effect transistor, thermistor-type, optoelectronic devices, photodiodes, fiber-optic devices, gas-sensing electrodes, piezoelectric crystals, and ion-selective electrode systems. Biosensors have advantages over physicochemical sensors, but they currently suffer from a short lifetime. Also, biosensors cannot be autoclaved.

biospecific adsorption chromatography \bī-ō-spi-'sif-ik ad-'sȯrp-shən ˌkrō-mə-'täg-rə-fē\ *See* affinity chromatography.

biostat \'bī-ō-ˌstat\ A continuous-culture vessel in which a parameter other than turbidity is used to monitor the biomass concentration.

biosurfactant \bī-ō-sər-'fak-tənt\ Any compound produced by a living organism that solubilizes compounds (e.g., oils) by reducing the surface tension between the compound and water.

biosynthesis \bī-ō-'sin(t)-thə-səs\ The synthesis of organic molecules by living organisms, i.e., an enzymatic synthesis.

biotin \'bī-ə-tən\ A water-soluble vitamin (MW 244) that is tightly bound by avidin (dissociation constant, $K_D = 10^{-15}$ M). This high affinity is used to amplify the detection of protein-protein or DNA–DNA interactions. For example, a DNA probe can be biotinylated by incorporating biotin derivatives of deoxyribonucleotides by nick translation. After hybridization the biotin can be detected by binding an avidin–enzyme complex. A substrate is then provided for the enzyme (usually peroxidase or alkaline phosphatase), which is converted to a colored product. The color indicates the position of the biotinylated probe.

biotinylation \bī-ə-tin-ə-'lā-shən\ The linking of biotin molecules to macromolecules such as DNA or protein. *See also* biotin.

biotransformation (bioconversion) \bī-ō-tran(t)s-fər-'mā-shən (bī-(‚)ō-'kən-'vər-zhən)\ The carrying out of a chemical reaction by biological systems. Reactions include reduction, oxidation, hydroxylation, epoxidation, esterification, and isomerization. Biotransformation is usually carried out by using enzymes, either in a purified or semipurified state, or whole cells (plant, animal, or microbial) that contain the relevant enzyme. For example, *Rhizopus nigricans* is used to hydroxylate progesterone at position 11 during the synthesis of the steroid cortisol.

biotrophic fungi \bī-ō-'trō-fik 'fən-jī\ Fungi that live as parasites on living host cells.

bis (bisacrylamide) \'bis (‚bis-ə-'kril-ə-mīd)\ An abbreviation used to refer to N,N'-methylenebisacrylamide, the cross-linking molecule used in the formation of polyacrylamide gels. *See also* polyacrylamide gel.

bispecific antibodies \bī-spi-'sif-ik 'ant-i-‚bäd-ēz\ Artificially produced antibodies that can react with and link two distinct antigens. One light and heavy chain is derived from one antibody, the second light and heavy chain is derived from another. Applications of such antibodies include diagnosis, where one arm detects the antigen of interest (e.g., a tumor cell marker) and the second arm is bound to an appropriate label. Alternatively, they may be used for targeting antitumor drugs, with one arm attached to the drug and the second arm reacting with a tumor cell marker. *See also* chimeric antibody.

bleed \'blēd\ Loss of material from a chromatography column as a result of high-temperature operation. These compounds may result in ghost peaks on the chromatogram, increased baseline offset, or detector noise.

bleed stream \'blēd 'strēm\ Low-volume flow from a continuous reaction process to allow removal of impurities that might impair the process. After removing impurities, the bleed may recycle to the reactor.

blinding \'blīnd-iŋ\ The wedging of particles that are not quite small enough to pass through the pores of a filter, so that an appreciable fraction of the filter surface becomes inactive.

blotting \'blät-iŋ\ A technique used to transfer proteins or nucleic acids from gels to another matrix while maintaining the physical separation developed in the gel. *See also* protein blotting, Southern blotting.

blue-green algae \'blü-'grēn 'al-jē\ Photosynthetic, oxygen-evolving prokaryotes. Although referred to as blue-green algae, they are not true algae and are more correctly classified as *Cyanobacteria*.

blue shift \'blü 'shift\ *See* hypsochromic shift.

blunt end (flush end) \'blənt 'end ('fləsh 'end)\ The end of a DNA molecule where both strands terminate at the same nucleotide position with no single-stranded extension. *Compare with* sticky ends. *See also* restriction endonucleases.

BOD \'bē 'ō 'dē\ *See* biological oxygen demand.

boronates \'bȯr-ə-ˌnāts\ Boron derivatives used as group-specific affinity ligands for the purification of sugars. Boronates are similar to lectins in their ability to bind sugars. Small diol-containing molecules can be chromatographed on immobilized boron columns. Because boronates will bind NAD via the ribose ring, immobilized phenylboronates have been used to bind nucleotides such as NAD sufficiently tightly to permit the subsequent biospecific chromatography of dehydrogenases and other NAD-binding enzymes.

bottom yeasts \'bät-em 'yēsts\ *See* brewing yeasts.

boundary layer \'baun-ˌd(ə-)rē 'lā-ər\ A layer of fluid in the immediate vicinity of a surface that forms a boundary between the surface and the bulk phase of the liquid. Flow in the boundary layer is laminar and thus provides the major resistance to heat and mass transfer because transfer can occur only by molecular diffusion.

bovine papilloma virus (BPV) \'bō-ˌvīn ˌpap-ə-'lō-mə 'vī-rəs ('bē 'pē 'vē)\ A DNA virus that causes warts on cattle. It has potential as a vector for transforming mammalian cells. The virus does not lyse its host cell but replicates as a plasmid with a copy number of 10–200 per cell.

Bowman–Birk inhibitors \'bō-mən-'birk in-'hib-ət-ərz\ A group of protease inhibitors found in soybeans.

bp \'bē 'pē\ An abbreviation for base pair. It is used as a measure of the size of double-stranded DNA. *See also* base-pairing.

Braun homogenizer \'braun hə-'mäj-ə-ˌnīz-ər\ *See* bead mill.

breakthrough profile \'brāk-ˌthrü 'prō-ˌfīl\ A plot of the concentration of compounds present in the effluent of the absorption column as a function of the effluent volume. Breakthrough profiles are used to show column efficiency in the separation of mixtures of components and to assess the optimum conditions for column operation. The shape of the profile is ideally a step function, indicating that the compound being studied is passing down the column of adsorbent as a discrete band. However, because of the adsorption of compounds onto the column, partitioning is not ideal and depends on the relevant adsorption isotherm. In addition, axial mixing occurs because of inconsistencies in column packing. An S-shaped curve is normally found for a breakthrough profile because of these problems. *See also* BET isotherm, Freundlich isotherm, Langmuir adsorption isotherm.

brewing yeasts \'brü-iŋ 'yēsts\ The two main groups of brewing yeasts are known as *top yeasts* (strains of *Saccharomyces cerevisiae*) and *bottom yeasts* (*Saccharomyces carlsbergensis*). Top yeasts, used for brewing traditional beers, tend to be carried to the surface of a vessel during fermentation and flocculate on the surface as a thick, scumlike layer. Bottom yeasts are used for brewing lagers. Comparatively longer fermentations are carried out at lower temperatures, and the yeast cells sediment to the bottom of the vessel at the end of the brewing cycle.

bridge plasmids \'brij 'plaz-mədz\ *See* shuttle vector.

broad-host-range \'bród 'hōst 'rānj\ Describing a plasmid or phage that replicates in a range of host species or strains, rather than being specific to one host species or strain.

bromelain (E.C. 3.4.22.4) \'brō-mə-lən\ A thiol protease (MW 33,500) isolated from pineapple juice and stems. The enzyme is used as a digestive aid, as a meat tenderizer, in chillproofing, and in animal feeds to improve their nutritional value.

Brookhaven Protein Data Bank \'brük-ˌhāv-ən 'prō-tēn 'dāt-ə 'baŋk\ A structural database of proteins, nucleic acids, viruses, and polysaccharides, including sets of atomic coordinates. It is distributed in Europe by the Cambridge Crystallographic Data Centre. *See also* Protein Identification Resource.

broth \'bróth\ The contents of a fermentor, comprising nutrients, microorganism, substrates, primary and secondary metabolites, and supporting fluid.

broth conditioning \'bróth kən-'dish-(ə-)niŋ\ The initial stage in downstream processing that occurs after the contents of a fermentor are harvested. It consists of a number of different processes designed to simplify and facilitate downstream processing and may include reduction of viscosity, aggregation of solids, and increase of particle size or density. These processes lead to easier control of solids and reduction of fouling. Processes that may be involved include addition of flocculants, filtration, pH variation, and heat treatment.

bubble-cap column \'bəb-əl-'kap 'käl-əm\ A contacting device using bubble-cap plates to ensure intimate contact between gas or vapor and liquid phases. *See* bubble-cap plate.

bubble-cap plate \'bəb-əl-'kap 'plāt\ A device for contacting vapor or gas and liquid in distillation or adsorption columns. It consists of a series of hollow mushroomlike protrusions on a plate that direct the vapor through the absorbing liquid.

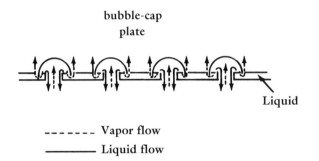

bubble-cap
plate

Liquid

------- Vapor flow
————— Liquid flow

bubble column \\'bəb-əl 'käl-əm\\ Any type of gas–liquid contactor in which the sparge gas is the only means of agitation. *See also* airlift fermentors.

bubble column fermentor \\'bəb-əl 'käl-əm fər-'ment-ər\\ A high-aspect-ratio fermentor with gas introduced at the base through a nozzle or porous plate. Gas bubbles rise through the medium in the fermentor and cause mixing. The bubbles may also be redispersed by a series of horizontal baffle plates set at intervals. This type of fermentor produces little shear effect on the organism. Also known as a sparger. *See also* airlift fermentors.

bubble-point test \\'bəb-əl-'pȯint 'test\\ A test to establish the integrity and pore size of membranes. The test depends on the pressure required to blow a liquid that wets the membrane material out of the pores of the membrane. The pressure (P) is proportional to the reciprocal of the pore diameter (d_p) as shown by the Cantor equation:

$$P = \frac{4\gamma(\cos\psi)}{d_p}$$

where γ is surface tension and ψ is the liquid–solid contact angle.

budding \\'bəd-iŋ\\ The production of a small budlike outgrowth from a parent cell. It is a method of asexual reproduction common among yeasts.

bursting disk \\'bərst-iŋ 'disk\\ *See* pressure release devices.

butanol \\'byüt-ᵊn-ȯl\\ *See* acetone–butanol fermentation.

C* \\'sē\\ The saturation concentration $(mmol/dm^3)$ of dissolved oxygen in a fermentation broth for any given set of conditions.

C_crit \\'sē 'krit \\ *See* critical dissolved oxygen concentration.

C_L \\'sē 'el\\ Concentration $(mmol/dm^3)$ of dissolved oxygen in a fermentation broth.

C-terminus \\'sē 'tər-mə-nəs\\ *See* carboxy terminal.

calcitonin \\ˌkal-sə-'tō-nən\\ A peptide hormone with 32 amino acids, secreted by both the thyroid and the parathyroid glands, that lowers the level of calcium in the blood. It is used in the treatment of Paget's disease and osteoporosis.

calcofluor white \\'kal-kō-flür 'wīt\\ A fluorescent reagent used to stain plant cell walls. It is used in particular to test whether protoplasts have regenerated cell walls.

calliclone \\'kal-ə-ˌklōn\\ A discrete colony that derives from a single cell when plant cells in suspension culture are plated out in semisolid medium.

callus \\'kal-əs\\ Plant tissue consisting of undifferentiated cells, produced either as a result of tissue wounding in the plant or by the growth of plant tissue explants on semisolid media. *See also* callus culture.

callus culture \\'kal-əs 'kəl-chər\\ The growth of plant tissue explants on semisolid agar, giving rise to an amorphous mass of undifferentiated cells with no regular form (callus). In many cases this callus material can be induced to differentiate into whole plants (via either organogenesis or embryogenesis) by the inclusion of appropriate growth regulators (e.g., auxins or cytokinins) in the culture medium. *See also* nurse callus, paper raft nurse technique.

calomel electrode \\'kal-ə-məl i-'lek-ˌtrōd\\ A type of reference electrode used with a glass electrode to measure hydrogen ion activity or concentration (i.e., pH). It consists of mercury in contact with solid mercury(I) chloride and a solution of potassium chloride, usually a saturated solution. A combined electrode is used with most modern pH meters; the glass electrode is surrounded by a concentric cylinder that contains the calomel electrode. A fiber or porous plug in the outer wall of the cylinder provides for electrical contact between the reference electrode and the surrounding liquid. These combined electrodes are capable of steam sterilization and so can be used as in-line sensors for fermentors.

cane molasses \\'kān mə-'las-əz\\ The concentrated liquor remaining after crystallization of the sucrose from cane sugar solutions. It contains approximately 33% wt/vol sucrose and 21% wt/vol invert sugar. The remainder is noncarbohydrate.

cane sugar \\'kān 'shȯg-ər\\ Sucrose extracted from sugar cane. It is not different from sucrose obtained from any other source, such as sugar beet.

Cantor equation \\'kan-tər i-'kwā-zhən\\ *See* bubble-point test.

capillary column \\'kap-ə-ˌler-ē 'käl-əm\\ Chromatographic or electrophoretic columns with a small diameter. These columns may contain a packing of support material (packed capillary columns), or have the stationary phase deposited on the inner wall (open-tubular columns), or be filled with a fluid as in capillary electrophoresis. *See* capillary electrophoresis; porous-layer, open-tubular column; support-coated, open-tubular column; wall-coated, open-tubular column.

capillary electrophoresis (CE) \\'kap-ə-ˌler-ē i-ˌlek-trə-fə-'rē-səs ('sē 'ē)\\ An analytical electrophoretic technique that involves the application of a potential difference across a length of capillary tube filled with an electrolyte. The silica capillaries are extremely narrow (typically 50 μm i.d. and 300 μm o.d.), 10–100 cm long, and may be coated for particular applications. For example, biomolecules tend to absorb strongly to ionized sites on the silica tube, but this problem can be alleviated by applying a polymer coating to the tube. Advantages over conventional electrophoresis include the ease and rapidity of capillary separation with low volumes of sample (nanoliters) and on-line detection during analysis, thereby removing the need for any post-separation identification processes. On-line detection is possible because the bulk electrolyte moves through the capillary under the action of electroendosmosis and so, regardless of the charge on the separated components, they all exit the capillary in the same direction and pass across the sensor. Another advantage of using capillaries is the large surface-to-volume ratio, which gives enhanced heat dissipation. This ratio helps eliminate both convection currents and zone broadening due to increased diffusion. Therefore, it is not necessary to include a stabilizing medium in the tube; hence, electrophoresis is by free flow. The method requires the use of high voltage (typically 10–20 kV) and can be used to separate a wide spectrum of biological molecules including amino acids, peptides, proteins, DNA fragments (e.g., synthetic oligonucleotides), and nucleic acids, as well as many small organic molecules, such as drugs or even metal ions. The method has also been applied to the problem of chiral separations. Although *capillary electrophoresis* is currently the accepted term, in the past the technique has been referred to as high-performance capillary electrophoresis (HPCE), capillary zone electrophoresis (CZE), and free-solution capillary electrophoresis (FSCE). *See also* capillary gel electrophoresis, capillary zone electrophoresis.

capillary gel electrophoresis \\'kap-ə-ˌler-ē 'jel i-ˌlek-trə-fə-'rē-səs\\ A form of capillary electrophoresis in which the capillary is filled with a gel to facilitate separation of compounds with similar mass-to-charge ratio as in conventional gel electrophoresis or SDS-PAGE techniques. The manipulation of cast gels within the capillary can cause problems, so linear polymers mixed with the buffer solution have practical advantages. Also known as sieving.

capillary membrane module \'kap-ə-ˌler-ē 'mem-ˌbrān 'mäj-ü(ə)l\ The arrangement of a membrane in the form of a bundle of capillaries installed in an outer shell. The mass separation layer of the polymeric membrane is on the inside of the capillary, and the feed solution is fed into the lumen. The permeate leaves the module from the outer shell. This type of module construction is used in ultrafiltration and pervaporation because it gives a large membrane area per volume with low capital and operating costs, together with good control of polarization concentration and fouling. A disadvantage is the relatively low bursting pressure of the capillaries (about 5×10^5 Pa). Also known as a hollow-fiber module.

capillary zone electrophoresis (CZE) \'kap-ə-ˌler-ē 'zōn i-ˌlek-trə-fə-'rē-səs ('sē 'zē 'ē)\ A capillary electrophoretic separation technique that uses a single buffer as the electrolyte with the separation of the sample components based on their mass-to-charge ratio. It is particularly suited to peptide separation because the eluate can be monitored with a UV detector using the 200-nm absorbance band.

capsid \'kap-səd\ The external protein coat or shell of a virus particle; it surrounds the core (RNA or DNA) of the virus.

capsomere \'kap-sə-mir\ Any of the individual protein units that form the capsid of a virus.

capsule \'kap-səl\ A compact gumlike layer of polysaccharide or polypeptide, exterior to the cell wall in some bacteria. Some of the gums are produced on a large scale for use as gelling agents or thickeners. *See also* glyocalyx, *Leuconostoc*, slime layer.

Carbbank \'kär-ˌbaŋk\ *See* Carbohydrate Data Bank.

Carbohydrate Data Bank (Carbbank) \ˌkär-bō-'hī-ˌdrāt 'dāt-ə 'baŋk ('kär-ˌbaŋk)\ A collection of primary structure data for complex carbohydrates produced by the Complex Carbohydrate Research Center at the University of Georgia.

carbohydrates (saccharides) \ˌkär-bō-'hī-ˌdrāts ('sak-ə-ˌridz)\ Generally, compounds with the empirical formula $(CH_2O)_n$. However, this formula is an oversimplification because although it applies to many of the more common carbohydrates, some carbohydrates also contain sulfate, phosphate, or amino groups. The smallest carbohydrates (in which $n = 3$) are referred to as trioses. Sugars with 4, 5, 6, and 7 carbon atoms are called tetroses, pentoses, hexoses, and heptoses, respectively. Monosaccharides are small monomeric molecules such as glucose (in which $n = 6$), ribose (in which $n = 5$), and fructose (in which $n = 6$). Many of the most important carbohydrates are formed by linking monosaccharide units together. If only a few monomer units are linked, we call the molecule an oligosaccharide; longer polymers are called polysaccharides. *See,* for example, disaccharides, polysaccharides, starch.

carbon dioxide gas analyzer \'kär-bən dī-'äk-sīd 'gas 'an-əl-ˌī-zər\ An instrument for measuring the concentration of carbon dioxide in a gas stream, often the exhaust gas from a fermentor. The detector in the instrument may be based on infrared spectrophotometry, thermal conductivity, or mass spectrometry.

carbon dioxide production rate (CPR) \'kär-bən dī-'äk-sīd prə-'dək-shən 'rāt ('sē 'pē 'är)\ The change in carbon dioxide concentration in the exhaust gas from a fermentor. CPR can be used to measure the growth of cells.

carbon source \'kär-bən 'sō(ə)rs\ A substance that provides the major part of the carbon required for a growing organism.

carboxy terminal \kär-'bäk-sē 'tərm-ən-ᵊl\ The end of a polypeptide chain (peptide or protein) that has a free carboxyl group. Also known as the C-terminus. *Compare with* amino terminal.

cardiac glycoside \'kärd-ē-ak 'glī-kə-ˌsīd\ Any one of a group of steroid glycosides that act on the heart muscle, increasing the force of contraction. Examples are ouabain and digitalis.

β-carotene \'bāt-ə 'kar-ə-ˌtēn\ A coloring agent used in the food industry. *See also* carotenoids.

carotenoids \kə-'rät-ən-ȯidz\ Plant pigments, usually red or yellow, based on a tetraterpene structure. Their color results from the presence of a long series of conjugated double bonds. They are precursors of vitamin A and are thus used as food supplements. The traditional source of carotenoids is the carrot, which provides a mixture of carotenoids, with β-carotene as the major component (>90%). β-Carotene is produced commercially from a range of algae, bacteria, fungi, and yeasts. It is used as a coloring material in foods and as a source of vitamin A in vegetarian margarines.

carrageenans \kar-ə-'gē-nənz\ A mixture of heterogeneous polysaccharides, comprising mainly α-D-galactopyranosyl sulfate esters, isolated from seaweed. κ-Carrageenan, used to immobilize cells, is the insoluble fraction obtained when potassium ions are added to an aqueous solution of carrageenan. Carrageenan is also used as a food additive and as an absorbent for metal ions.

cascade \kas-'kād\ A type of continuous processing in which a series of identical operations, such as liquid–liquid extraction or solid–liquid leaching, are operated so that the feed for a particular stage is the product from the previous stage. Several different configurations are possible; the choice depends on the particular requirements of the flowsheet. *See also* cocurrent flow, countercurrent flow, crosscurrent flow.

cascade fermentation \kas-'kād ˌfər-men-'tā-shən\ An arrangement of a series of fermentors such that the fermenting liquor is fed through the system as a cascade. Modifications are possible to retain the cells in intermediate fermentors (tanks) while the liquor is allowed to continue.

casein \kā-'sēn\ The major protein component of milk (2.6% of whole milk). Casein is not a single protein but a heterogeneous group of phosphoproteins. The nomenclature of the casein complex has changed continuously as research progresses but essentially consists of three major components: α-, β-, and κ-casein, representing approximately 40%, 35%, and 15% of the total casein, respectively. Minor casein components make up the remaining 10%.

Casson body \\'ka-sen 'bäd-ē\ A rheological term originally used by Casson to describe a type of non-Newtonian fluid that behaves as a pseudoplastic. The apparent viscosity decreases with increasing shear rate, but it also displays a yield stress and therefore resembles a Bingham plastic.

Casson fluid \\'ka-sen 'flü-əd\ A fluid of non-Newtonian rheological behavior shown by some fermentation broths that can be represented by the equation:

$$\tau^{0.5} = \tau_0^{0.5} + k\gamma^{0.5}$$

where τ is applied stress, τ_0 is the yield stress, γ is the shear rate, and k is the apparent viscosity.

CAT \\'sē 'ā 'tē\ Abbreviation for chloramphenicol acetyl transferase. *See* reporter gene.

catabolic pathway \kat-ə-'bäl-ik 'path-ˌwā\ Any metabolic pathway related to the degradation of organic molecules, with concomitant generation of energy. *Contrast with* anabolic pathway.

catabolism \kə-'tab-ə-ˌliz-əm\ Reactions that provide chemical energy by the oxidation of organic substrates or that generate metabolic intermediates subsequently used in anabolic reactions.

catabolite repression \kə-'tab-ə-ˌlīt ri-'presh-ən\ The inhibition of enzyme synthesis by increased concentrations of certain metabolic products or substrates (e.g., glucose repression of the *lac* operon).

catalytic antibody \kat-ᵊl-'it-ik 'ant-i-ˌbäd-ē\ A monoclonal antibody that has been designed to catalyze a particular organic reaction. Enzymes function by reducing the energy barrier along a chemical reaction pathway and use binding energy to reduce the chemical activation energy. By analogy, catalytic antibodies have been obtained by raising an immune response against a small molecule, or hapten, rationally designed to resemble the transition state for the reaction of interest. The binding energy of the antibody stabilizes the transition state of the required reaction and thus catalyzes the forward reaction, i.e., the binding site of the antibody simulates the environment of the enzyme-active site. Rate enhancements of more than a millionfold have been achieved using catalytic antibodies. Also known as mabzymes (mab means monoclonal antibody).

catenated DNA \\'kat-ə-ˌnāt-əd 'dē 'en 'ā\ Two interlocking strands of DNA, frequently generated in circular DNA replication. It arises as a result of rapid replication of a closed circular DNA molecule and the slow nicking of one of the closed circular DNA molecules to release it from the other. *See also* covalently closed circular DNA.

cation exchange \\'kat-ī-ən iks-'chānj\ The replacement of one positively charged ion for another, often involving the use of ion-exchange resins (e.g., in water softening, calcium and magnesium ions are replaced by sodium).

cationic surfactant \kat-ī-'än-ik sər-'fak-tənt\ A compound that, in aqueous solution, forms positively charged surface-active species. These surfactants are often derived from quaternary alkylammonium salts or derivatives of fatty acid amides (e.g., cetyltrimethylammonium chloride).

cauliflower mosaic virus (CaMV) \'käl-i-flau(-ə)r mō-'zā-ik 'vī-rəs ('sē 'ā 'em 'vē)\ A member of the caulimovirus family of viruses. CaMV is of interest as a possible cloning vector in plants. It has the advantage that a single infection of a plant results in the virus being transmitted throughout the plant, with every cell becoming infected. Regeneration of plants from transformed cell cultures is therefore not necessary. However, the use of CaMV is limited in that only small inserts can be added to the CaMV genome without interfering with the packing of the DNA into the viral protein coat. CaMV also has a very limited host range, infecting only a small number of species, primarily the brassicas.

caulimoviruses \käl-i-mō-'vī-rəs-əz\ One of the two classes of DNA viruses (the other is geminiviruses) that infect plants. They have potential as cloning vectors in plants. *See also* cauliflower mosaic virus.

cDNA \'sē 'dē 'en 'ā\ *See* complementary DNA.

cecropin \sə-'krō-pən\ A peptide fragment with antibacterial activity, produced from a giant silkworm hemolymph protein. The peptide acts as an ionophore against a broad spectrum of bacterial species. The gene has been introduced and expressed constitutively in transgenic potato plants. Field tests showed that the expression of cecropin gave significant resistance to the bacterial diseases soft rot and black leg (stem rot).

cell culture \'sel 'kəl-chər\ *See* tissue culture.

cell cybrids \'sel 'sī-brədz\ A population of cells produced by combining the cytoplasm of one cell type with the nucleus of another cell type.

cell cycle (mitotic cycle) \'sel-'sī-kəl (mī-'tät-ik 'sī-kəl)\ The various stages during nuclear and cell division in a mitotically dividing eukaryotic cell. For cultured cells, the length of the cell cycle varies from one cell line to another, but the majority have a cycle time of $18-24$ hours. The cell cycle is divided into four stages, which occur in the following order: M phase (mitosis), G_1 phase (the period before DNA synthesis), S phase (the period of DNA synthesis), and G_2 phase (the period between DNA synthesis and mitosis). The G_2 phase is then followed by another mitotic phase and so on. A fifth phase, G_o, is the period when cells have ceased replication and are in a resting or nondividing state.

cell disruption \'sel dis-'rəp-shən\ *See* lysis.

cell-free translation system \'sel-'frē tran(t)s-'lā-shən 'sis-təm\ A crude cell extract (which thus contains ribosomes) plus added factors such as tRNAs, amino acids, creatine phosphate, and creatine phosphokinase, giving a solution that allows the translation of added mRNA molecules. Cell extracts (lysates) from rabbit reticulocytes or wheat germ are most commonly used. *See also* in vitro protein synthesis.

cell hybrids \'sel 'hī-brədz\ Cells produced by the fusion of two different cell lines. *See also* hybridoma cell.

cell immobilization \\'sel im-ِō-bə-lə-'zā-shən\\ The conversion of cells from the free mobile state to the immobilized state, either by attachment to a solid support or by entrapment in an appropriate matrix. The three major methods of cell immobilization are *aggregation* (e.g., by cross-linking with glutaraldehyde); *adsorption* (carrier binding) onto materials such as glass, plastic, brick, or ion-exchange materials; and *entrapment* in gels made from materials such as polyacrylamide, calcium alginate, or κ-carrageenan. The supports can exist in a range of shapes and sizes, including sheets, tubes, fibers, cylinders, and spheres. Whole cells are increasingly used as biocatalysts because they avoid the need for expensive and lengthy extraction and purification of the enzymes of interest. The use of immobilized cells, rather than free cells, allows reuse or continuous use of the biocatalyst. If free cells are used, their recovery and reuse is not feasible because of the expense involved, and they contaminate the product or the waste stream. Immobilization of cells does not necessarily require the retention of cell viability, as long as the enzyme of interest remains active. Although generally used to catalyze single-enzyme reactions, immobilized cells can be used to catalyze multienzyme reactions. The most successful use of immobilized cells to date has been in providing glucose isomerase for the production of high-fructose syrups. For example, ruptured *Bacillus coagulans* cells aggregated by cross-linking with glutaraldehyde have been used, as have *Actinoplanes missouriensis* cells entrapped in gelatin cross-linked by using glutaraldehyde and then granulated. *See also* gas flooding.

cell line \\'sel 'līn\\ *See* secondary cell cultures.

cell recycle reactor \\'sel rē-'sī-kəl rē-'ak-tər\\ A reactor (fermentor) in which cells are removed from the product and recycled to the tank. This reaction is used in particular for recycling activated sludge in wastewater treatment.

cell sorter \\'sel 'sō(ə)rt-ər\\ A device designed to separate and analyze individual classes of cells, microorganisms, or organelles from a mixed population. *See also* flow cytometry.

cellobiase \\ِsel-ə-'bī-ās\\ An enzyme that hydrolyzes cellobiose to glucose. *See also* cellulase.

cellobiose \\ِsel-ə-'bī-ōs\\ A disaccharide comprising two β-(1,4)-linked glucose molecules, produced by the hydrolysis of cellulose. Cellobiose is the basic structural unit of cellulose but only exists free in nature in trace concentrations. *See also* cellulase.

cellulase \\'sel-yə-ِlās\\ An inducible enzyme system that degrades cellulose to glucose. Cellulase is present in the digestive juices of snails and wood-boring insects. Cellulase-producing microorganisms in the stomach allow ruminants to partially digest cellulose-containing material such as straw. Commercially used cellulases are produced by strains of *Aspergillus niger*, *Trichoderma reesei*, *Penicillium funiculosum*, and *Rhizopus* species. Three main types of cellulase enzymes are produced by fungi: an endoglucanase that cleaves cellulose randomly; *exo*-1,4-glucanases that produce cellobiose; and cellobiase (β-glucosidase), which breaks down cellobiose to glucose. Cellulase has considerable potential for the large-scale production of glucose and alcohol from cellulose, once the problems of releasing cellulose from its protected environment in lignocellulose have been solved (*see* lignin). Cellulases are currently used in cereal processing to speed up mash filtration and to increase extract yield; in brewing as a supplement to starch-degrading enzymes to increase the yield of alcohol by degrading cellobiose to a fermentable product; in fruit processing to speed up color extraction from the skins of fruits and to increase liquefaction or maceration of fruits (and vegetables); in waste treatment to speed up the process by causing the release of fermentable sugars; and in the recovery of alginate from seaweed.

cellulose \'sel-yə-ˌlōs\ The major polysaccharide of plants. It has a structural rather than nutritional role. Cellulose comprises linear polymers of β-1,4-linked D-glucose units. Cellulose has considerable potential for the large-scale production of glucose and alcohol, but this potential has yet to be realized because most cellulose is found as lignocellulose and is thus resistant to enzymic or microbial attack (*see* lignin). However, the degradation of cellulose by cellulase is used in a number of industrial processes (*see* cellulase). Beaded cellulose can be prepared by solidification of liquid particles and is used as a support matrix for affinity and ion-exchange chromatography.

cellulose ion exchanger \'sel-yə-ˌlōs 'ī-än iks-'chān-jər\ A beaded form of cellulose that has been derivatized to form an ionic grouping at the surface. Both cation exchanges (with carboxymethyl groups) and anionic exchanges (with diethylaminoethyl substituents) are commercially available.

centimorgan \'sent-ə-'mȯrg-ən\ A unit for expressing the relative distance between genes on a chromosome. The frequency of crossing over (crossover) between any two genes on a chromosome is directly proportional to their distance from each other. The closer together, the lower their recombination frequency; the farther apart, the higher. A map unit, or measure of distance between genes, was therefore assigned. The map unit is named in honor of the geneticist T. H. Morgan. A centimorgan (cM) is a crossover value of 1%; a decimorgan (dM) is a crossover value of 10%; one morgan (M) equals a crossover value of 100%.

centrifugal concentrator (vacuum centrifuge) \sen-'trif-(y)ə-gəl 'kän(t)-sən-ˌtrāt-ər ('vak-yü-əm 'sen-trə-fyüj)\ A bench-scale centrifuge in which samples in solution can be placed under a vacuum and dried. Samples are centrifuged before and during evacuation to prevent bumping (i.e., samples jumping out of the tube during rapid degassing). The centrifuge bowl can also be heated. Heating prevents freezing of the sample and increases the rate of evaporation.

centrifugal contactor \sen-'trif-(y)ə-gəl 'kän-ˌtakt-ər\ *See* Alfa Laval extractor, Podbielniak extractor.

centrifugal evaporator \sen-'trif-(y)ə-gəl i-'vap-ə-ˌrāt-ər\ A mechanical device for the large-scale concentration of solutions. It usually consists of a rotating disk or cone heated on one side, with the feed solution fed to the axis of the disk or cone on the other side. The condenser is placed close to this rotor so that the evolved vapor is withdrawn rapidly from the system. Costs are high and the capacity low, making this device suitable only for high-value products.

centrifuge \'sen-trǝ-ˌfyüj\ A mechanical device that may be used in downstream processing as a means of separating systems according to their specific gravity. It operates on the principle that the contents of a rotating container exert a centrifugal force toward the vertical walls of the cylinder. This force causes rapid sedimentation of heavy solid particles through a layer of liquid or of a heavier liquid through a less dense liquid. The centrifugal force is many times greater than gravity and thus allows the rapid separation of solids and liquids with very small density differences. Samples may be placed in discrete tubes attached to the rotor or, in the case of continuous centrifuge, the feed flows through a rotating cylinder or a series of concentric plates. Many different sizes and designs are available to separate either solids and liquids or two immiscible liquids, with and without the presence of solids. The choice depends on the requirements of the process. *See also* basket centrifuge, disk centrifuge, multichamber centrifuge, solid-bowl scroll centrifuge, tubular bowl centrifuge, ultracentrifuge.

centromere \'sen-trǝ-ˌmi(ǝ)r\ A sequence within the chromosome that is essential for the division of chromosomes during mitosis. Before metaphase, all the chromosomal DNA replicates except the centromere, resulting in two identical sister chromatids held together at the centromere. The resultant short arm is designated *p*, and the long arm, *q*. The centromere is the point of attachment of the chromosome to the spindle apparatus. At late metaphase, the centromere divides and the two chromatids separate to become two chromosomes, which move away from each other along the microtubule spindle to opposite poles of the cell.

cephalosporins \ˌsef-ǝ-lǝ-'spōr-ǝnz\ A group of closely related β-lactam antibiotics (i.e., they contain the β-lactam ring common to penicillins) isolated from the mold *Cephalosporium*. Although they have a core structure in common with penicillins, they are sufficiently different to be used with patients who are allergic to penicillin.

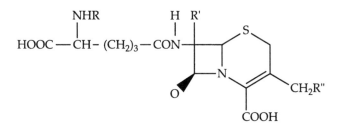

in cephalosporin C, R, and R' = H, R" = –OCOCH$_3$

chain termination method \'chān ˌtǝr-mǝ-'nā-shǝn 'meth-ǝd\ A DNA sequencing method. *See* dideoxy sequencing.

chaotropic ions \ˌkā-ō-'trō-pik 'ī-änz\ Ions that disrupt the water structure. Because they reduce hydrophobic interactions, they can be used as desorbing agents for affinity chromatography. Ions can be arranged in order of increasing chaotropicity (e.g., $Cl^- < I^- < ClO_4^- < SCN^- < CCl_3COO^-$).

chaperones \\'shap-ə-ˌrōnz\\ Cellular proteins that assist in the folding of other proteins into their biologically active forms but do not become part of the final structure. Also known as molecular chaperones.

Charon vector \\'ka-rən 'vek-tər\\ A cloning vector constructed from λ-phage. A range of Charon vectors have been constructed, most of which are replacement vectors. In Greek mythology, Charon ferried the souls of the dead over the river Styx.

cheese \\'chēz\\ Foodstuff produced from milk. The basic process involves the inoculation of milk with lactic acid, producing bacteria (usually *Streptococcus* or *Lactobacillus* species). The milk is curdled by the addition of rennet, and the curd is separated from the liquid (whey). The resulting curd is incubated, pressed, and left to ripen. *See also* curd, rennet, rennin, secondary fermentations.

CHEF \\'sē 'āch 'ē 'ef\\ *See* pulse-field gel electrophoresis.

chelating agent \\'kē-lāt-iŋ 'ā-jənt\\ A chemical capable of bonding via two or more donor atoms to a metal ion (e.g., glycine, H_2NCH_2COOH, which can donate via both the nitrogen and one of the oxygen atoms; also EDTA, which has the possibility of six donor atoms). Chelating agents may be used to solubilize and stabilize metal ions in, for example, the preparation of culture media.

chelation \\kē-'lā-shən\\ The process by which a molecule (ligand) is attached to a metal ion by two or more donor atoms to provide a chelate ring. The name is derived from *chelos* (Greek), meaning crab's claw.

ChemFET (chemical field effect transistor) \\'kem-fet ('kem-i-kəl 'fē(ə)ld i-'fekt tranz-is-tər)\\ A field effect transistor that has been modified by depositing onto the gate sites chemicals that can respond reversibly to changes in the activity of ions in solution. Thus ChemFETs act as sensors, with the advantage of very small size and the ability to react to changes in activity of a number of ions because each gate can be made to respond to a different ion. However, currently they suffer from lack of reliability and reproducibility. Also known as ISFET (ion-selective field effect transistor).

ChemFET

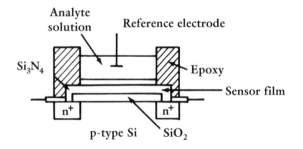

chemical fusogen \\'kem-i-kəl 'fyü-zō-jen\\ Any chemical that can be used to fuse together two cells or protoplasts. For example, a suspension of plant protoplasts in sodium nitrate solution induces rapid fusion. Polyethylene glycol (PEG) is used to fuse both protoplasts and animal cells (e.g., in hybridoma formation).

chemical ionization (CI)   \\'kem-i-kəl ˌī-ə-nə-'zā-shən ('sē 'ī)\\ A method of producing ions for analysis by mass spectrophotometry. It involves a reaction between molecules of the sample and ions produced in a reagent gas by electron-impact ionization in the source. Sample molecules are in much lower concentration than the reagent gas, and a number of different ion–molecule reactions are possible. CI is a much gentler process than electron impact. As a result, the fragmentation patterns are intermediate between simple field-ionization spectra and complex electron-impact spectra. The choice of reagent gas depends on the type of system to be studied. Thus, ammonia is useful for polyhydroxy compounds like sugars, isobutane when the molecular mass is required, and deuterium oxide for active hydrogens. *See also* fast-atom bombardment.

chemical oxygen demand (COD)   \\'kem-i-kəl 'äk-si-jən di-'mand ('sē 'ō 'dē)\\ A measure of the amount of oxygen required (milligrams per cubic decimeters) for the oxidation of organic matter in water with a strong oxidizing agent at a raised temperature. Samples are treated with a known amount of boiling acidic potassium dichromate solution for 2.5–4 hours. The organic matter is estimated as proportional to the potassium dichromate used. Such correlations depend on both the nature and the strength of the effluent. Most organic compounds are oxidized to completion by the dichromate, but benzene, toluene, and pyridine are not oxidized at all. Certain inorganic sulfur-containing compounds, nitrites, and ferrous ions are also oxidized by the treatment, and chlorides may interfere with the reaction. When sewage effluents are tested for COD and biochemical oxygen demand (BOD), the BOD : COD ratios are usually between 0.2 : 1 and 0.5 : 1. *See also* biochemical oxygen demand.

chemiluminescence   \\ˌkem-i-ˌlü-mə-'nes-ᵊn(t)s\\ The emission of radiation, usually visible, caused by the decay of a chemical reaction product from an electronic excited state to ground state. For reasons of both safety and increased sensitivity there has been considerable interest in replacing the use of radiolabeled proteins and other molecules, particularly those used in diagnostic kits, with reactions that have a luminescent end product. *See also* reporter gene.

chemisorption   \\ˌkem-i-'sȯrp-shən\\ The adsorption of a substance at a surface, involving the formation of chemical bonds between the adsorbate and the adsorbing surface.

chemoautotroph   \\ˌkē-mō-'ȯt-ə-ˌtrōf\\ An autotroph that synthesizes biological molecules from carbon dioxide and derives its energy from the oxidation of reduced forms of various elements present in the biosphere, such as NH_3 and NO_2^-.

chemolithotroph   \\ˌkē-mō-'li-thō-trōf\\ An organism that can obtain its energy from the oxidation of inorganic compounds, such as hydrogen, reduced-sulfur compounds, nitrite, and iron(II). Also known as lithotroph.

chemometrics \ke-mō-'me-triks\ The use of mathematics and statistics to improve analytical instrumentation and the manipulation of data to increase the information available from such measurements.

chemostat \'kē-mə-stat\ A continuous-culture system in which the growth rate of the culture is controlled by the availability of one limiting component in the medium. *See also* continuous culture.

chill haze \'chil 'hāz\ A precipitation of protein (probably in association with other compounds, such as tannins and polysaccharides) formed during the cold storage of beer. This problem is overcome in the brewing industry by the use of proteases, such as papain and bromelain, to degrade and solubilize the protein component of the haze. This enzymatic treatment is known as chillproofing.

chillproofing \'chil 'prüf-iŋ\ *See* chill haze.

chimera \kī-'mir-ə\ Any recombinant DNA molecule produced by joining fragments of DNA from more than one organism (e.g., a bacterial plasmid containing an inserted eukaryotic gene). It is named after a mythological beast with a lion's head, a goat's body, and a serpent's tail.

chimeric antibody \kī-'mir-ik 'ant-i-bäd-ē\ An antibody in which part of the molecule is derived from one species and part from a second species. The majority of monoclonal antibodies are currently produced in rodents. Their use in human therapy is, however, restricted, particularly when multiple doses are needed because of the problem of allergic reactions and the generation of an antibody response to the rodent IgG. Monoclonal antibodies have been constructed in which the antigen-binding variable region derived from a rodent source is genetically combined to a human constant region. When used in vivo, these molecules are less antigenic than their pure rodent counterparts because the constant domains of mouse IgGs are probably the most antigenic part of the mouse IgG molecule. This process is sometimes referred to as *humanizing* the antibody molecule. However, in practice the process used is not really humanization of rodent antibodies but site-directed mutagenesis of human antibodies. Synthetic oligonucleotides are used to provide recombinant human immunoglobulins with the hypervariable (complementarity-determining) region sequences of rodent antibodies. A further possibility is to replace the F_c region of an antibody with another active protein, e.g., an enzyme or a toxic protein, which could then be used in diagnosis or therapy. All these strategies have proved necessary because of the inherent difficulty of producing human monoclonal antibodies in large amounts. However, the recent introduction of repertoire cloning, which allows the production of human monoclonal antibodies in microorganisms, may well make the above strategies redundant in the future. *See also* bispecific antibodies, immunoglobulin G, repertoire cloning.

chirality \kī-'ral-əd-ē\ Nonidentity of a molecule with its mirror image (i.e., left- and right-handedness). This property, often found in carbon compounds, is demonstrated by the ability in solution to rotate the plane of polarized light to the 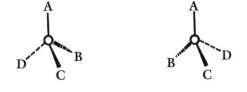 left or right. These compounds are said to display optical activity. A carbon atom that is attached to four different atoms (groups) is commonly called a chiral atom, but more strictly the term *chiral* refers to the environment in which the atom is found (i.e., a chiral center). *See also* dextrorotatory, levorotatory.

chloramphenicol acetyl transferase (CAT) \klōr-ˌam-'fen-i-ˌkȯl ə-'sēt-ᵊl 'tran(ts)-fər-ˌās ('sē 'ā 'tē)\ *See* reporter gene.

chloramphenicol amplification \ˌklōr-ˌam-'fen-i-kȯl ˌam-plə-fə-'kā-shən\ *See* amplification of plasmids.

chloroplasts \'klōr-ə-ˌplasts\ Cytoplasmic organelles, found in plants and algae, in which photosynthesis takes place.

chorionic villus sampling (CVS) \ˌkōr-ē-'än-ik 'vil-əs 'sam-pliŋ ('sē 'vē 'es)\ Removal of part of the chorionic villi that surround the fetus in the uterus. It is carried out via the cervix using a needle. The tissue obtained is diploid and fetal in origin and may be used for prenatal diagnosis. The advantage of this method is that it can be carried out in the first trimester of pregnancy, but the method carries a risk of spontaneous abortion after the procedure. Also known as chorionic biopsy. *Contrast with* amniocentesis, fetoscopy.

chromatin \'krō-mət-ən\ The complex of DNA and protein found in the eukaryotic cell nucleus. *See also* chromosome.

chromatofocusing \krō-'mat-ə-ˌfō-kəs-iŋ\ A liquid chromatographic technique that separates biomolecules on the basis of their respective isoelectric points (pI). Chromatofocusing is a chromatographic analogue of isoelectric focusing and is particularly useful for separating proteins. The method is based on the interaction of an ion-exchange resin and a suitable buffer; this interaction results in the formation of a pH gradient without the use of any gradient mixer. Proteins that are bound to the ion exchange elute in order of their pI values. *See also* isoelectric focusing, isoelectric point.

chromatography \ˌkrō-mə-'täg-rə-fē\ Generally, several analytical and preparative separation techniques based on the partitioning of a substance between a mobile and a stationary phase. Because the degree of partitioning is closely related to the structure of the substance and the nature of the mobile and stationary phases, the technique is very powerful; separations of homologous series, stereoisomers, and even chiral species are possible. Classification of the different techniques can be based on the nature of the stationary phase, e.g., gel, paper, ion-exchange, thin-layer; or on the type of mobile phase, e.g., gas, liquid, supercritical fluid. Individual techniques are defined under specific headings.

chromatophore \krō-'mat-ə-ˌfō(ə)r\ *See* chromophore.

chromogenic substrate \krō-mə-'jen-ik 'səb-ˌstrāt\ Any enzyme substrate that is converted to a colored product by the enzyme. For example, alkaline phosphatase converts colorless *p*-nitrophenol phosphate to colored (yellow) *p*-nitrophenol. Chromogenic substrates are used particularly to detect enzyme-linked antibodies in techniques such as ELISA and immunoblotting. *See also* enzyme-linked immunosorbent assay, immunoblotting.

chromophore \'krō-mə-ˌfō(ə)r\ An atom or group of atoms that conveys a particular spectral band (color) to a substance. Also known as a chromatophore.

chromoplasts \'krō-mə-ˌplasts\ *See* plastid.

chromosome \'krō-mə-ˌsōm\ A highly condensed form of chromatin identifiable in the eukaryotic cell nucleus at the time of cell division.

chromosome sorting \'krō-mə-ˌsōm 'sȯ(ə)rt-iŋ\ *See* flow cytometry.

chromosome walking \'krō-mə-ˌsōm 'wȯ-kiŋ\ A technique that allows the sequential isolation of clones carrying overlapping sequences of DNA, thus allowing large regions of the chromosome to be mapped and sequenced. Starting with a particular cloned sequence, sequences obtained from one end of this starting clone are used to probe a genomic library. The new set of clones isolated from this library are themselves sequenced; sequences that are located furthest away from the starting point are then used as probes for a further screening of the library. Proceeding in this way, one can "walk" along the chromosome, obtaining large stretches of chromosome sequence. For example, if the library was constructed in a λ-vector, each clone should contain about 15 kilobase of chromosomal DNA and thus one would "walk" the chromosome in steps of about 15 kilobase.

chymosin \'kī-mə-sən\ *See* rennin.

Cibacron Blue 3G-A \'sē-bə-krȯn 'blü 'thrē 'jē 'ā\ A Ciba-Geigy trade name for a triazine dye used in dye–ligand chromatography for the purification of proteins. It has been designated the color index (C.I.) number 61211. Also known as Cibacron Blue F3G-A, Procion Blue H-B, Reactive Blue 2. *See also* dye–ligand chromatography.

Cibacron Blue F3G-A \'sē-bə-krȯn 'blü 'ef 'thrē 'jē 'ā\ *See* Cibacron Blue 3G-A.

2-μm circle \'tü-'mī-krō-ˌmēt-ər 'sər-kəl\ A 6-kilobase pair plasmid found in the yeast *Saccharomyces cerevisiae*. It is used as a cloning vector and exists in the yeast cell at a copy number between 50 and 200. The plasmid contains the gene *leu* 2, which codes for one of the enzymes involved in the conversion of pyruvic acid to leucine. If a *leu* 2⁻ host is used, transformants can be identified by their ability to grow in the absence of leucine. *See also* complementation.

cis-acting region \'sis-ˌakt-iŋ 'rē-jən\ Any region of DNA that affects the activity only of DNA sequences of its own molecule of DNA. For example, an operator sequence is cis-acting because it can only control genes that are adjacent to it. If these same genes are introduced into the cell on a plasmid the genomic operator cannot affect the expression of the plasmid genes, which need their own operator within the plasmid. In general, the term *cis configuration* is used in molecular biology to describe two sites on the same molecule of DNA. A cis-acting protein has the property of acting only on the molecule of DNA from which it was expressed. *Contrast with* trans-acting region.

cis isomers \\'sis 'ī-sə-mərs\\ A configuration of geometric isomers in which two atoms or groups are on the same side of the molecule or atom. *Contrast with* trans isomers. *See also* Z-isomers (the IUPAC preferred term).

cistron \\'sis-ˌträn\\ A region of DNA that codes for a particular polypeptide.

citric acid \\'si-trik 'as-əd\\ An industrially useful organic acid. It has been traditionally produced by submerged fungal fermentation of carbohydrate sources such as molasses and glucose hydrolysate. *Aspergillus niger* is the fungus predominantly used, although more recently developed processes preferentially use yeasts. It is also produced by solid-state fermentation, by the traditional Koji process. Citric acid has extensive use in the food, confectionary, and beverage industries as an acidulant. Its esters are used in the plastics industry.

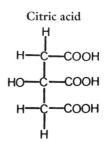

Citric acid

Cladosporium resinae \\klad-ō-'spōr-ē-em 'rez-ən-ē\\ A fungus commonly found as a contaminant in aviation and diesel fuels. Such fuels invariably contain a small amount of water from seepage during storage. The fungus grows at the fuel-water interface. If unchecked, the fungus can spread into the fuel supply system and cause blockage of filters and fuel lines. This is a particular problem in airplane jet engines and ship diesel engines.

clarification \\klar-ə-fə-'kā-shən\\ A process by which suspended matter is removed from solutions. Several techniques are available, including filtration, centrifugation, and sedimentation.

clarifier \\'klar-ə-fī(-ə)r\\ A large tank with continuous feed and outlet, in which suspended matter is allowed to settle under gravity. The term may also be used to describe liquid–solid centrifuges. *See also* thickening.

clavulanic acid \\klav-yü-'lan-ik 'as-əd\\ A β-lactam antibiotic produced by *Streptomyces clavuligerus* and a few other *Streptomyces* species. Of particular interest because it is a potent inhibitor of bacterial β-lactamases. It is available together with the antibiotic amoxicillin as a commercial preparation. Clavulanic acid is included to inhibit the degradation of amoxicillin by β-lactamases produced from penicillin-resistant strains.

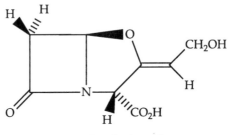

clavulanic acid

cleared lysate \\'kli(ə)rd 'lī-ˌsāt\\ A cell extract that has been centrifuged to remove cell debris, subcellular particles, and most chromosomal DNA.

climacteric fruit \\klī-'mak-tər-ik 'früt\\ Fruit, such as apples, bananas, and tomatoes, in which ripening is associated with a marked increase in respiration.

climbing-film evaporator \\'klīm-iŋ-ˌfilm i-'vap-ə-ˌrāt-ər\\ *See* rising-film evaporator.

clonal propagation \\'klōn-ᵊl ˌpräp-ə-'gā-shən\\ The generation of large numbers of identical plants from a single plant cell or protoplast, or by the growth of plant cuttings. *See also* clone.

clone \\'klōn\\ A group of organisms or cells, all derived asexually from a single ancestral organism or cell and therefore genetically identical. However, a somewhat looser interpretation of the word *clone* must be used when considering the clonal propagation of plants. Current techniques for plant single-cell and protoplast culture enable thousands of plants to be ultimately derived from a single cell. These plant products should, by definition, each be considered a single clone. However, the product of callus and cell-suspension cultures consists of many abnormal genotypes (somatic variants), so 100% true-to-type clonal propagation does not actually occur.

cloning \\'klōn-iŋ\\ *See* gene cloning.

cloning vector \\'klōn-iŋ 'vek-tər\\ *See* vector.

cloning vehicle \\'klōn-iŋ 'vē-(h)ik-əl\\ *See* vector.

clonogenic \\ˌklō-nō-'jen-ik\\ Capable of generating a clone of cells.

Clostridium \\klä-'strid-ē-əm\\ A genus of Gram-positive, endospore-forming, rod-shaped bacteria. Most strains are strictly anaerobic and obtain energy from fermentation of sugars, starch, pectin, cellulose, amino acids, purines, and other organic compounds. The main habitat of *Clostridia* is the soil, although some are found in the mammalian intestinal tract. A few *Clostridia* can cause human diseases by toxin production: Botulism is caused by *C. botulinum*, tetanus by *C. tetani*, and gas gangrene by *C. perfringens* and other species. *See also* acetone–butanol fermentation.

cmc \\'sē 'em 'sē\\ *See* critical micelle concentration. Also an abbreviation used for carboxymethyl cellulose.

CoA (CoASH) \kō-ˈā (kō-ˈā ˈes ˈāch)\ *See* coenzyme A.

coacervation \kō-ˌas-ər-ˈvā-shən\ Separation of lyophilic colloids into two immiscible phases, each of which has a different concentration of the dispersed phase.

cocurrent flow \kō-ˈkər-ənt ˈflō\ A type of flow configuration in continuous processing whereby two process streams flowing in the same direction are brought into contact for the purpose of carrying out heat or mass transfer. This particular configuration is inefficient because once equilibrium between the phases has been reached, no further transfer will occur. *See also* countercurrent flow, crosscurrent flow.

code-blocker therapeutics \ˈkōd-ˌbläk-ər ˌther-ə-ˈpyüt-iks\ The use of artificially constructed strands of nucleic acids that target either DNA or RNA inside a cell to inhibit gene expression. *See also* antisense RNA, aptamer, ribozymes, triple helixes.

codeine \ˈkō-ˌdēn\ An alkaloid plant product from *Papaver somniferum*, used as an analgesic.

codon \ˈkō-ˌdän\ In a molecule of DNA or RNA, the sequence of three adjacent bases that provides the code for a specific amino acid. For example, in mRNA, the sequence adenine–cytosine–cytosine (usually abbreviated ACC) codes for the amino acid threonine. Thus, given any sequence of DNA or mRNA one can read along the sequence to determine the amino acid sequence for which the polynucleotide codes. Because both DNA and RNA are linear polymers of four different nucleotides, there are 4^3 (64) possible codon triplets. (The sequence of bases in a codon is important. GAC codes for a different amino acid than GCA). Because only 20 amino acids are found in proteins, most amino acids are specified by more than one codon. The genetic code is thus said to be degenerate. However, in highly expressed genes of several species only a subset of the potential codons are used. For example, although arginine has six codons, arginine residues of abundant proteins are encoded in *Escherichia coli* most frequently by CGU and CGC, whereas yeast preferentially uses AGA. The amounts of specific tRNAs in the cell have been shown to correlate with codon usage. Knowledge of codon usage is important in helping to optimize heterologous gene expression. For example, when designing synthetic genes one should use codons that are optimal for the host organism.

codon usage \ˈkō-ˌdän ˈyü-sij\ *See* codon.

coenocytic fungal hypha \sē-nə-ˈsit-ik ˈfəŋ-gəl ˈhī-fə\ A fungal hypha that has no cross walls (septa) and in which the nuclei are distributed fairly uniformly throughout the cytoplasm. The majority of the *Mastigomycotina* and *Zygomycotina* are coenocytic.

coenzyme \kō-ˈen-ˌzīm\ *See* cofactor.

coenzyme A \kō-ˈen-ˌzīm ˈā\ A molecule that functions as a carrier of acyl groups. It is formed by the linkage of adenosine diphosphate via the 5′-phosphate to the phosphate of pantetheine 4′-phosphate. Because the thiol (—SH) group of pantetheine phosphate is responsible for the biological activity of CoA, uncombined CoA is often represented as CoASH. CoA functions as a carrier of acyl groups by forming thioesters (CoASCOR) and provides the acyl group for a range of biosynthetic steps. Also known as CoA, CoASH.

cofactor (coenzyme) \\'kō-ˌfak-tər (kō-'en-ˌzīm)\\ A small, nonprotein organic molecule that associates with the protein portion (apoenzyme) of an enzyme and is essential for enzyme function. The complete enzyme (apoenzyme and cofactor–coenzyme) is called a *holoenzyme*. In the course of the catalytic reaction the coenzyme may be changed chemically and dissociate from the apoenzyme, but it is regenerated in associated reactions and can combine again with the apoenzyme. Coenzymes include nicotinamide adenine dinucleotide (NAD$^+$), nicotinamide adenine dinucleotide phosphate (NADP$^+$), coenzyme A, and flavine adenine dinucleotide (FAD). *See also* cofactor recycling.

cofactor recycling \\'kō-ˌfak-tər rē-'sī-kliŋ\\ Regeneration of cofactors. About a third of the known enzymes require coenzyme A, NAD, NADP, FAD, or adenosine triphosphate as a cofactor. For any of these enzymes to be commercially useful, methods must be available for the regeneration of these cofactors after their oxidation or reduction during the catalytic process. Regeneration is part of the normal integrated metabolism of the cell. However, when isolated enzymes and disrupted cells are used, cofactor regeneration is a major problem because cofactors are too expensive to supply continuously. Methods must be developed to regenerate the cofactors in vitro. Methods that have potential are the enzymatic regeneration of cofactors by other enzymes that use cheap substrates and the electrochemical oxidation–reduction of reduced–oxidized cofactors.

cohesive ends \\kō-'hē-siv 'endz\\ *See* sticky ends.

coimmobilized enzymes \\kō-im-'ō-bə-ˌlīzd 'en-ˌzīmz\\ Complementary immobilized enzymes. Usually only one enzyme is immobilized to catalyze a specific reaction. However, it can be advantageous to immobilize two complementary enzymes. For example, glucose oxidase can be coimmobilized with catalase when hydrogen peroxide produced by glucose oxidase has been degraded by catalase before it can inactivate the glucose oxidase. Other potentially useful coimmobilizations include glucoamylase and glucose isomerase for use in the formation of high-fructose syrups from dextrin. *See* enzyme immobilization.

col plasmids \\'kōl 'plaz-mədz\\ Plasmids that code for the antibiotic proteins known as colicins.

cold sterilization \\'kōld ˌster-ə-lə-'zā-shən\\ Sterility achieved by ionizing radiation. *Compare with* autoclave.

colicins \\'kō-lə-sənz\\ Antibiotic proteins produced by bacteria carrying col plasmids. Colicins kill bacteria by a range of mechanisms, including the inhibition of active transport, inhibition of protein synthesis, and promotion of DNA degradation. Bacteria carrying col plasmids are obviously immune to the effects of the colicin specified by the col plasmid they are carrying, and this immunity forms the basis of a selection system for cells transformed by col plasmids.

coliform \\'kō-lə-ˌfȯrm\\ A group of bacteria, primarily strains of the genera *Escherichia*, *Citrobacter*, *Klebsiella*, and *Enterobacter*, most frequently used as indicators of fecal pollution in assessment of water quality. In the United States, coliforms are defined as aerobic and facultatively anaerobic, Gram-negative, non-spore-forming rod-shaped bacteria that ferment lactose with gas formation within 48 h at 35 °C. In

Great Britain, coliforms are defined as Gram-negative, non-spore-forming, oxidase-negative rod-shaped bacteria that can grow aerobically on agar media containing bile salts and ferment lactose with the production of both acid and gas within 48 h at 37 °C.

colloid \\'käl-óid\\ A stable suspension of microscopic particles, size range 1 nm to 1 µm, dispersed in a continuous medium. Two types of colloid may be formed. In *hydrophilic* colloids, the substance consists of macromolecules whose size is in the colloid range. Therefore they exhibit properties characteristic of dispersed systems, such as light scattering, but are soluble in water. These colloids (e.g., gelatin and starch) are thermodynamically stable and are kept in the disperse state by their affinity for water. They can be precipitated only by removal of the water. *Hydrophobic* colloids consist of insoluble particles in a finely divided state suspended in water. These colloids are thermodynamically unstable, but often have a kinetic stability as a result of surface-charge repulsion among the particles.

colloid mill \\'käl-óid 'mil\\ A device for grinding matter into microscopic particles such that, on suspension in a liquid, a colloid is formed.

colony \\'käl-ə-nē\\ A group of cells produced from an individual cell when grown on a solid medium such as an agar plate.

colony hybridization (Grunstein–Hogness method) \\'käl-ə-nē ˌhī-brəd-ə-'zā-shən ('grün-stīn 'hòg-nes 'meth-əd)\\ A method for identifying a clone or clones containing a particular DNA sequence. The colonies are partially transferred to a nitrocellulose membrane by briefly placing the membrane, with light pressure, on the surface of the nutrient plate containing the organisms. The membrane is then treated to lyse the cells and denature the DNA so that the hydrogen bonds between the DNA strands are broken. Baking them fixes the single-strand DNA to the membrane. The membrane is then washed in a solution containing a radiolabeled DNA probe for the sequence of interest; then the filter is washed and autoradiographed. The position where the probe has bound to the DNA (by complementary base-pairing) is detected as a darkening of the film. This position can then be related to the original nutrient plate, and the required clone can be identified.

colony stimulating factor \\'käl-ə-nē 'stim-yə-ˌlāt-iŋ 'fak-tər\\ *See* cytokines.

color index (C.I.) \\'kəl-ər 'in-deks ('sē 'ī)\\ An international index listing all coloring matter, both natural and synthetic. If the structure of the coloring is known, it is given a five-digit index number. For example, the C.I. number for Cibaron Blue 3G-A is 61211. Each entry also includes data on fastness and brand names.

column reactor–extractor \\'käl-əm rē-'ak-tər-ik-'strak-tər\\ A general name given to equipment in which the main feature is a long vertical cylinder that may be fitted with devices to promote mixing or coalescence. The simplest design is the spray column, in which one reactant is distributed as a fine spray into the other without any additional means of mixing. Because this is not very efficient, mixing devices are often fitted to the equipment. These devices can be turbines mounted on a vertical shaft or rotating disks, both of which may also use a pack of coalescent material between the mixing devices to aid reaction efficiency. *See also* Oldshue–Rushton contactor, perforated-plate column, pulsed-plate column, rotating-disk contactor, Scheibel contactor.

column volume \\'käl-əm 'väl-yəm\\ *See* bed volume.

combined electrode \\kəm-'bīnd i-'lek-ˌtrōd\\ An electrochemical device that consists of a glass electrode, or ion-selective electrode, and a reference electrode in a single assembly. This design avoids the need for a separate reference electrode.

cometabolism \\ˌkō-mə-'tab-ə-ˌliz-əm\\ The process whereby a substrate is modified but not used for growth by an organism that is growing on or metabolizing another substrate.

commensalism \\kə-'men(t)-sə-ˌliz-əm\\ A system in a culture containing two microbial species in which one species derives benefits, e.g., detoxification, as a result of interactions with the other species. *See also* amensalism, mutualism, neutralism.

commensals \\kə-'men(t)-səlz\\ Organisms of different species that live together, neither being detectably harmed.

comminution \\ˌkäm-ə-'n(y)ü-shən\\ The process of size reduction of materials by crushing and grinding.

communities \\kə-'myü-nət-ēz\\ Collections of different species of organisms that exist together and can between them carry out reactions that cannot be achieved by any of the single component species. The presence of communities is commonplace in natural ecosystems in waste treatment processes, and in microbial leaching. Also known as consortia.

compatibility \\kəm-ˌpat-ə-'bil-ət-ē\\ The ability of two or more types of plasmid to coexist in the same cell. Several types of plasmids can be found in a single cell at any time. If two plasmids are incompatible, then one or the other is rapidly lost from the cell. Plasmids can be categorized in different *incompatibility groups* on the basis of whether or not they can coexist.

competent \\'käm-pət-ənt\\ Having enhanced ability to take up DNA molecules, used to describe bacteria. For example, *Escherichia coli* cells are treated with $CaCl_2$ to make them competent before uptake of plasmid DNA in transformation experiments.

competitive inhibitor \\kəm-'pet-ət-iv in-'hib-ət-ər\\ Any substance that inhibits an enzyme reaction by complexing with the enzyme at the active site.

complement \\'käm-plə-mənt\\ A group of nine serum proteins essential for antibody-mediated immune hemolysis.

complementarity-determining region (CDR) \\ˌkäm-plə-men-'tar-əd-ē di-'term-(ə-)niŋ rē-jən ('sē 'dē 'är)\\ The portion of an antibody variable region that binds to antigen.

complementary base-pairing \\ˌkäm-plə-'ment-ə-rē 'bās-'pa(ə)r-iŋ\\ The ability of two polynucleotide sequences (DNA or RNA) to form a double-stranded structure by hydrogen bonding between bases in the two sequences. The two sequences may be on different strands (e.g., double-stranded DNA) or the same strand (e.g., transfer RNA). Adenine base pairs with thymine (or uracil in RNA) and cytosine, with guanine.

complementary DNA (cDNA) \ˌkäm-plə-'ment-ə-rē 'dē 'en 'ā ('sē 'dē 'en 'ā)\ DNA synthesized from an mRNA template with the enzyme reverse transcriptase. The RNA specifies the base sequence of the DNA by complementary base-pairing. Once the cDNA strand has been synthesized, the RNA chain in this hybrid molecule can be degraded by mild alkali treatment. The single-stranded cDNA can then be converted into double-stranded cDNA by using the Klenow fragment of DNA polymerase I. The resultant DNA fragment can be ligated into a vector and cloned.

complementation \ˌkäm-plə-men-'tā-shən\ A method used to identify transformants in which one of the incoming genes complements a defective copy of the same gene on the chromosome of the recipient. For example, the *lacz*⁺ gene (coding for β-galactosidase) will complement a *lacz*-deficient host to give transformants that produce β-galactosidase. These transformants are selected by their ability to produce red colonies on MacConkey's agar.

complex medium \'käm-'pleks 'mēd-ē-əm\ *See* undefined medium.

composting \'käm-pōst-iŋ\ The aerobic decomposition of solid organic matter by a consortium of bacteria, fungi, and some invertebrates.

con A \'kän 'ā\ Abbreviation for concanavalin A.

concanavalin A (con A) \ˌkän-kə-'nav-ə-lən 'ā (ˌkän-'ā)\ A lectin isolated from the plant *Canavalia ensiformis* (jack bean). Concanavalin A binds specifically to residues of α-D-mannopyranose and α-D-glucopyranose with unmodified hydroxyl groups at C-3, C-4, and C-6. Concanavalin A is used to purify glycoproteins by affinity chromatography and to agglutinate cells by cross-linking cell-surface glycoproteins. The protein also stimulates resting lymphocytes to undergo mitosis and multiply (i.e., it is mitogenic). *See also* lectins.

concatamer \kän-'kat-ə-mər\ A DNA sequence consisting of a repeating series of genomes joined end to end. The entire genome is often repeated several times. For example, in the replication of both lambda and T4 phage, concatameric DNA is first formed. The concatameric DNA is inserted into a preformed viral head and is cleaved off when a "genome's worth" of DNA has been inserted.

concentration polarization \ˌkän(t)-sən-'trā-shən ˌpō-lə-rə-'zā-shən\ The increase in concentration of a solute that arises in a fluid boundary layer adjacent to a membrane surface. This increase in solute concentration affects the flux across the membrane and the rejection coefficient of particles excluded from the membrane. In certain cases a gel layer is built up at the membrane surface as a result of concentration polarization. An example is found in desalination, where the transport of water creates a salt buildup close to the membrane surface.

concentration
polarization

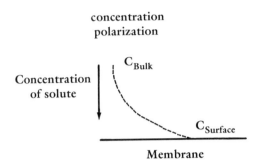

Concentration of solute

C_{Bulk}

$C_{Surface}$

Membrane

concentric draft tube fermentor \kən-'sen-trik 'draft 't(y)üb fər-'ment-ər\ *See* airlift fermentor.

conditioned medium \kən-'dish-ənd 'mēd-ē-əm\ *See* conditioning.

conditioning \kən-'dish-(ə-)niŋ\ The process whereby a cell culture medium is partially used by cells. Although thus depleted of some components, the medium also now contains some cell-derived material, in particular growth factors. This conditioned medium, either alone or supplemented with fresh medium, supports the growth of cells at low cell densities. Growth at these low cell densities cannot be achieved in unconditioned (fresh) media because of the lack of appropriate growth factors. Conditioned medium is therefore particularly useful for use in cell cloning experiments.

conductimetric sensor \kən-dək-tə-'me-trik 'sen(t)-sər\ A sensor with a transducer system that measures localized conductivity changes, produced in the proximity of a membrane-linked enzyme by the presence of enzyme substrate. Conductivity changes are proportional to substrate concentrations.

confluent growth \'kän-flü-ənt 'grōth\ The growth of bacterial cells on a solid medium such that the entire surface of the medium is covered by cells.

congener \'kän-jə-nər\ One of a family of related chemical substances, such as derivatives of a compound. Congeners are also elements belonging to the same group in the periodic table.

conidiophore \kə-'nid-ē-ə-fō(ə)r\ A specialized fungal hypha, produced by members of the *Ascomycotina* and *Deuteromycotina*, bearing conidia either at the tip or, rarely, along its length.

conidium \kə-'nid-ē-əm\ A spore formed asexually on a fungal hypha, normally at the tip or, more rarely, along its length.

conjugation \ˌkän-jə-'gā-shən\ The process whereby genetic information is transferred from one bacterial cell to another by cell-to-cell contact.

consensus sequence \kən-'sen(t)-səs 'sē-kwən(t)s\ The sequence that gives the most common nucleotide at each position in cases where there are a number of minor variations between nucleotide sequences that have the same function (e.g., promoters). **consortia:** Groups of microorganisms that coexist or interact when growing together (e.g., in microbial leaching). *See also* communities.

constant boiling mixture \'kän(t)-stənt 'bȯi(ə)l-iŋ 'miks-chər\ *See* azeotropic mixture.

constitutive \'kän-stə-ˌt(y)üt-iv\ Of or relating to the continuous expression of a gene. Constitutive genes are continuously expressed, being neither inducible nor repressible. Many of the "housekeeping" genes, such as those for the enzymes involved in the Krebs cycle, are constitutive.

contact angle \'kän-ˌtakt 'aŋ-gəl\ The angle between a solid and fluid interface used to determine interfacial tension. It may also be used to assess the surface properties of cells (e.g., bacteria or blood cells).

contact inhibition \'kän-ˌtakt ˌin-(h)ə-'bish-ən\ The phenomenon whereby healthy eukaryotic cells growing in culture stop dividing and become immobilized after they have formed a contiguous monolayer covering the surface on which they grow.

contact stabilization \'kän-ˌtakt ˌstā-bə-lə-'zā-shən\ *See* activated sludge process.

containment \kən-'tān-mənt\ The prevention of the distribution of an organism, compound, or material that is normally hazardous outside a defined boundary (containment facility). Codes of practice or legislation are often used to restrict the movement and define the zone in which pathogens, biological substances, radioactive compounds, and other hazardous materials can be kept and used.

containment facility \kən-'tān-mənt fə-'sil-ət-ē\ A building, room, cabinet, or vessel designed to prevent the escape of hazardous organisms or materials. The codes of practice, handling procedures, and containment structure depend on the potential danger if the material or organism were released or escaped.

contamination \kən-ˌtam-ə-'nā-shən\ The entry of an undesirable or unwanted organism or contaminant into a culture vessel.

contig \kən-'tig\ A group of cloned DNA sequences that are contiguous.

continuous cell line \kən-'tin-yə-wəs 'sel 'līn\ *See* secondary cell cultures.

continuous chromatography \kən-'tin-yə-wəs ˌkrō-mə-'täg-rə-fē\ The operation of chromatographic separations in a continous regime. Several techniques have been devised, and some examples follow:

(1) The chromatographic support is packed into the annular space between two concentric cylinders, and the carrier fluid is fed axially while the whole bed is slowly rotated. The feed solution is fed at a point on the cylinder circumference. As the feed rotates, the components in the feed are separated at the base of the cylinder at a series of points, depending on their individual R_f values.

(2) Another version uses two rotating disks placed close together, with the space filled with the support phase. In this device, the carrier fluid and feed material are fed at the center of the disks. Separation results from the rotation of the disks and radial flow of the carrier fluid. Again the separated components emerge at points on the circumference.

(3) Instead of moving the support phase, the direction of flow of the carrier fluid can be varied. Here a rectangular plate is used, and the feed mixture is fed at one corner. The carrier fluid is allowed to flow across the plate in alternate directions at right angles to one another; it carries the separated components up and across the plate in a series of steps.

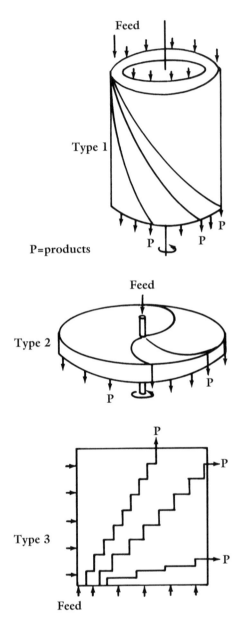

continuous culture \kən-'tin-yə-wəs 'kəl-chər\   A system that is used to maintain a cell culture at a steady growth rate. This growth rate is usually achieved in a chemostat by pumping medium that contains a growth-limiting substrate into the culture vessel at a constant rate and removing spent medium by an overflow at the same rate. The growth rate of the cells is controlled by the dilution rate

$$\frac{\text{flow rate of medium}}{\text{vessel volume}}$$

into the vessel. In turbidostat systems, an optical sensor is used to measure the cell concentration in the growth vessel. Signals from this sensor control the rate of addition of fresh medium to the vessel and maintain the cell density within defined limits. *Compare with* batch culture.

continuous-flow FAB \kən-'tin-yə-wəs-'flō 'ef 'ā 'bē\ *See* fast-atom bombardment.

continuous-flow reactor \kən-'tin-yə-wəs-'flō rē-'ak-tər\ A reactor used for the continuous processing of chemical and biological reactions, all possessing the basic facilities for continuous supply of reactants and withdrawal of products. Many designs have been produced, based on a simple tank or column with or without facilities for removal of products and recycling of unreacted material. *See also* fluidized-bed reactor, loop reactor, stirred-tank reactor.

continuous process \kən-'tin-yə-wəs 'präs-es\ A process whereby raw materials are supplied and products are removed continuously at volumetrically equal rates to maintain the process under the designed operating conditions.

continuously stirred tank reactor (CSTR) \kən-'tin-yə-wəs 'stərd 'taŋk rē-'ak-tər ('sē 'es 'tē 'är)\ One of the most commonly used reactor designs. A CSTR is usually an upright baffled cylinder with a vertical rotating shaft, with flat-bladed impellers at regular intervals. The impellers create agitation and mixing within the bioreactor and facilitate aeration. Aeration is achieved by introducing air at the base of the vessel via an open pipe or ring sparger.

controlled-pore glass \kən-'trōld-ˌpō(ə)r 'glas\ Glass beads that contain a network of extremely small tunnels and pores (25–70 Å). The pore-size distribution is very narrow, giving the most uniform porosity of any material available. Derivatization of the glass with an alkylsilane provides a suitable support matrix for affinity chromatography.

convective mass transfer \kən-'vek-tiv 'mas 'tran(t)s-fər\ Mass transfer produced by simultaneous convection and molecular diffusion. The term is used to describe mass transfer between a moving fluid and a boundary surface, which may be a solid or an immiscible liquid.

convective transfer \kən-'vek-tiv 'tran(t)s-fər\ Transfer of mass, heat, or momentum in a medium with a nonhomogeneous distribution of concentration, temperature, or velocities. The process is accompanied by a displacement of elements within the system.

conversion \kən-'vər-zhən\ *See* embryogenesis.

cooperativity \kō-'äp-ə-rət-ˌiv-ə-tē\ The interaction between substrate binding sites of an allosteric enzyme. Thus the binding of a substrate to one binding site of an enzyme affects the conformation and therefore the affinity of other binding sites. Cooperative enzymes typically give an S-shaped plot of reaction rate versus substrate concentration. *See* allosteric enzymes.

copy number \\'käp-ē 'nəm-bər\\ The number of molecules of an individual plasmid present in a single bacterial cell. Each plasmid has a characteristic value ranging from 1 to more than 50. Copy number should be considered before a gene is cloned. A high copy number is an advantage because this number will result in a higher yield of the protein encoded by the plasmid gene; plasmids with high copy numbers are less likely to be lost from bacterial cultures. However, if the protein product is toxic to the bacterial cell, then a low-copy-number plasmid is preferred. *See also* runaway plasmid.

corn steep liquor \\'kȯ(ə)rn 'stēp 'lik-ər\\ A component of many fermentation media. It is a byproduct of starch extraction from corn (maize). Although primarily used as a nitrogen source because of its amino acid content, it also contains vitamins, lactic acid, and small amounts of reducing sugars and polysaccharides.

corn syrup \\'kȯ(ə)rn 'sər-əp\\ A mixture of glucose, sugars, and dextrins. Corn syrup is produced by the hydrolysis of starch and is often referred to commercially as *glucose*, whereas pure glucose is called *dextrose.*

cos sites \\'sē 'ō 'es 'sīts\\ The "sticky" or "cohesive" ends found at each end of bacteriophage λ-DNA. They are 12 nucleotides long and are complementary so that they can base-pair with one another. They are produced by restriction endonuclease digestion of λ-DNA, which is initially synthesized as long lengths of repeat units joined together at the cos sites.

cosmid \\'cäz-mid\\ A cloning vector that is a hybrid between a phage DNA molecule and a bacterial plasmid. The cos sites of λ-DNA are inserted into a plasmid to form a molecule that can be packaged (*see* in vitro packaging) into the λ-phage head through the presence of the cos sites. These λ-phages can then be used to infect bacterial cells. When the cells are plated out, plaques are not formed because most of the DNA is missing. Transformants can be detected by their ability to grow on a selection medium because of the presence of antibiotic resistance genes encoded on the plasmid. Cosmids can be used to clone fragments of DNA up to about 47 kilobase pairs long.

Coulter counter \\'kōl-tər 'kaunt-ər\\ Trade name for a device manufactured by Coulter that automatically counts cells by measuring the changes in resistance that occur when a suspension of microorganisms in saline solution is drawn through a small glass orifice. A transient increase in resistance occurs when a cell passes through this orifice, the size of which is proportional to the cell volume. The pulses within a defined volume of solution are sorted electronically to give both concentration and size fraction of the microorganisms. Any particles with suitable dimensions may be counted by this device (e.g., bacteria, algae, and blood cells).

countercurrent chromatography \\'kaunt-ər-ˌkər-ənt ˌkrō-mə-'täg-rə-fē\\ Chromatographic methods, developed over the past 20 years, with the common feature that a solid support is not used. Two liquid phases of different densities are contained in a coil that slowly rotates about its own axis. This rotation causes the two phases to undergo a countercurrent movement across the original interface, and new interfaces are introduced every half rotation of the coil. Introduction of a solute into one of the phases allows partition between the two liquids as the segments move through the coil. The two phases are collected at either end of the coil, and the partitioned solutes are collected and analyzed. Improved performance can be obtained by reducing the internal diameter of the tubing and the helical diameter of the coil and subjecting the whole system to centrifugal motion. This improved performance allows the system to be used in both analytical and preparative modes. Various patterns of centrifugal force fields have been tested; for example, synchronous planetary motion, in which the coil rotates about its own axis while revolving about a central axis. This scheme has been shown to produce effective analytical separation, whereas the toroidal coil planetary arrangement gives efficient preparative separations with, for example, macromolecules and cells in two-phase aqueous partitioning systems.

countercurrent flow \'kaunt-ər-ˌkər-ənt 'flō\ A type of flow pattern encountered in continuous processing such that one of the process streams is contacted, with the second stream flowing toward (counter to) the first stream. This configuration conveys high efficiency to the resulting process and is commonly found in heat exchangers, where the two streams are usually separated by glass or metal plates, and in continuous extraction equipment operating in the cascade mode.

coupling efficiency of immobilized enzymes \'kəp-liŋ i-'fish-ən-sē əv im-'ō-bə-ˌlīzd 'en-ˌzīmz\ *See* effectiveness factor.

covalently closed circular (ccc) DNA
\kō-'vā-lənt-lē 'klōzd 'sər-kyə-lər ('sē 'sē 'sē) 'dē 'en 'ā\ A completely double-stranded circular DNA molecule with no nicks or discontinuities (e.g., a plasmid). Most plasmids exist in the cell as *supercoiled* molecules: the ccc DNA folds up in the conformation shown in Figure 1. However, if one of the polynucleotide strands is nicked, the torsional strain is removed and the double helix reverts to the *relaxed* (nonsupercoiled) state referred to as the open circular (oc) DNA (Figure 2).

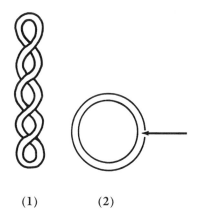

(1)   (2)

cp DNA \'sē 'pē 'dē 'en 'ā\ DNA found in chloroplasts.

Crabtree effect \'krab-ˌtrē i-'fekt\ A type of substrate inhibition seen in yeasts, e.g., *Saccharomyces cerevisiae*, whereby glucose concentration over 5% inhibits the formation of respiratory enzymes even in the presence of air.

creep flow \'krēp 'flō\ A flow condition in which the Reynolds number is less than one. Under such conditions, inertial forces can be neglected relative to the viscous forces. Also known as slow flow. *See* Reynolds number.

crenotes \\'kren-ōtz\\ One branch of the archaeal phylogenetic tree. This group comprises only thermophilic and hyperthermophilic organisms, many of which were formerly referred to as S-dependent *Archaebacteria* or thermoacidophiles. *See also* archae, euryotes.

critical dilution rate (D_{crit}) \\'krit-i-kəl də-'lü-shən 'rāt ('dē ˌkrit)\\ The dilution rate when washout of cells occurs from a culture vessel. Under these conditions the dilution rate approaches the maximum specific growth rate (m_{max}). *See* dilution rate, Monod kinetics, washout.

critical dissolved oxygen concentration (C_{crit}) \\'krit-i-kəl diz-'älvd 'äk-sə-jen ˌkän(t)-sən-'trā-shən ('sē ˌkrit)\\ The concentration of dissolved oxygen in a submerged culture when oxygen becomes the limiting substrate. Below this value, often in the range 0.1–0.5 ppm, cells may be metabolically disturbed, with subsequent lower yields of biomass or metabolites. Under normal operating conditions, the air supply to a fermentor must be adequate to maintain an oxygen concentration well above C_{crit}, especially during conditions of high oxygen demand. Monitoring the dissolved oxygen concentration is achieved with an oxygen electrode.

critical fluid \\'krit-i-kəl 'flü-əd\\ A fluid that can be defined in terms of the critical temperature (T_c) and critical pressure (P_c), as shown in the pressure–density–temperature diagram for carbon dioxide. (In this diagram, reduced units have been used in which the actual parameter is divided by the respective critical value.) The critical temperature is the temperature above which it is not possible to liquefy a gas regardless of the applied pressure, and the critical pressure may be defined as the pressure required to liquefy a gas at the critical temperature. These critical parameters meet on the diagram at the critical point (C_p).

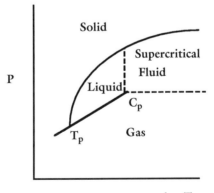

T_p is the Triple Point

C_p is the Critical Point

The region for which the temperature (T) is greater than T_c and the pressure (P) is greater than P_c is called the supercritical fluid region.

The ability of a supercritical fluid to extract a component is called the solvent power and depends on the pressure and temperature. In the fractionation region of the diagram, the solubility gradient is at a maximum and the solvent loading can be varied widely by relatively small changes in pressure and temperature. Mixtures of solutes can therefore be fractionated by variation of these properties over small ranges.

critical micelle concentration (cmc) \\'krit-i-kəl mī-'sel ˌkän(t)-sən-'trā-shən ('sē 'em 'sē)\\ The lowest concentration of a surfactant at which micelles are formed.

cross hybridization \\'kròs hī-ˌbrəd-ə-'zā-shən\\ Hybridization of a DNA probe to an imperfectly matching (less than 100% complementary) DNA molecule.

crosscurrent flow \\'kròs-kər-ənt 'flō\\ A type of flow pattern found in continuous processing where the main reactant stream (A) is contacted with a second stream (B), for example, for the purpose of mass transfer. After this initial contact, feed stream A is brought into contact with another fresh stream B. This succession of contacts with fresh B streams continues until the required process conditions are met in terms of, for example, residual material in stream A. Crosscurrent flow processing gives a number of product streams equal to the number of contacts made, with decreasing concentration of solute as the number of contacts increases. The recovery of the extracted solute then generally requires more concentration than that found for countercurrent flow processing.

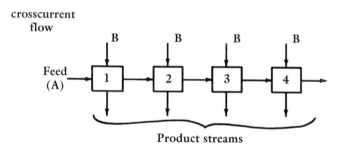

crosscurrent flow

Product streams

crossflow chromatography \\'kròs-'flō ˌkrō-mə-'täg-rə-fē\\ *See* continuous chromatography.

crossflow field-flow fractionation \\'kròs-'flō 'fē(ə)ld-'flō ˌfrak-shə-'nā-shən\\ *See* field-flow fractionation.

crossflow filtration (tangential-flow filtration) \\'kròs-'flō fil-'trā-shən (tan-'jen-chəl-'flō fil-'trā-shən)\\ An operating regime for a filtering device in which the main fluid flow is parallel to the filter, such that the fluid passes through the filter at right angles to the main flow. This regime minimizes the buildup of filter cake and also the consequential reduction in filtration rate. It allows rapid filtration without the need for filter aids or flocculants. The disadvantage is that a higher pumping energy is required to move the slurry, in addition to that required for filtration itself. *Compare with* dead-end filtration.

crossflow
filtration

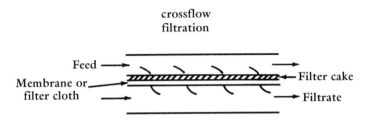

Feed ⟶

Membrane or
filter cloth ⟶

⟵ Filter cake

⟶ Filtrate

crossflow membrane \\'krȯs-'flō 'mem-ˌbrān\\ A membrane used in a crossflow regime, as in crossflow filtration.

crossover \\'krȯ-sō-vər\\ The reciprocal exchange of homologous genes between chromosomes that occurs during mitosis and meiosis. Crossover is responsible for genetic recombination. *See also* centimorgan.

crown gall disease \\'kraun 'gȯl diz-'ēz\\ *See Agrobacterium tumefaciens.*

crud \\'krəd\\ A solid or semisolid deposit formed at the interface between two immiscible liquids. Also known as insoluble interface.

crude extract \\'krüd ik-'strakt\\ A fluid derived from a cell preparation after the cells are disrupted and cell debris removed but before any further purification has been carried out. Also known as crude lysate.

crude lysate \\'krüd 'lī-ˌsāt\\ *See* crude extract.

cryogen \\'krī-ə-jən\\ A substance capable of providing low temperatures, such as a freezing mixture or liquid gases (e.g., nitrogen or air).

cryogenics \\krī-ə-'jən-iks\\ The study of the behavior of matter and materials at low temperatures. The application of low-temperature techniques.

cryopreservation \\krī-ə-ˌprez-ər-'vā-shən\\ Literally, preservation in the frozen state. In practice, this means storage over solid carbon dioxide (-79 °C), in low-temperature deep freezers (-30 °C or lower), or in liquid nitrogen (-196 °C). The term is applied to the preservation of cells, protoplasts, tissues, organs, and embryos. At low temperatures, metabolic processes and biological deterioration are minimal, and the preserved material remains genetically stable. Cryopreservation is an ideal way to store important cells safely, such as strains of microorganisms and hybridoma cell lines. *See also* cryoprotectants.

cryoprotectants \\krī-ə-prə-'tek-tənts\\ Chemicals added to suspensions of cells before cryopreservation to enhance the survival of cryopreserved cells. Cryoprotectants bring about changes in cell permeability, freezing point, and response to the stresses of freezing and thawing. Typically, 5–10% wt/vol or vol/vol dimethylsulfoxide (DMSO), glycerol, or sucrose are used.

cryptic plasmid \\'krip-tik 'plaz-mid\\ A plasmid that does not contain any defined functions other than those required for replication and transfer.

CSTR \\'sē 'es 'tē är\\ *See* continuously stirred tank reactor.

ctDNA \\'sē 'tē 'dē 'en 'ā\\ DNA found in chloroplasts.

culture \\'kəl-chər\\ A population of cells.

culture maintenance \\'kəl-chər 'mānt-ᵊn-ən(t)s\\ Any method used to store or preserve a microorganism with minimum degeneration of its genetic capabilities. These methods include subculturing on agar medium, drying, freeze-drying, and cryopreservation.

curd \\'kərd\\ A precipitated form of casein produced by the proteolytic action of rennin on milk (curdling). Curd production is encouraged by the presence of bacteria in the milk. Bacteria reduce the pH by lactic acid production to a value close to the pI of casein. The resulting curd is separated from the water component (whey), incubated and pressed, then left to ripen and produce cheese. *See* cheese, rennet, rennin.

curdlan \\'kərd-lən\\ A microbial polysaccharide produced by *Alcaligenes faecalis* var. *myxogenes* 10C3. Curdlan is an α-1,3-glucan. Heating aqueous solutions of curdlan above 54 °C results in the formation of a nonreversible resilient gel. Curdlan has considerable potential as a gelling agent in cooked foods.

CVS \\'sē 'vē 'es\\ *See* chorionic villus sampling.

cyanobacteria \\ˌsī-ə-nō-ˌbak-'tir-ē-ə\\ Photosynthetic, oxygen-evolving prokaryotes. Cyanobacteria use water as the hydrogen donor and evolve oxygen in the light..

$$H_2O + CO_2 \xrightarrow{\text{light}} \text{cell substance} + O_2$$

Also referred to as blue-green algae, although they are not true algae, which are eukaryotic. *Compare with* purple bacteria.

cyanogen bromide \\ˌsī-'an-ə-jən 'brō-ˌmīd\\ CNBr. Used in protein chemistry, it generates relatively large peptides from proteins because of its ability to cause specific cleavage at methionine residues. However, cleavage is rarely 100% at each residue. It is also used to cross-link proteins to various matrixes to form an affinity matrix. Sepharose, agarose, and cellulose can be "CNBr-activated" through the reaction of CNBr with vicinal diols, forming a reactive matrix. This matrix can then be reacted directly with ligands, such as proteins or spacer arms, that contain unprotonated primary amines.

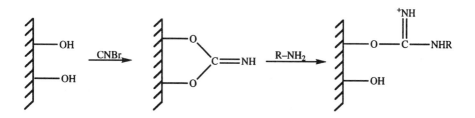

cybrids \'sī-brədz\ *See* cell cybrids.

cyclodextrins \sī-klō-'dek-strənz\ Cyclic compounds composed of six, seven, or eight α-1,4-linked anhydroglucose units: α-, β-, and γ-cyclodextrins, respectively. Uses include the enrichment of unsaturated fatty acids and the removal of organic solvent vapors from air, toxic substances from industrial wastes, and caffeine from coffee. These compounds are able to form inclusion complexes with a variety of hydrophobic substances, and this property has led to applications in the foods, cosmetics, toiletry, pharmaceutical, and pesticide industries. In aqueous solution, their macrorings are cylindrical with a hydrophilic outer surface and an apolar internal cavity large enough to accommodate guest molecules. Cyclodextrins can be produced from starch by using the enzyme cyclodextrin glycosyl transferase. An enzyme preparation that produces predominantly β-cyclodextrins has been isolated from a strain of *Bacillus circulans*.

cytochrome P-450 (also P450, P$_{450}$) \'sīd-ə-ˌkrōm 'pē-'fō(ə)r 'fift-ē \ A widely distributed monooxygenase, active in many biological hydroxylation reactions. The enzyme has considerable potential as a catalyst in the specific hydroxylation of chemicals. The mass production of the enzyme by cloning in yeast is being investigated.

cytofectins \sīt-ō-'fek-tenz\ A class of cationic lipid formulations that can be used as gene delivery vehicles. Lipofectin is a commercially available cytofectin. These liposomes condense with DNA to form complexes in which all the DNA is entrapped. Such positively charged lipid vesicles interact spontaneously with negative charges on cell surfaces, thus fusing with the cell membrane and delivering the associated polynucleotide into the cell in a manner that allows the entrapped DNA to avoid degradation in the lysosomal compartment. If conventional liposomes are used, these particles are taken up by phagocytosis but do not easily escape the lysosomal compartment of the cell, which completely degrades the liposomes and the enclosed polynucleotide.

cytokines \'sīt-ə-kīnz\ A heterogeneous group of human proteins that are active at extremely low concentrations and that regulate cell growth, differentiation, and function. Cytokines are produced by many cell types, including lymphocytes (where the cytokines they produce are called *lymphokines*), mononuclear phagocytes (producing *monokines*), endothelial cells, and fibroblasts. A number of cytokines have been cloned and are either in clinical use or undergoing clinical evaluation. These cytokines include interferon-α (for antitumor activity), interferon-β (for treatment of viral diseases such as common cold and warts), interferon-γ (for antitumor activity and treatment of arthritis and asthma), interleukins-1, -2, and -4 (for antitumor activity), interleukin-3 (for restoration of white blood cells following chemotherapy), granulocyte-colony-stimulating factor (for restoration of white blood cells), and tumor necrosis factor (TNFα) (for antitumor activity due to its ability to cause necrosis in solid tumors).

cytokinins \sīt-ō-'kī-nənz\ Plant-growth substances derived from the purine adenine. Cytokinins are characterized by their ability to promote cell division and cell and shoot differentiation in cultures of plant cells and tissues. Among the cytokinins commonly used in plant-tissue culture are benzylaminopurine (BAP), 2-isopentenyladenine, kinetin, zeatin.

cytosol   \\'sīt-ə-säl\\   The part of the cytoplasm that remains when organelles and internal membranes are removed. Cytosol is basically the soluble components of the cytoplasm. It used to be known as cell sap.

D

D_{crit} \'dē 'krit\ *See* critical dilution rate.

Dalton \'dȯl-tən\ The unit of atomic mass equal to one twelfth of the mass of an atom of carbon-12, the most common isotope of carbon. It is equal to 1.66×10^{-27} kg. Abbreviated as D or Da. The molecular masses of compounds are usually given in Daltons.

Darcy law (Hagen–Poiseuille law) \'där-sē 'lȯ (‚häg-ən-pwä-'zȯi 'lȯ)\ A law concerning the relationship between flux force, V_w, and resistance. The law, applicable to membrane processes, is expressed as

$$V_w = L_P(\delta P - \sigma/\delta\pi)$$

where σ is the Staverman reflection coefficient with a range from 0 (no solute rejection) to 1 (total rejection of solute); L_P is hydrodynamic permeability; P is the applied pressure; and π is the osmotic pressure.

DBT \'dē 'bē 'tē\ *See* dibenzothiophene.

DE \'dē 'ē\ *See* dextrose equivalent.

dead-end filtration \'ded-'end fil-'trā-shən\ An operating regime for filtration in which the fluid passes perpendicularly to the plane of the filter. This configuration is limited in total throughput because the retained material remains close to the upstream surface of the filter, with consequent decrease in the rate of filtration. *Compare with* crossflow filtration.

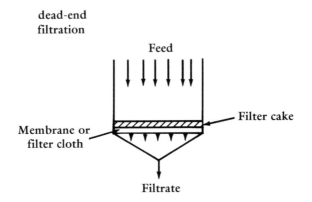

dead-end
filtration

Feed

Filter cake

Membrane or
filter cloth

Filtrate

dead volume \\'ded 'väl-yəm\\ The noncolumn volume through which solutes pass in a chromatographic system as a result of empty space above column packing and the presence of distributors and filters. The presence of excess dead volume is one cause of peak broadening.

death phase \\'deth 'fāz\\ The period of time that follows the stationary phase of a microbial culture, when the number of viable cells declines. *See* stationary phase.

debranching enzyme \\di-'bran-chiŋ 'en-zīm\\ An endoglucosidase that hydrolyzes the 1,6-α-D-glucosidic linkages found in polysaccharides such as amylopectin. Two debranching enzymes are known. These are α-dextrin-*endo*-1,6-α-glucosidase (E.C. 3.2.1.41) and isoamylase (E.C. 3.2.1.68). They are differentiated by the inability of isoamylase to attack pullulan.

decanter centrifuge \\di-'kant-ər 'sen-trə-fyüj\\ *See* solid-bowl scroll centrifuge.

decarboxylases \\dē-kär-'bäk-sə-ḷās-əz\\ *See* lyases.

deceleration phase \\dē-ṣel-ə-'rā-shən 'fāz\\ A period of growth slowdown before the stationary phase in a microbial culture. As log phase proceeds in a microbial culture, toxic byproducts tend to accumulate, and the substrate (nutrient) becomes increasingly depleted until it is exhausted. Growth of the microorganism therefore slows down and finally ceases, at which point the culture is said to have reached the stationary phase. This period of slowdown in growth is the deceleration phase.

decimorgan \\'des-ə-ṃȯrg-en\\ *See* centimorgan.

dedifferentiation \\dē-ḍif-ə-ṛen-chē-'ā-shən\\ A reversal of the process of differentiation, in which cells that have been committed to a specialized function lose this specialized character and revert to a relatively unspecialized structure and function. A good example of dedifferentiation is the formation of callus tissue in plants. Dedifferentiation is rare in mammalian tissue.

deep-jet fermentor \'dēp-'jet fər-'ment-ər\ A reactor design in which mixing of contents is achieved by injecting high-pressure air into the bottom of the reactor or by injecting a recycle stream back into the main reactor under pressure to cause agitation. It is useful for viscous media.

deep-shaft airlift fermentor \'dēp-'shaft 'a(ə)r-ˌlift fər-'ment-ər\ A fermentor that increases gaseous transfer at high pressures; high levels of dissolved oxygen result. It is used in processes with high oxygen demand, such as the activated-sludge process for waste treatment. The introduction of the fermentor has made the activated-sludge process more economical than the conventional process because it reduces residence time, has lower running costs, and takes up less space. Incoming waste, returning activated sludge, and compressed air are introduced into the downflow sections of a fermentor shaft. As the mixture rises through the upflow section, the pressure in the system decreases and gas bubbles are released, which increases biomass mixing.

deep-shaft process \'dēp-'shaft 'präs-ˌes\ *See* activated sludge process.

defined medium \di-'fīnd 'mēd-ē-əm\ A medium used for microbial or other cell culture that is of known composition and can be duplicated whenever it is required. All the components are pure chemicals at defined concentrations. Also known as synthetic medium.

degeneration \di-jen-ə-'rā-shən\ The phenomenon whereby a microbial strain decreases in productivity of an excessive metabolite after repeated transfer in culture media. It is essential that an industrial microbial strain should retain the desirable characteristics that led to its selection (e.g., antibiotic potency). The culture should be stored in such a way as to eliminate genetic change, and repeated subculture should be reduced to a minimum. In fact, 10–20 serial subcultures may lead to degeneration and a marked loss of antibiotic potency or to the production of sterile mycelia at the expense of sporing structures.

degeneration of media \di-jen-ə-'rā-shən 'əv 'mēd-ē-ə\ A loss in quality of media nutrient during sterilization. This loss may be caused by interactions between nutrient components of the medium or by degradation of heat-labile components, such as vitamins or amino acids.

degree of polymerization \di-'grē 'əv pə-ˌlim-ə-rə-'zā-shən\ A measure of the amount of polymerization of a monomeric compound. It is given as an average number of monomeric units of a molecule in a polymer. The method of averaging must be stated (i.e., number average or weight average).

dehydratases \dē-'hī-drə-ˌtās-əz\ *See* lyases.

dehydrogenases \dē-hī-'dräj-ə-ˌnās-əz\ *See* oxidoreductases.

deionized water \dē-'ī-ə-ˌnīzd 'wät-ər\ Water with a very low (zero) content of ionic species, formed by passage down a mixed-bed ion-exchange column, containing both anionic and cationic resins. Water purified by this process may, however, still contain nonionized (i.e., organic) matter. Also known as demineralized water.

del factor \\'del 'fak-tər\\ A measure of the fractional reduction in viable organism count produced by a certain heat and time regime during a sterilization process. Thus

$$\text{del factor} = \frac{N_o}{N_t}$$

where N_o is the number of viable organisms present at the start of the sterilization treatment and N_t is the number of viable organisms present after a treatment period t. Also known as nabla factor, sterilization criterion.

deletion \\di-'lē-shən\\ The loss of a segment of a chromosome or a gene.

deletion mutation \\di-'lē-shən myü-'tā-shən\\ Any event that results in the loss of a part of a DNA molecule. Deletions may range from individual nucleotides to large chromosomal segments.

delignification \\dē-ˌlig-nə-fə-'kā-shən\\ *See* lignin.

demineralized water \\dē-'min-ə-rə-ˌlīzd 'wät-ər\\ *See* deionized water.

denaturation of proteins \\dē-ˌnā-chə-'rā-shən 'əv 'prō-ˌtēnz\\ Any treatment that results in the loss of the tertiary (three-dimensional) structure of a protein. Denaturation leads to the loss of biological activity and often results in the precipitation of proteins. Commonly used protein denaturants include urea (8 M), guanidinium chloride (6 M), detergents (e.g., sodium dodecyl sulfate, SDS), heat, and phenol (used to remove protein when preparing DNA or RNA).

denaturation of DNA \\dē-ˌnā-chə-'rā-shən 'əv 'dē 'en 'ā\\ Any treatment that disrupts the hydrogen bonds in a DNA sample, causing the DNA double helix to separate into two distinct strands. This disruption is usually achieved by alkali treatment (*see* Southern blotting) or by heat treatment (*see* melting of DNA).

denitrification \\dē-ˌnī-trə-fə-'kā-shən\\ Microbial conversion of nitrate to nitrite and nitrogen under anaerobic conditions by strains of a few genera, including *Bacillus* and *Pseudomonas.* This main biological process for the production of N_2 is detrimental in ecosystems. Nitrates can be converted to nitrites by human or animal gut flora. The nitrite that is formed combines with hemoglobin in the blood to form methemoglobin. This reaction can be physiologically dangerous and may cause death. The denitrification of rivers used as a source of drinking water is of considerable interest because of the excess discharge of nitrates into rivers from agricultural fertilizers.

density-gradient centrifugation \\'den(t)-sət-ē-'grād-ē-ənt ˌsen-trə-f(y)ə-'gā-shən\\ The separation of molecules or particles (e.g., phages or subcellular particles) on the basis of their buoyant density, usually carried out in a gradient of sucrose or cesium chloride. As molecules or particles are centrifuged through an appropriate density gradient, they reach a point where the density of the gradient is equal to their own buoyant density and band at this position. *See also* ethidium bromide, ultracentrifuge.

deoxyribonuclease (DNase) \\dē-ˌäk-si-rī-bō-'n(y)ü-klē-ˌās ('dē 'en-āse)\\ *See* nuclease.

dermatophyte \\dər-'mat-ə-fīt\\ A fungus that causes skin diseases.

desalting \dē-'sȯl-tiŋ\ The removal of ionic compounds (salts) from a product solution, often by passage across a membrane (as in reverse osmosis, microfiltration, or dialysis) or by gel filtration.

detergency \di-'tər-jən-sē\ The process by which water-soluble, surface-active chemicals are able to wet the surfaces of substances and materials. *See also* anionic surfactant, cationic surfactant, nonionic surfactant.

Deuteromycotina (Fungi imperfecti) \d(y)üt-ə-ˌrō-mī-kō-'tin-ə ('fən-jī im-pər-'fek-tī)\ The group of fungi that includes those that have no sexual sporing stage and reproduce only by means of asexual spores (conidia) or by fragmentation of mycelium. It includes genera such as *Aspergillus, Botrytis, Cephalosporium, Cladosporium, Fusarium, Penicillium,* and *Trichoderma.*

dextran \'dek-stran\ A branched-chain polymer of glucose residues, linked mainly in 1,6-linkages. Occasionally branches are formed by 1,2-, 1,3-, or 1,4- linkages, depending on the species from which the polymer is derived. Dextran is produced by a wide range of bacteria but is obtained commercially by the growth of *Leuconostoc mesenteroides,* with sucrose as a substrate. Dextran is synthesized extra-cellularly by the enzyme dextransucrase (E.C. 2.4.1.5). The molecular weight of the dextran produced depends on the sucrose concentrations of the substrate used. At 70% sucrose concentration, low-molecular-weight (10,000–25,000) dextrans are produced. With low-molecular-weight dextran as a primer and 10% sucrose, high-molecular-weight (50,000–100,000) dextran represents more than 56% of the product. High-molecular-weight dextrans are used clinically as plasma substitutes. Cross-linked dextrans are used as gel filtration media and, with functional groups attached, as ion-exchange resins. For these applications raw dextran is purified, partially hydrolyzed, and fractionated by ethanol precipitation to give a product suitable for forming gel filtration media. The soluble polymer chains are cross-linked by glycerin ether bonds by the reaction of dextran with epichlorohydrin in alkali solution to yield a solidified three-dimensional gel (dextran gel). The gels are sold commercially as Sephadex or Sephacryl (Pharmacia). Dextran has also proved useful in forming aqueous two-phase systems with polyethylene glycol. *See also* dextranase.

dextran gels \'dek-stran 'jelz\ *See* dextran.

dextranase (E.C. 3.2.1.11) \'dek-strə-ˌnās\ An *endo*-α-1,6-glucanase that hydrolyzes dextran, mainly to isomaltose and isomaltotriose. It is produced commercially from strains of *Penicillium lilacinum* and *Penicillium funiculosum.* Dextrans, produced by *Leuconostoc* species in damaged sugar cane and sugar cane juices, increase viscosity. The viscosity causes problems in clarification and in the evaporation and concentration steps. Addition of dextranase overcomes this problem.

dextransucrase \ˌdek-stran-'sü-krās\ *See* dextran.

dextrins \'dek-strənz\ Mixture of short-chained polysaccharides formed during the breakdown of starch to maltose and glucose.

dextrorotatory \ˌdek-strə-'rōt-ə-ˌtōr-ē\ Able to rotate the plane of polarized light to the right, said of a chiral compound. Denoted by the prefix D-, e.g., D-glucose. The converse of levorotatory. *See also* chirality.

dextrose \'dek-ṣtrōs\ An alternative name for glucose. The term *glucose* is often used commercially to mean corn syrup (a mixture of glucose, sugars, and dextrins produced by starch hydrolysis), and pure glucose is called dextrose.

dextrose equivalent (DE) \'dek-ṣtrōs i-'kwiv(-ə)-lənt ('dē 'ē)\ A measure of the extent to which dextrose (glucose) has been produced from starch. The greater the degree of hydrolysis, the higher the DE value. Starch has a DE of 0, whereas complete conversion to dextrose gives a DE of 100. The DE of a solution is determined by measuring the amount of reducing sugars present. *See also* starch.

diafiltration \dī-ə-fil-'trā-shən\ A process whereby a solution is continuously recycled through a membrane filtration device so that the process stream containing the permeating species is removed. Concurrent with this removal of material, fresh solvent is added to the reactor. Thus within a short time the reactor contents are free of membrane-permeating species.

dialysis \dī-'al-ə-səs\ The separation of macromolecules from low-molecular-weight compounds by using a semipermeable membrane. These compounds pass across the membrane into another fluid (water) stream while the macromolecules are retained in the original solution. The process often uses a tubular membrane surrounded by a flowing or stirred stream of pure water to allow rapid diffusion from the membrane surface. Dialysis may be defined as the transfer of solute molecules across a membrane by simultaneous diffusion of solvent molecules across the membrane in the opposite direction. *See also* osmosis.

dialysis fermentor \dī-'al-ə-səs fər-'ment-ər\ A fermentor fitted with a semipermeable membrane that allows the transport of substrate and product across the membrane while retaining the biomass within the reactor.

dialysis fermentor

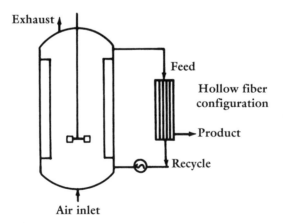

dialyzable   \'dī-ə-ˌlī-zə-bəl\ Capable of being removed through a semipermeable membrane, used to describe a solute.

dialyzate   \dī-'al-ə-ˌzāt\ Fluid flowing through a semipermeable membrane by the process of dialysis.

diamagnetism   \dī-ə-'mag-nə-ˌtiz-əm\ Magnetic behavior of substances that are repelled from areas of high magnetic field. This feature is classically explained by the presence of electron pairs in molecules. Diamagnetic materials (e.g., water and most organic compounds) are considered magnetically unresponsive. Paramagnetic and ferromagnetic effects, if present, normally outweigh the diamagnetic effects. *See also* ferromagnetism, magnetic separation techniques, paramagnetism.

diatomaceous earth   \dī-ət-ə-'mā-shəs 'ərth\ The skeletal remains of microscopic plants deposited on oceans and lake bottoms 20 million years ago. It is mined from chalklike deposits, ground to a powder, sterilized, and calcined at 800–900 °C. Used as an adsorbent. *See also* blinding, filter aids, kieselguhr.

diauxic growth   \dī-'äk-sik 'grōth\
Microbial growth when multiple log phases are observed because of the presence in the fermentation broth of multiple carbon sources. For example, when *Escherichia coli* is grown on a medium containing glucose and xylose, glucose is used first, then there follows a lag phase during which the cells rearrange their metabolic function and begin to use the xylose.

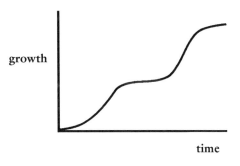

dibenzothiophene (DBT)   \dī-ˌben-zō-'thī-ə-fēn ('dē 'bē 'tē)\ The generally accepted model compound representative of the forms of organic sulfur found in coal. It is used as a substrate in enrichment culture experiments designed to identify microorganisms that can remove organic sulfur from coal.

dideoxy sequencing   \'dī-dē-äk-sē 'sē-kwən(t)s-iŋ\ A method for sequencing DNA by using dideoxynucleotides. Single-stranded (ss) DNA is required for the method and is produced by cloning in bacteriophage M13. This ss DNA is used, in four separate experiments, as a primer for the synthesis of the complementary strand of DNA. The process uses the Klenow fragment of DNA polymerase in the presence of low levels of one of the dideoxynucleotides and the four "normal" deoxynucleotides. Complementary strands are produced, varying in length and each ending with a dideoxynucleotide. The strands are analyzed by polyacrylamide gel electrophoresis. Examination of the gel pattern produced allows the DNA sequence to be read directly. Also known as the chain termination method.

dideoxynucleotides \dī-dē-ˌäk-si-'n(y)ü-clē-ə-ˌtidz\ Nucleotides that lack the 3′ hydroxyl group and therefore prevent further chain elongation when incorporated into a growing polynucleotide. *See also* dideoxy sequencing.

dielectrophoresis \dī-ə-ˌlek-trō-fə-'rē-səs\ A separation technique involving the movement of electrically charged or neutral polarizable particles in a nonuniform electrical field. It has been shown to offer two major advantages. The motion of the particle is toward the region of highest field strength, regardless of electrode polarity. Therefore alternating current may be used; this advantage reduces problems associated with the use of direct current in aqueous solution. The system can be very selective; the force applied to the particles is a function of their polarizability, which is a specific function for nonhomogeneous particles such as cells.

differential thermal analysis \dif-ə-'ren-chəl 'thər-məl ə-'nal-ə-səs\ *See* thermogravimetric analysis.

differentiation \dif-ə-ˌren-chē-'ā-shən\ A period of cellular development during which cells acquire special characteristics that fit them for the specialized tasks for which they are destined.

diffuser \dif-'yü-zər\ A device to facilitate the distribution of one fluid in another to produce a uniform mixture, used especially for gas–liquid mixing.

diffusion \dif-'yü-zhən\ A process by which molecules or ions mix as a result of thermal motion. The process is dependent on the ease of relative movement of molecules of the components and thus is most rapid in gases and slowest in solids. Ease of diffusion depends on molecular size. Thus, each species can be assigned a molecular diffusivity, which is the rate of diffusion under defined conditions.

diffusion dialysis \dif-'yü-zhən dī-'al-ə-səs\ *See* Donnan dialysis.

diffusion limitation \dif-'yü-zhən ˌlim-ə-'tā-shən\ Factors affecting the diffusion of solvents or solutes. For example, when considering an enzyme immobilized on a solid support, the term relates to factors affecting the diffusion of substrate and products to and from the enzyme. An external diffusion barrier is caused by the thin layer of unstirred solvent that surrounds any particle in a stirred solution (Nernst layer). Second, internal diffusion limitation reflects limitations on free diffusion within the particle matrix caused by the matrix itself.

diffusivity \dif-yü-'siv-ət-ē\ The proportionality coefficient of Fick's first law of diffusion, which relates molecular flux and concentration gradient and thus is a measure of the ease of diffusion under defined conditions. *See also* Fick's law of diffusion.

digoxin \dij-'äk-sən\ A cardiac glycoside used therapeutically as a heart stimulant. Digoxin is produced from the plant *Digitalis lanata* (foxglove) and by the 12β-hydroxylation of digitoxin, either by suspended cell cultures or by immobilized cells of *Digitalis lanata*.

dilatant fluid \dī-'lāt-ᵊnt 'flü-əd\ A fluid for which the viscosity increases as the shear force increases (i.e, the fluid thickens as it is sheared). The behavior of the shear force (τ) with shear rate (γ) follows a power law:

$$\tau = k\gamma^n$$

where $n > 1$. *See also* pseudoplastic fluid.

dilution cloning \də-'lü-shən 'klō-niŋ\ A method for obtaining clones of cells. Increasingly dilute cell suspensions are plated into wells (often containing nondividing support cells such as primary cultures of macrophages, which produce mitogenic factors for the added cells). Colonies derived from the most dilute samples are considered to be derived from a single cell. By repeating the cloning procedure a number of times, true clones can be established.

dilution rate (*D*) \də-'lü-shən 'rāt ('dē)\ The ratio of the flow rate of the in-going medium to the volume of the culture. Thus, in a continuous culture the flow rate of medium is related to the vessel volume by the dilution rate:

$$D = \frac{F}{V}$$

where V is the volume of the vessel (dm^3) and F is the flow rate of the medium (dm^3/h). Under steady-state conditions, the dilution rate is equal to the specific growth rate. *See* critical dilution rate.

diosgenin \dī-əs-'jen-in\ A steroid plant product isolated from the yam (*Dioscorea deltoidea*) and used in the synthesis of steroids for use as antifertility agents.

diphtheria toxin \dif-'thēr-ē-ə 'täk-sən\ A toxin protein produced by the microorganism *Corynebacterium diphtheriae*. It is toxic to mammalian cells because of its ability to interfere with the process of translation. It is one of a number of proteins being investigated as part of an antibody-toxin conjugate for use in chemotherapy. *See also* bispecific antibodies, chimeric antibody.

diploid cells \'dip-lóid 'selz\ Cells having chromosomes in pairs.

direct digital control \də-'rekt 'dij-ət-ᵊl kən-'trōl\ Control of a process by computer. This control is achieved when the computer receives information directly from a sensor and, after manipulation of the data via control loops in the computer software, returns a signal to adjust a control device.

direct injection \də-'rekt in-'jek-shən\ The injection of a sample into the carrier fluid stream, which sweeps it onto a chromatography column. The whole sample is transferred onto the column and no splitting or venting of the sample occurs. *See* split injection, splitless injection.

disaccharide \dī-'sak-ə-ˌrīd\ A compound formed by joining two sugars by an *O*-glycosidic bond. The three most abundant disaccharides are sucrose (glucose-α-(1 → 2)-fructose), lactose (galactose-β-(1 → 4)-glucose), and maltose (glucose–α-(1 → 4)-glucose). Sucrose is common table sugar and is hydrolyzed to glucose and fructose by invertase (sucrase). Lactose, the disaccharide of milk, is hydrolyzed to its component sugars by lactase in humans and by β-galactosidase in bacteria. Maltose is formed by the hydrolysis of starch and is hydrolyzed to glucose by maltase.

disk centrifuge \'disk 'sen-trə-ˌfyüj\ A centrifuge commonly used to clarify solutions. A typical design consists of a stack of perforated disks (actually truncated cones) spaced 5–12 mm apart, with a half angle of the disk with the vertical of 35–50°. This stack is surrounded by a bowl to contain the solids. The feed enters the bowl near the base and rises up through the stack of disks. The purpose of the disks is to reduce the sedimentation distance because particles have to travel only a short distance before striking the underside of a disk. Once there, they are removed from the liquid and flow down the disk until they are deposited on the walls of the bowl. Liquid is discharged from the top of the centrifuge via a discharge pipe. Bowl diameter sizes range from 10 to 75 cm, with a centrifugal force of up to 10,000 *g*. Efficiency of separation is about the same as that of a tubular centrifuge, in spite of the lower *g* force. Fluid flows in the range of 20–900 dm³ per minute. In addition to their use for solution clarification, they may also be used for liquid–liquid separation and emulsion breaking.

disk centrifuge

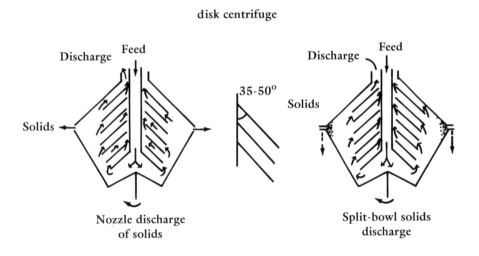

Nozzle discharge of solids

Split-bowl solids discharge

disk centrifuge with solids discharge \'disk 'sen-trə-ˌfyüj 'with 'säl-əds 'dis(h)-chärj\ Disk centrifuges that allow the continuous or intermittent discharge of solids from the centrifuge bowl. *See also* intermittent-discharge centrifuge, nozzle-discharge centrifuge.

Dispase \'dis-pāz\ Trade name for a commercially available enzyme preparation used to break down connective tissue and thus release viable cells for cell culture.

displacement chromatography \dis-'plā-smənt ˌkrō-mə-'täg-rə-fē\ A chromatographic procedure in which the sample components are fed into the system as a discrete slug of solution, adsorbed on the column, and displaced from the column by a compound (the displacer) in the mobile (elution) phase. The displacer is more strongly retained by the stationary phase. Advantages of the technique include the focusing of minor components into narrow bands and high resolution and the simplicity of a stepwise elution process. *See also* gradient elution, stepwise elution.

displacer \dis-'plā-sər\ A compound added to a mobile chromatographic phase to displace solutes retained on a column by virtue of its greater affinity for the stationary column phase.

dissociation constant (K_d) \dis-ˌō-sē-'ā-shən 'kän(t)-stənt ('kā 'dē)\ A measure of the extent of a reversible association between two molecular species at equilibrium. In a chemical equilibrium of the form A + B → AB, at equilibrium the concentrations of the reactants are related such that

$$\frac{[A][B]}{[AB]}$$

is a constant. This number is the dissociation constant, K_d. The smaller the value of K_d, the greater the association between the two molecules because [AB] is large and both [A] and [B] are low. For example, lectin–carbohydrate interactions have K_d values of about 10^{-4} M; antigen–antibody interactions have K_d values of 10^{-8}–10^{-10} M, and the avidin–biotin interaction has a K_d of 10^{-15} M. The reciprocal of association constant.

dissolved oxygen concentration \diz-'älvd 'äk-si-jen ˌkän-sen-'trā-shən\ A measure of the concentration of dissolved oxygen in a medium (mmol/dm^3). It is usually determined in a fermentation broth with sterilizable galvanic or polarographic electrodes. *See also* critical dissolved oxygen concentration.

dissolved oxygen electrode \diz-'älvd 'äk-si-jen i-'lek-trōd\ A device for measuring oxygen in solution. Various types are available, but the majority rely on a sensor consisting of a semipermeable membrane through which oxygen diffuses at a rate proportional to the partial pressure of oxygen outside the membrane. Once inside the sensor, the oxygen is reduced at a cathode to provide a current that is proportional to the partial pressure of oxygen, providing that the oxygen diffusion is the limiting step. The partial pressure (or activity) of oxygen is measured, not the concentration. The thin oxygen-permeable membrane protects the electrodes from both chemical and biological contamination. *See also* Phillips and Johnson tube.

distillation column \ˌdis-tə-'lā-shən 'käl-əm\ A device to facilitate the separation of components as a function of their relative volatilities or boiling points. The device usually consists of a cylindrical column with various forms of packing or coalescence devices to aid the establishment of equilibrium between the vapor and liquid–condensate phases within the column. *See also* bubble-cap column, packed column, plate column, theoretical plate.

distributed digital control \dis-'trib-yüt-əd 'dij-ət-ᵊl kən-'trōl\ A type of control for a process in which each unit operation is individually controlled by its own computer.

distribution coefficient \dis-trə-'byü-shən kō-ə-'fish-ənt\ The ratio of the total analytical concentration of the substance in one phase to its total analytical concentration in the other. Under defined conditions, the distribution coefficient describes the distribution of a substance between two phases in a heterogeneous system of two (liquid) phases in equilibrium. This description distinguishes the distribution coefficient from the partition coefficient, which requires a single identical form of the solute in the two phases. For example, in an oil–water system, the distribution of a solute A between these solvents is defined by

$$D = \frac{C_A(\text{oil})}{C_A(\text{water})}$$

Compare with distribution constant, partition coefficient.

distribution constant \dis-trə-'byü-shən 'kän(t)-stənt\ The same as distribution co-efficient, with concentrations replaced by activities. *See also* distribution coefficient, partition constant.

λ-DNA \'lam-də 'dē 'en 'ā \ A single linear DNA molecule (49 kilobase pairs) found in the head of λ-bacteriophage. The molecule has sticky ends (cos sites) of 12 nucleotides at each end, which are complementary. This structure allows the DNA to circularize by complementary base-pairing, which is a prerequisite for replication and for insertion into the bacterial genome when λ-phage infects bacteria.

DNA duplex \'dē 'en 'ā 'd(y)ü-pleks\ The normal double-stranded form of DNA.

DNA fingerprinting \'dē 'en 'ā 'fiŋ-gər-'print-iŋ\ The generation of a complex pattern of DNA fragments from the DNA of an individual, the pattern being unique for any given individual. Only identical twins give identical DNA fingerprints, although related individuals can be identified because of similarities in their DNA profiles. DNA fingerprinting has therefore proved useful in paternity, immigration, and forensic cases, as well as having applications in transplantation biology and gene linkage studies. DNA fingerprinting is based on the observation that the human genome contains a large number of tandemly repeated short sequence units. This tandem repeat is referred to as a *minisatellite,* and similar regions are *hypervariable* (the number of tandem repeats is variable in different regions of the chromosome). DNA is isolated, cleaved with a specific restriction enzyme, and separated by electrophoresis; then a Southern blot is hybridized under low-stringency conditions with a probe consisting of the core repeat; a complex ladder of DNA fragments is detected. The restriction fragment length polymorphism detected by the probe constitutes a DNA fingerprint for an individual.

DNA ligase \'dē 'en 'ā 'lī-ˌgās\ An enzyme that has the physiological role of catalyzing the synthesis of a phosphodiester bond between the 3′ hydroxyl group and the 5′ phosphoryl group at a single-strand break in double-stranded DNA. Because such nicks are generated in important processes such as DNA replication and DNA repair, DNA ligases are found in all living cells. However, two forms of the enzyme have found particular use in genetic engineering. These forms are DNA ligase (MW 74,000) from *Escherichia coli*, which is NAD-dependent, and T4 ligase (MW 68,000) from bacteriophage T4, which is ATP-dependent and also requires magnesium ions. Both enzymes are used extensively for covalently joining restriction fragments that have been joined by the base-pairing of their mutually cohesive sticky ends. The T4 enzyme, purified from T4-infected *E. coli*, is used most because it is easier to prepare. The T4 enzyme has the additional advantage that at high concentrations it can join DNA fragments that have blunt ends. In this case the molecules to be joined are not initially held together by hydrogen bonds between mutually cohesive termini (sticky ends). The *E. coli* enzyme cannot join DNA fragments that have blunt ends. The T4 enzyme is now prepared by cloning its gene in phage vectors.

DNA polymerase I (DNA pol. I) (E.C. 2.7.7.7) \'dē 'en 'ā 'päl-ə-mə-ˌrās 'wən ('dē 'en 'ā 'päl 'wən)\ An enzyme isolated from *Escherichia coli* that synthesizes double-stranded DNA by using single-stranded DNA as a template. A primer is required. The enzyme also contains both 5′ → 3′ and 3′ → 5′ exonucleolytic activities. It is used to radiolabel DNA by nick translation. *See also* Klenow fragment.

DNA polymerases \'dē 'en 'ā 'päl-ə-mə-ˌrās-əz\ Enzymes that synthesize a new strand of DNA complementary to an existing DNA or RNA template. *See* DNA polymerase I, Klenow fragment, *Micrococcus luteus* polymerase, reverse transcriptase, *Taq* polymerase, T4-DNA polymerase.

DNA sequencing \'dē 'en 'ā 'sē-kwən(t)s-iŋ\ Determination of the order of nucleotides in a DNA sample. *See also* dideoxy sequencing, Maxam–Gilbert method.

DO₂ \'dē 'ō 'tü\ *See* dissolved oxygen concentration.

Donnan dialysis \'dän-ən dī-'al-ə-səs\ Dialysis in which two solutions of different concentrations are separated by an ion-exchange membrane, named for F. G. Donnan (1870–1956). Ions of the appropriate sign migrate across the membrane from the higher concentration to the lower. This migration causes the reverse movement of the other ions to maintain electroneutrality. Migration occurs until the relative concentrations of ions on both sides of the membrane are constant. Also known as diffusion dialysis.

Donnan dialysis

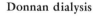
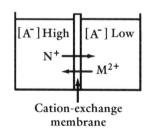

Cation-exchange
membrane

Donnan membrane equilibrium \'dō-nən 'mem-ˌbrān ˌē-kwə-'lib-rē-əm\ A variation of concentration. When two solutions, a polyelectrolyte and a salt, are separated by a membrane such that the ions in the salt solution are able to diffuse across the membrane but the polyelectrolyte is contained, then the distribution of ions will be influenced by the presence of the polyelectrolyte. Hence their concentrations will be unequal on both sides of the membrane. In turn, it generates a potential difference across the membrane, the Donnan membrane potential. Also known as Gibbs–Donnan equilibrium.

Donnan membrane potential \\'dō-nən 'mem-ˌbrān pə-'ten-chəl\\ *See* Donnan equilibrium.

Dorn effect \\'dorn i-'fekt\\ *See* sedimentation potential.

dot–blot \\'dät-'blät\\ A method for rapidly screening a large number of DNA or RNA samples to detect and quantify the presence of a particular base sequence. The aqueous samples are applied to a nitrocellulose filter, either by dotting them on with a pipet or by adsorbing them onto the paper by means of suction through a custom-built apparatus. This dot–blot apparatus comprises a series of wells (10×10) drilled in a plastic block. Directly beneath the block is a sheet of nitrocellulose paper, which in turn is connected to a vacuum source. Individual samples are placed in separate wells. When a mild vacuum is applied, the liquid in each well passes through the membrane, leaving the DNA bound as individual dots on the nitrocellulose sheet. The filter is then incubated in a solution containing a radiolabeled probe for the required sequence. It is washed, dried, and then subjected to autoradiography to detect samples that have hybridized to the probe. The extent of blackening of the film can be taken as a semiquantitative measure of the amount of sample present.

double digestion \\'dəb-əl dī-'jes(h)-chən\\ The cleavage of a given DNA molecule with two different restriction endonucleases, either consecutively or together.

downflow \\'daún-'flō\\ The passage of a mobile phase under gravity. Downflow is the converse of upflow as a regime for the operation of columns in, for example, chromatography, ion exchange, or extraction processes.

downstream processing \\'daún-'strēm 'präs-es-iŋ\\ The unit operations required to separate, concentrate, and purify a product arising from a reactor or fermentor. The converse of upstream processing.

driselase \\'dris-ə-ˌlāz\\ A mixture of cellulase and a pectinase derived from a basidiomycete, used to degrade cell walls and produce protoplasts.

drug targeting \\'drəg 'tär-gət-iŋ\\ Any strategy whereby a drug is selectively delivered to a particular target cell-type or tissue, e.g., directing a cytotoxic agent to a tumor. For examples of such strategies, *see* bispecific antibodies, chimeric antibody, immunotoxin, liposomes.

drum filter \\'drəm 'fil-tər\\ *See* rotary vacuum filter.

Dunaliella bardawil \\ˌdün-ə-lē-'el-ə 'bar-də-wēl\\ A unicellular green alga with potential as a source of glycerol. The alga grows in sunlight and in high saline concentrations, accumulating intracellular free glycerol to counterbalance the salt concentration of its external environment. It can produce up to 85% of its dry weight as glycerol.

dye exclusion tests \\'dī iks-'klü-zhən 'tests\\ Tests used to determine the viability of cells. One drop of stain (e.g., trypan blue) is mixed with one drop of cell suspension.

After 5 minutes the cells are washed with culture medium and observed on a light microscope with white-light illumination. Living cells remain unstained, but dead cells accumulate the stain. *See also* fluorescein diacetate.

dye–ligand chromatography \'dī-'lig-ənd ̩krō-mə-'täg-rə-fē\ Affinity chromatography using dyes as the affinity ligand. The affinity ligands most widely used for protein purification are the triazine-based reactive textile dyes. These dyes can bind representatives of virtually every protein class. Reactive dyes are polyaromatic molecules that consist of a chromophore (azophthalocyanin or anthraquinone) linked to a reactive group (usually a mono- or dichlorotriazine ring). They are made water-soluble by the incorporation of sulfonic or carboxylic acid groups. The Procion dyes (of ICI Chemical and Polymers) are among the most common groups used for this purpose.

dynamic light scattering \dī-'nam-ik 'līt 'skat-ər-iŋ\ *See* photocorrelation spectroscopy.

E

E-isomers \\'ē 'ī-sə-ˌmərz\\ A configuration of geometric isomers in which similar groups are on opposite sides of a double bond. This term is preferred by IUPAC; it is derived from *entgegen*, German for opposite. *Contrast with* Z-isomers. Also known as trans isomers.

E number \\'ē 'nəm-bər\\ A reference number for a food additive that has been approved for use at the European level, e.g., riboflavin is E101. The use of colorings in food in the European Community (EC) is controlled by EC Directive 2645/62.

Eadie–Hofstee plot \\'ē-dē 'hȯf-stā 'plät\\ A mathematical treatment of enzyme kinetic data used in the determination of maximum reaction rate and the Michaelis constant. Rearrangement of the Michaelis–Menton equation gives:

$$V = V_{max} - \frac{K_m V}{[S]}$$

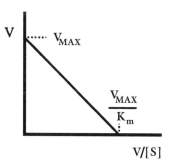

where V is the rate of reaction, $[S]$ is the substrate concentration, V_{max} is the maximum rate of reaction, and K_m is the Michaelis constant. A plot of V versus $V/[S]$ gives a straight line with a gradient of $-K_m$, a y-intercept of V_{max}, and an x-intercept of V_{max}/K_m. *See also* Hanes–Wilkinson plot, Lineweaver–Burke plot, Michaelis–Menton equation.

EBV \\'ē 'bē 'vē\\ *See* Epstein–Barr virus.

E.C. number \\'ē 'sē 'nəm-bər\\ *See* Enzyme Commission number.

ecosystem \\'ē-kō-ˌsis-təm\\ A community of interdependent organisms together with the environment they inhabit.

ectomycorrhiza \\ˌek-tō-ˌmī-kə-'rī-zə\\ *See* mycorrhiza.

eddy diffusion \\'ed-ē dif-'yü-zhən\\ Diffusion that results from eddy formation in a fluid and thus produces rapid mixing and high diffusivities. An eddy is a macroscopic element of a fluid possessing rotational movement and is a feature of turbulent flow.

eddy length \\'ed-ē 'leŋ(k)th\\ *See* turbulent shear.

Edman degradation \\'ed-mən ˌdeg-rə-'dā-shən\\ A series of chemical reactions (together referred to as one cycle of the Edman degradation) that selectively removes amino acids one at a time from the N-terminus of a peptide or protein. The released amino acids are recovered as a phenylthiohydantoin (PTH) derivative, which can be identified by high-performance liquid chromatography. Thus the sequence of amino acids from the N-terminus can be identified. Since the introduction of the method by Per Edman in 1953, the technique has remained the major method for protein sequence determination, although considerable success has also been achieved in recent years using mass spectrometry to obtain sequence data. Attempts to increase both the efficiency and the sensitivity of the method have resulted in a number of developments in recent years. Automated machines are available for carrying out the Edman degradation. The most recent developments are the gas phase protein sequencer, in which the reagents are supplied in the gas phase to the protein immobilized on a membrane, and the pulsed liquid protein sequencer, in which reagents are provided as a minute pulse in liquid form. About 30–40 residues can be determined using 50–100 pmol of protein, and up to 10 residues can be determined on as little as 5–10 pmol. The N-terminal sequencing of protein that has been separated by SDS polyacrylamide gel electrophoresis and blotted onto a PVDF membrane is routine. *See also* polyvinylidenedifluoride membrane.

EDTA (ethylenediaminetetraacetic acid) \\'ē 'dē 'tē 'ā (ˌeth-ə-lēn-ˌdī-ə-mīn-ˌte-trə-ˌə-sēd-ik 'as-əd)\\ A multidentate chelating agent used extensively as a titrant in analytical chemistry for determining metals in solution and as a means of solubilizing metals in the preparation of culture media.

effective kinetics of an enzyme \\i-'fek-tiv kə-'net-iks 'əv 'an 'en-ˌzīm\\ *See* intrinsic kinetics.

effective yield \\i-'fek-tiv 'yē(ə)ld\\ The quantity of cells produced from unit weight of substrate in a continuous-culture system.

effectiveness factor \\i-'fek-tiv-nes 'fak-tər\\ For immobilizing cells or enzymes, the activity of the immobilized enzyme (or cells) divided by the activity of an equivalent quantity of free enzyme (or cells). The activity of an immobilized enzyme is defined as the number of grams of product formed per gram of immobilized enzyme (or cells) per hour, or as the number of grams of product formed per milliliter of reactor volume per unit of time. Also known as activity yield, coupling efficiency.

effector \\i-'fek-tər\\ A small molecule that can either increase (activate) or decrease (inhibit) the activity of a protein by binding to the protein at a regulatory site. *See* allosteric enzymes, allostery.

effluent \\'ef-lü-ənt\\ The liquid and solid waste materials of a manufacturing process. Typical wastes can include unconsumed inorganic and organic media components; microbial cells; suspended solids; filter aids; waste wash water from cleansing operations; and water contaminated with solvents, acids, alkalis, and human sewage. Emissions are governed by legislation. *See also* biochemical oxygen demand, chemical oxygen demand.

EGF \\'ē 'jē 'ef\\ *See* epidermal growth factor.

elasticity \\i-ˌlas-'tis-ət-ē\\ One of the fundamental parameters of rheology. Elasticity is the energy retained in a fluid such that some flow reversal occurs when shear stress, the force causing fluid flow, is removed.

electrical field-flow fractionation \\i-'lek-tri-kəl 'fē(ə)ld-'flō ˌfrak-shə-'nā-shən\\ *See* field-flow fractionation.

electroblotting \\i-'lek-trō-ˌblät-tiŋ\\ The electrophoretic transfer of separated protein molecules from a gel matrix onto a nitrocellulose sheet, where they bind. *See also* protein blotting.

electrochemical reactors \\i-ˌlek-trō-'kem-i-kəl rē-'ak-tərz\\ *See* biochemical fuel cells.

electrochemical sensor \\i-ˌlek-trō-'kem-i-kəl 'sen(t)-sər\\ A monitoring device based on the production of a change in an electrical property (such as variation in current, potential, or resistance) proportional to changes in the analyte concentration.

electrochromatography \\i-ˌlek-trō-ˌkrō-mə-'täg-rə-fē\\ A separation technique in which a high electrical potential is applied along the axis of a liquid chromatographic column. The effect of the potential is to vary the retention times of charged molecules, thus allowing their separation from uncharged species. In this technique the separation of solutes depends on two processes, partition with the stationary phase and electromobility.

    A problem with all electrochromatographic methods is that the heat generated in the system tends to disperse the sample bands, so fine-bore capillaries are used to dissipate this heat rapidly. *See* capillary electrophoresis.

electrodialysis \\i-ˌlek-trō-dī-'al-ə-səs\\ The application of a potential difference across a membrane to facilitate the transport of ions. The membranes contain ion-exchange groups and carry a fixed electrical charge. By suitable choice of the membrane, either cations or anions can be transported. The technique is used for desalting and concentration of proteins from fermentation broths, as well as their fractionation and purification. Other applications include desalting of cheese whey and recovery of reagents used in downstream processing.

electroendosmosis \\i-ˌlek-trō-en-däs-'mō-səs\\ The movement of electrophoresis buffer through a supporting medium (e.g., paper, agarose) during electrophoresis. Ideally, all support media used for electrophoresis should have zero charge. In practice, even the best media contain a small number of attached charged groups (e.g., agarose contains some sulfate groups). Such charges are balanced by the presence of counterions, and in an electric field these cations and associated water molecules migrate toward the cathode. This solvent flow, termed *electroendosmosis*, tends to carry sample molecules with it. Its movement, in the opposite direction to that of most biological molecules, is directly related to the content of ionic groups in the medium.

Electroendosmosis occurs in capillary electrophoresis. The fused silica capillary acquires a surface charge in fluids above pH 3, which attracts cations in the electrolyte to form an electrical double layer. The double layer, under the high applied voltage, tends to move the bulk fluid in the direction of the negative electrode. Also known as electroosmosis. *See also* capillary electrophoresis.

electrofusion \i-ˌlek-trō-'fyü-zhən\ The fusing of cells (e.g., in the production of hybridomas) or protoplasts by the use of an electric field. A high-frequency alternating field (~1 MHz) is applied. This field induces dipoles on the cells, which cause the cells to align. Cells are then fused by the application of a brief (μs) direct-current pulse (~1000 V/cm). The pulse induces structural and permeability changes in cell membranes (micropore formation), which allow intermixing of the cellular contents and fusion of adjacent cells.

electroinsertion \i-ˌlek-trō-in-'sər-shən\ *See* electroporesis.

electron-impact ionization \i-ˌlek-trōn im-'pakt ī-ə-nə-'zā-shən\ The formation of positive ions by electron bombardment of molecules. It is the most common form of ion source in mass spectrometry. The electrons come from an incandescent filament and cover the energy range 10–100 eV, most commonly 70 eV, and bombard the molecules in the sample in the ionization chamber of the mass spectrometer.

electroosmosis \i-ˌlek-trō-äz-'mō-səs\ *See* electroendosmosis.

electroosmotically driven electrochromatography \i-ˌlek-trō-äz-'mät-i-kə-lē 'driv-ən i-ˌlek-trō-ˌkrō-mə-'täg-rə-fē\ A technique for the separation of compounds by the application of electroendosmosis to liquid chromatography without a pressurized flow. The chromatographic separation of solutes in the sample proceeds by a combination of electroosmosis, electrophoretic mobility, and partition. The introduction of electroendosmosis improves the degree of separation over that obtainable by capillary zone electrophoresis. *See* capillary zone electrophoresis, electrochromatography.

electropermeabilization \i-ˌlek-trō-ˌpər-mē-ə-ˌbil-ə-'zā-shən\ *See* electroporesis.

electropherogram \i-ˌlek-trə-'fir-ə-gram\ A plot of solute concentration against time and distance generated by an electrochromatographic technique.

electrophoresis \i-ˌlek-trə-fə-'rē-səs\ A separation process whereby charged particles (ions) may be separated by their movement through a stationary fluid or gel under the influence of an electrical current. Ions can be separated because the rate of migration depends on size and charge of the ion. Because the extent of dissociation of a substance into ions may be pH-dependent, change of pH can provide an additional means of separation. *See also* capillary electrophoresis, isoelectric focusing, pulse-field gel electrophoresis, SDS polyacrylamide gel electrophoresis.

electroporation \i-ˌlek-trō-pō(ə)r-'ā-shən\ A method for introducing large molecules into cells, particularly to transform cells. Cells or protoplasts in solution are subjected to a brief (~1 ms) direct-current pulse (~1000 V/cm), which induces changes in the structure and permeability of the cell or protoplast membrane (formation of micropores). If carried out in the presence of molecules such as DNA, this pulse allows passage of DNA into the cell (transformation). Also known as electroinsertion, electropermeabilization, electroporesis, electrotransformation.

electroporesis \i-ˌlek-trō-pō-'rə-sis\ *See* electroporation.

electrospray mass spectrometry (ESMS) \i-'lek-trō-ˌsprā 'mas spek-'träm-ə-trē ('ē 'es 'em 'es)\ A mass spectrometric method that provides the ability to measure protein mass accurately and at high resolution up to 100 kDa. This method is particularly useful for detecting and identifying side-chain modifications in proteins and for confirming primary structure data. The electrospray method is an exceptionally mild process in which ionization takes place at temperatures and pressures close to ambient. The main feature of this method is the production of multiply charged, intact protein ions. Up to 100 charges can be produced on a large molecule. This feature greatly reduces the mass-to-charge ratio (m/e) for the molecule and allows molecular mass measurements to be made on relatively low-cost quadrupole instruments with a nominal m/e range of 100–4000 Da. Before the introduction of this method, intact proteins could not be analyzed by mass spectrometry. *See also* mass spectrometer.

electrotransformation \i-ˌlek-trō-tran(t)s-fər-'mā-shən\ *See* electroporesis.

ELISA \ə-'lī-sə\ *See* enzyme-linked immunosorbent assay.

eluant \'el-yə-wənt\ A solvent used for the elution of species from an adsorbent.

elution \ē-'lü-shən\ Removal of adsorbed material from an adsorbent by means of a solvent.

elution profile \ē-'lü-shən 'prō-ˌfīl\ A diagrammatic representation of the amount of substance removed from an adsorbent as a function of volume of eluant passed. Thus, the elution of a sequence of substances can be represented by a series of concentration peaks at varying volumes of eluant passed. *See also* breakthrough profile.

elution volume \ē-'lü-shən 'väl-yüm\ Volume of eluant necessary to remove a particular substance from an adsorbent.

EMBL \'ē 'em 'bē 'el\ Abbreviation for European Molecular Biology Laboratory, located in Heidelberg, Germany.

embryo culture \'em-brē-ō 'kəl-chər\ Plant regeneration by the aseptic culture of a zygote embryo. The embryo is excised from either the seed or ovule and planted on a nutrient medium. Subsequent embryo development and germination occur as they would from the seed.

embryogenesis \ˌem-brē-ō-'jen-ə-səs\ A pathway of differentiation in plants, induced in undifferentiated cell, tissue, or organ cultures by appropriate control of nutritional and hormonal conditions, that results in the formation of organized embryolike (embryoid) structures. Under appropriate cultural conditions, these organized structures can develop to form plantlets and eventually whole plants. The production of a whole plant (true leaves and roots) from a somatic embryo is referred to as conversion.

Eminase \'em-ə-ˌnās\ SmithKline Beecham trade name for anistreplase. It is a streptokinase derivative used as a thrombolytic agent. *See* streptokinase.

emulsifier \i-'məl-sə-ˌfī(-ə)r\ A substance that aids the formation and stability of an emulsion. Also known as a surfactant.

emulsion \i-'məl-shən\ A colloidal suspension of one liquid in another (e.g., oil and water). According to their relative concentrations, it is possible to form either an oil-in-water or a water-in-oil emulsion. The stability of an emulsion can be increased by the presence of surface-active agents (i.e., surfactants or emulsifiers). Emulsions are used, for example, in steroid transformations to get the steroid into the correct environment for enzymic reaction.

emulsion liquid membrane extraction \i-'məl-shən 'lik-wəd 'mem-ˌbrān ik-'strak-shən\ An extraction technique used to recover solutes from either water or oil phases. The process, in the case of recovery from an aqueous phase, relies on the formation of a stable water-in-oil emulsion, which is then distributed as droplets in the aqueous feed solution. The oil phase contains an appropriate extractant for the solute together with a surfactant to stabilize the emulsion. The aqueous strip phase within the emulsion may contain a reagent that reacts with the solute to remove it from the oil phase as in conventional liquid-liquid extraction (*see* diagram).

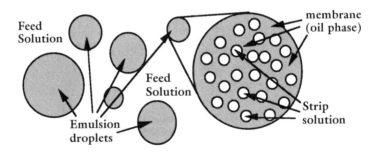

Thus, for the extraction of citric acid (H^+Cit^-) from an aqueous feed solution an appropriate extractant would be a quaternary ammonium salt ($R_4N^+OH^-$); the aqueous strip reagent could be sodium hydroxide, as shown by the following equations:

$$H^+Cit^- + [R_4N^+OH^-]_o \longrightarrow [R_4N^+Cit^-]_o + H_2O \quad \text{(extraction)}$$

$$[R_4N^+Cit^-]_o + NaOH \longrightarrow [R_4N^+OH^-]_o + H^+Cit^- \quad \text{(stripping)}$$

where the subscript o indicates the oil phase. The process can operate in either a batch or continuous mode using either mixer–settlers or column contactors. Gentle agitation of the feed and emulsion phases is necessary to avoid premature breakdown of the emulsion. After extraction is complete, the emulsion phase is separated from the aqueous feed raffinate. The emulsion is broken, usually by an electrostatic coalescer, to release the strip aqueous phase, which contains the extracted product. The oil phase containing the extractant and surfactant may be recycled to be mixed with fresh strip solution to reform an emulsion for continuous operation.

The advantages of this process over conventional liquid–liquid extraction include reduced reagent costs and a unified process combining extraction and stripping with fast kinetics resulting from the thin organic membrane. Disadvantages include a relatively complex flowsheet and problems with premature breakdown of the emulsion phase. Also known as liquid surfactant membrane extraction, surfactant liquid membrane extraction. *Compare with* supported liquid membrane extraction.

enantiomers \in-'ant-ē-ə-ˌmərz \ Isomers of a molecule that possess an asymmetric atom, such that the arrangement of groups (atoms) about this atom are different, with one isomer the mirror image of the other. *See also* chirality, optical activity.

enantioselective \in-ˌant-ē-ō-sə-'lek-tiv\ *See* stereoselective.

end-filling \'end-'fil-iŋ\ The conversion of a DNA fragment with sticky ends to one with blunt ends by the enzymatic synthesis of a strand complementary to the single-strand extension.

end product inhibition \'end 'präd-(ˌ)əkt ˌin-(h)ə-'bish-ən\ *See* feedback inhibition.

endo- \'en-dō\ Prefix meaning within or internal. For example, restriction endonucleases hydrolyze phosphodiester bonds at internal positions in a polynucleotide strand, rather than at the ends. *See* restriction endonucleases. *Contrast with* exo-.

endogenous \en-'däj-ə-nəs\ Originating within an organism or cell. *Contrast with* exogenous.

endomycorrhiza \ˌen-dō-ˌmī-kə-'rī-zə\ *See* mycorrhiza.

endonuclease \ˌen-dō-'n(y)ü-klē-ˌās\ An enzyme that cleaves phosphodiester bonds within a nucleic acid molecule. *Contrast with* exonuclease. *See also* nuclease, restriction endonucleases.

endospore \'en-də-ˌspō(ə)r\ A bacterial spore formed within the cell of species of *Bacillus*, *Clostridium*, *Desulfotomaculum*, *Sporolactobacillus*, *Sporosarcina*, and *Thermoactinomyces*. The spores may be extremely resistant to heat, UV radiation, X-rays, and chemical agents. *See also Bacillus, Clostridium.*

endothermic reaction \ˌen-də-'thər-mik rē-'ak-shən\ A chemical reaction during which energy is absorbed from the surroundings. Some input of energy to the system is required to achieve reaction. Endothermic reactions therefore have a positive enthalpy change and cannot occur spontaneously. *Compare with* exothermic reaction.

endotoxin \en-dō-'täk-sən\ The lipopolysaccharide of Gram-negative bacteria. *See* pyrogen.

enhancer \in-'han(t)-sər\ A nucleotide sequence that dramatically increases promoter efficiency, thus resulting in a large increase in expression of the gene under that promoter's control. These enhancer sequences can be located either upstream or downstream of the basic promoter structure, often substantial distances away. Their mode of action is still unclear, but they may provide sites where RNA pol. II can bind highly efficiently to DNA or they may be sites that enzymes can structurally alter to give DNA sequences in an active conformation for transcription.

enrichment culture \in-'rich-mənt 'kəl-chər\ A technique that leads to an increase in the number of a given organism relative to the numbers of other types of organism in the original inoculum in a liquid medium. The procedure involves taking a mixed culture and providing conditions either suitable for growth of the required organism or unsuitable for growth of the other organisms. This procedure is usually done by using particular substrates or including specific inhibitors. It may be necessary to subculture a number of times before the required organism can be isolated after streaking onto solid media.

ensiling \en-'sī(ə)l-iŋ\ The solid-substrate fermentation of agricultural vegetation, using indigenous microflora. It is carried out for 1–2 weeks at 25–30 °C, during which time *Lactobacillus* species become the dominant organisms. Lactic acid is produced, which inhibits the growth of putrefactive bacteria. The absence of oxygen prevents the growth of aerobic mold fungi.

enterotoxin-B \ent-ə-rō-'täk-sən 'bē\ A secreted, single-chain protein (MW 28,400) that is the causative agent of staphylococcal food poisoning. However, the compound is also mitogenic and may have uses in the treatment of cancer and in immunology.

entrapment \in-'trap-mənt\ A popular technique for immobilizing enzymes or cells. The enzyme or cell is entrapped within a polymeric mesh by carrying out the polymerization or cross-linking reaction in the presence of the enzymes or cells in question. Entrapment media that have been used successfully include polyacrylamide gels, calcium alginate gels, agar gels, and κ-carrageenan. The pores of the matrix are large enough to allow substrates, products, and nutrients to diffuse through but small enough to prevent loss of the higher-molecular-weight biological material. *See also* cell immobilization, enzyme immobilization, microencapsulation.

entrapment vector \in-'trap-mənt 'vek-tər\ A vector containing a reporter gene whose expression depends on the cis-acting regulatory sequences of an endogenous cellular gene. After integration of the vector near or within a gene, the expression of the reporter gene should reflect the usual expression of the endogenous gene. The system may be used, for example, to identify genes important in developmental processes by monitoring the pattern of reporter gene activity during embryogenesis.

ENV glycoproteins \'env glī-kō-'prō-ˌtēnz\ Glycoproteins found in the lipoprotein envelope of RNA tumor viruses (retroviruses). They are necessary for binding the virus to the surface of host cells on infection. ENV is the name given to the region of RNA in a retrovirus that codes for the glycoprotein.

enzyme \'en-ˌzīm\ A molecule, produced by living cells, that functions as a catalyst for biochemical reactions. An enzyme increases the rate of a reaction by decreasing the activation energy of the reaction, but it cannot alter the equilibrium constant for the reaction. Some enzymes require small organic molecular cofactors for their activity. It was originally thought that all enzymes were proteins (but not all proteins were enzymes). However, some RNA molecules have now been shown to have catalytic activity. Substances acted on by enzymes are known as substrates. Enzymes are highly specific in the substrates they will catalyze, often catalyzing reactions involving only one of a few closely related compounds. They can, for example, distinguish between stereoisomers. Catalytic reactions occur at a region known as the active site of the enzyme. Microbial enzymes in particular have considerable use in commercial processes. These enzymes include amylases, cellulases, glucose isomerase, invertase, pectic enzymes, restriction enzymes, and proteases. *See also* activity of an enzyme, cofactor, Enzyme Commission number, enzyme immobilization, ribozymes.

enzyme activity \\'en-ₓzīm ak-'tiv-ət-ē\\ *See* activity of an enzyme.

enzyme classification   \\'en-ₓzīm ₖklas-(ə-)fə-'kā-shən\\ *See* Enzyme Commission number.

Enzyme Commission (E.C.) number \\'en-ₓzīm kə-'mish-ən ('ē 'sē) 'nəm-bər\\ A code number, different for each enzyme, providing a systematic and logical nomenclature for enzymes. The system was introduced by the nomenclature committee of the International Union of Biochemistry. The code number, prefixed by E.C., contains four numbers separated by points. The four numbers are as follows:

(1) The first number shows to which of the six main divisions or classes the enzyme belongs. The six classes are oxidoreductases, transferases, hydrolases, lyases, isomerases, and ligases.
(2) The second figure indicates the subclass.
(3) The third figure gives the sub-subclass.
(4) The fourth figure is the serial number of the enzyme in its sub-subclass.

For example, papain is E.C. 3.4.22.2, and bromelain is E.C. 3.4.22.4. The 3 indicates that they both belong to the class of hydrolases. The sub class 4 indicates that they both act on peptide bonds. The sub-sub class 22 indicates that they are both cysteine proteases, and the fourth figure is a specific serial number identifying these two enzymes within the same sub-subclass. *See also* hydrolases, isomerases, ligases, lyases, oxidoreductases, transferases.

enzyme electrode \\'en-ₓzīm i-'lek-ₜtrōd\\ The combination of an immobilized enzyme with an ion-selective electrode sensor to provide a selective and sensitive method for detecting a given compound in a complex solution. The compound to be measured is a substrate of the enzyme. The enzyme is usually immobilized on the ion-selective electrode, either by chemical immobilization or by entrapment in an inert matrix. The substrate to be measured diffuses through an external membrane to the enzyme and is converted to a product that is measured at the surface of the ion-selective electrode. In some designs the change in activity (concentration) of one of the reactants is measured. The detection range of most enzyme electrodes is $10^{-2}-10^{-4}$ M. One of the first uses of enzyme electrodes to be described was that of glucose oxidase to measure glucose concentration. In this instance, oxygen uptake was measured with an oxygen electrode.

$$\text{glucose} + O_2 + H_2O \xrightleftharpoons{\text{glucose oxidase}} H_2O_2 + \text{gluconic acid}$$

More than 100 enzyme electrodes have now been described. They can determine a spectrum of compounds, including amino acids, steroids, sugars, alcohols, and inorganic ions. For example, penicillin levels in fermentation broths are routinely determined by using a pH probe coated with immobilized penicillinase. The pH electrode detects penicilloic acid produced by the enzyme.

$$\text{penicillin} \xrightarrow{\text{penicillinase}} \text{penicilloic acid}$$

See also biosensor.

enzyme engineering \\'en-ˌzīm ˌen-je-'ni(ə)-riŋ\\ *See* protein engineering.

enzyme immobilization \\'en-ˌzīm im-ˌō-bə-lə-'zā-shən\\ The conversion of enzymes from the free mobile state in solution to the immobilized state. The use of immobilized enzymes, rather than the free enzyme, has particular advantages for industrial processes because it allows the reuse or continuous use of the enzyme, which in many cases may be expensive to produce. Immobilization can also increase the stability of the enzyme. If the enzyme is used in free solution, it cannot be recovered and also results in contamination of the product or waste stream. Various methods have been used to immobilize enzymes. These methods include

- *entrapment*, either in cross-linked insoluble gels (e.g., acrylamide), within the microcavities of synthetic films, or microencapsulation within a semiper-meable polymer membrane;
- *covalent binding* to an insoluble carrier polymer (e.g., cellulose);
- *adsorption* onto an insoluble matrix (e.g., polystyrene or glass); and
- *copolymerization* of the enzyme with a cross-linking agent, such as glutaralde-hyde, to produce an insoluble matrix of enzyme.

See also gas flooding.

enzyme induction \\'en-ˌzīm in-'dək-shən\\ The synthesis of an enzyme or enzymes in response to an inducer molecule. In bacteria many catabolic enzymes are synthesized only when the appropriate substrate is available in the medium. In this case the substrate is the inducer molecule and initiates expression of the relevant gene. When more than one enzyme is required for a particular catabolic pathway, the different enzymes are often coordinately produced in response to one inducer. This growth is achieved by the inducer activating a group of genes, known as an operon, for the enzymes required.

enzyme-linked immunosorbent assay (ELISA) \\'en-ˌzīm-ˌliŋ(k)t ˌim-yə-nō-'sȯr-bənt 'as-ā (ə-'lī-sə)\\ A highly sensitive assay technique for detection and mea-surement of antigens or antibodies in solution. ELISA is particularly useful in immunodiagnostic applications such as serodiagnosis to detect antigens from a wide range of specific viruses, bacteria, fungi, and parasites, and to measure the presence of antibodies against these various microorganisms. It is also used to measure vaccine responses and to monitor factors involved in noninfectious diseases such as hormone levels, hematological factors, serum tumor markers, drug levels, and antibodies. The technique is particularly useful for screening hybridoma clones for the production of monoclonal antibodies. The name derives from the fact that the assay uses enzyme-linked antigens or antibodies to amplify an antigen–antibody reaction. A single enzyme-linked antibody–antigen complex can convert orders of magnitude more of colorless substrate molecules into detectable colored products, thus considerably amplifying the original antigen–antibody reaction. The antigen or antibody is adsorbed onto the surface of the well of a microtiter plate, and all the relevant reactions take place in solution inside the well.

enzyme nomenclature \\'en-ˌzīm 'nō-mən-ˌklā-chər\\ *See* Enzyme Commission number, hydrolases, isomerases, ligases, lyases, oxidoreductases, transferases.

enzyme reactor \\'en-ˌzīm rē-'ak-tər\\ *See* bioreactor.

epidermal growth factor (EGF) \\ep-ə-'dər-məl 'grōth 'fak-tər ('ē 'jē 'ef)\\ A 53-residue polypeptide of human origin that stimulates the growth of epithelial cells and may also play a role in nerve regeneration, tendon healing, and blood vessel formation (angiogenesis). It is mitogenic for a large number of cell types. The cloned product has shown considerable potential for improving wound healing.

epigenetic factors \\ep-ə-jə-'net-ik 'fak-tərz\\ Factors that influence the phenotype but do not arise in the genotype.

epiphase \\'ep-i-ˌfāz\\ The less dense phase in a two-phase system, particularly where the two phases may be similar in properties (e.g., two-phase aqueous systems). *Compare with* hypophase.

episomal \\ep-ə-'sō-məl\\ Existing as an independent, autonomously replicating, genetic element not associated with cellular chromosomes.

episomes \\'ep-ə-ˌsōmz\\ Plasmids that replicate by inserting themselves into the bacterial chromosome. They are maintained in this form through numerous cell divisions but at some stage exist as independent elements. Also known as integrative plasmids.

epitope \\'ep-ə-ˌtōp\\ The region of an antigen to which the variable region of an antibody binds. Most antigens have a large number of epitopes, and therefore a polyvalent antiserum to the antigen contains many different antibodies, each antibody capable of binding to a different epitope on the antigen.

Eppendorf \\'ep-ən-ˌdȯ(ə)rf\\ A German manufacturing company. However, the word is used in laboratory jargon to describe small ($0.5-1.5$ cm^3) plastic centrifuge tubes (Eppendorf tubes) used in microcentrifuges and to describe disposable plastic tips used on micropipets (Eppendorf tips).

Epstein–Barr virus (EBV) \\'ep-stīn 'bär 'vī-rəs ('ē 'bē 'vē)\\ A herpes virus that naturally infects B-lymphocytes and certain human epithelial cells and is the causative agent of infectious mononucleosis. The virus has a linear double-stranded genome of ~172,000 kilobase pairs. EBV is used to immortalize human B-lymphocytes. This immortalization allows, for example, the establishment of long-term culture of B-lymphocytes that secrete a specific antibody. Vectors that contain elements of the EBV genome are used to maintain cloned DNA inserts as plasmids in mammalian cells.

equivalent weight   \i-'kwiv(-ə)-lənt 'wāt\ The weight of an element or compound that will combine with or displace eight parts by weight of oxygen or one part by weight of hydrogen. It is commonly used to determine the amount of an ion that can be exchanged in ion-exchange processes because the capacity of a resin is quoted in milliequivalents or equivalents per gram. Thus, equivalent weight of sodium (Na^+) is the same as the atomic mass because it is "equivalent to" H^+. Calcium equivalent weight is half the atomic mass because Ca^{2+} is "equivalent to" $2H^+$. Equivalent weights of anions can be calculated in a similar way, e.g., equivalent weight of chloride is the same as the atomic mass of chlorine because Cl^- is "equivalent to" H^+. Sulfate (SO_4^{2-}) has an equivalent weight half of the ionic mass. The equivalent weight depends on the ionic state of a species; thus for ferric iron (Fe^{3+}) the equivalent weight is a third of atomic mass, but ferrous iron (Fe^{2+}) has an equivalent weight half of the atomic mass.

ergot alkaloids   \'ər-gət 'al-kə-lȯidz\ A group of pharmacologically active compounds, mainly produced by species of *Claviceps* while they are growing as fungal parasites of grasses and cereals. Historically, the alkaloids were obtained by extraction from sclerotia formed on rye infected with the fungi. Ergot alkaloids are now produced by submerged culture of *Claviceps* rather than by the systematic infection of rye. Ergot alkaloids have a wide range of uses, including treatment of postpartum hemorrhage, migraine, senile cerebral insufficiency, and Parkinson's disease. The biotransformation of simple ergot alkaloids has also been used to generate a further range of ergot alkaloids.

erythromycins   \i-ˌrith-rə-'mīs-ᵊnz\ A number of closely related nonpolyene macrolide antibiotics that are produced by *Streptomyces erythraeus*. Although broad-spectrum, they have strong activity against *Staphylococcus aureus* and *Streptococcus faecalis* and are used mainly to treat patients sensitive to penicillin. These antibiotics inhibit bacterial protein synthesis.

erythropoiesis   \i-ˌrith-rō-pȯi-'ē-səs\ *See* erythropoietin.

erythropoietin   \i-ˌrith-rə-'pȯi-ət-ᵊn\ A heavily glycosylated protein (MW 35,000), produced by the kidney in human adults, that stimulates red blood cell production (erythropoiesis). It therefore has potential for use in the treatment of anemia.

Escherichia coli   \ˌesh-ə-'rik-ē-ə 'kō-ˌlī\ A species of Gram-negative, facultatively anaerobic, rod-shaped bacteria. First isolated as a cause of diarrhea and named by Theodore Escherich in 1885. It is normally found in the lower intestinal tract of warm-blooded animals. Many strains may show opportunistic pathogenicity. It is used as an indicator organism for fecal contamination of water. It grows readily on simple nutrient media and has been used extensively as a model microorganism in biochemical and genetic investigations. Because a range of vectors may be used to introduce new genetic material into *E. coli*, it has been used extensively in genetic engineering studies. *See also* coliform.

essential oil \i-'sen-chəl 'ȯi(ə)l\ An oil extracted from biological material, such as plants or animals, that possesses some property of the material (e.g., smell or taste). Used in production of perfumes and flavorings (e.g., jasmine oil, musk oil, peppermint oil).

established cell line \i-'stab-lisht 'sel 'līn\ Any cell line that appears to be capable of unlimited growth through in vitro propagation.

ethanoic acid \eth-ə-'nō-ik 'as-əd\ *See* acetic acid.

ethanol (ethyl alcohol) \'eth-ə-nȯl ('eth-əl 'al-kə-ˌhȯl)\ C_2H_5OH. An alcohol with a wide range of industrial uses, mainly as a solvent, as a chemical intermediate, and as a fuel (*see* gasohol). About 40% of the world production of ethanol relies on fermentative processes; the rest is prepared chemically from ethylene gas. Production by fermentation is more prevalent in less industrialized nations. The economic viability of the fermentative production of alcohol depends on the current price of crude oil. Yeasts are the only organisms currently used for large-scale industrial ethanol production. Alcohol is produced by the fermentation of saccharine raw materials such as sugar cane juice, sugar beet juice, molasses, and sweet sorghum. Starchy raw materials such as cereal grains and starch root plants are also used but cannot be fermented directly by yeast. Such material must first be broken down to fermentable sugars by the use of enzymes or weak acids. These preliminary treatments add about 20% to the cost of the alcohol plant. Alcohol recovery, invariably by distillation, is energy-intensive, normally accounting for more than 50% of the ethanol plant energy consumption.

ethidium bromide \e-'thid-ē-əm 'brō-ˌmīd\ A planar, polycyclic molecule used to detect DNA (e.g., in gels) by staining. Ethidium bromide binds to DNA by intercalating between adjacent base pairs and can be identified by its fluorescence under UV light. Ethidium bromide is also used to separate supercoiled DNA (e.g., plasmids) from linear and open-circular DNA. Ethidium bromide binding to linear and open-circular DNA causes partial unwinding of the double helix, with a resultant decrease in the buoyant density. Supercoiled DNA, however, can bind only limited amounts of DNA because of its reduced ability to unwind, and therefore its buoyant density is decreased by a small amount. These forms of DNA can therefore be separated on the basis of their different buoyant densities by density-gradient centrifugation.

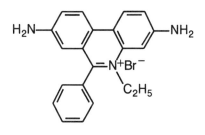

ethylene \'eth-ə-ˌlēn\ A gas, C_2H_4. Ethylene is particularly important in plant sciences because it functions as a plant-growth substance. It is involved in many auxin-induced growth responses, and it plays a part in leaf aging and the ripening of some fruits, such as tomatoes.

ethylenediaminetetraacetic acid \eth-ə-lēn-ˌdī-ə-mēn-ˌte-trə-ə-'sēd-ik 'as-əd\ *See* EDTA.

eukaryote \yü-'kär-ē-ōt\ Any organism that is made up of eukaryotic cells (e.g., animals, plants, fungi, and yeasts). *See also* eukaryotic cell.

eukaryotic cell \yü-ˌkar-ē-'ät-ik 'sel\ A cell whose genetic material is organized into a well-defined compartment (the nucleus). Plant and animal cells, fungi, and yeast are all eukaryotic. *Compare with* prokaryotic cell.

Eupergit C \'yü-pər-jit 'sē\ Trade name for oxirane acrylic beads obtained by the copolymerization of methacrylamide, methylene bis(methacrylamide glycidyl) methacrylate, and allyl glycidyl ether. It is used as a support matrix for affinity chromatography in downstream processing.

euploid \'yü-ˌplȯid\ *See* ploidy.

European Molecular Biology Laboratory (EMBL) nucleotide sequence data library \yur-ə-'pē-ən mə-'lek-yə-lər bī-'äl-ə-jē 'lab-ə-rə-ˌtōr-ē ('ē 'em 'bē 'el)\ A compilation of all published nucleotide sequences. The September 1994 issue contained 209,352 entries, with a total of 211,071,104 nucleotides. The rate at which DNA sequence data is currently being produced can be seen by comparing these data with a 1986 issue, which contained only 7,630 entries, with 7,813,214 nucleotides. Although originally based in Germany, all activities and services of the EMBL data library are now run by the European Bioinformatics Institute (EBI), which is located at Hinxton Hall, Cambridge, United Kingdom.

euryotes \yür-ē-'ōts\ One branch of the archaeal phylogenetic tree, comprising the most primitive of the thermophiles, the methanogens, and the extreme halophiles. *See also* archae, crenotes.

eutrophication \yü-ˌtrō-fə-'kā-shən\ The enrichment of a body of water, for example, by the input of organic material or by surface runoff containing nitrates and phosphate. Eutrophication may occur naturally but is often caused by pollution or effluent disposal. It leads to an increase in the growth of photosynthetic algae, which results in the depletion of the dissolved oxygen content of the water.

evaporation \i-ˌvap-ə-'rā-shən\ The removal of a liquid by conversion into its vapor by the application of heat or vacuum.

evaporator \i-'vap-ə-ˌrāt-ər\ An apparatus designed to carry out evaporation of a liquid. Evaporators are sometimes operated at a reduced pressure to reduce heating costs. Various designs are available to meet the particular requirements of the process and product, especially product heat stability. *See also* centrifugal evaporator, falling-film evaporator, flash evaporator, forced-circulation evaporator, natural-circulation evaporator, rising-film evaporator, wiped-film evaporator.

exchange capacity (ion exchange) \iks-'chānj kə-'pas-ət-ē (ī-än iks-'chānj)\ The number of ions (generally measured in milliequivalents per gram of dry resin) that can be exchanged by a particular resin. Capacity of a resin depends on the number of exchange sites and therefore on the type of resin. *See also* gel resin, macroporous resin.

exclusion chromatography \iks-'klü-zhən ˌkrō-mə-'täg-rə-fē\ *See* gel filtration.

exo- \\'ek-sō\ Prefix meaning outside or external. For example, an exonuclease catalyzes the sequential hydrolysis of nucleotides from one end of a polynucleotide strand. *See*, for example, exonuclease III. *Contrast with* endo-.

exogenous \ek-'säj-ə-nəs\ Originating outside an organism or cell. *Contrast with* endogenous.

exon \\'ek-sän\ The part of the sequence of a eukaryotic gene that codes for the final protein product. In most eukaryotic genes, noncoding regions known as introns separate the coding region into a number of distinct exons. *See also* intron.

exonuclease \ek-sō-'n(y)ü-klē-ās\ An enzyme that sequentially removes nucleotides from the ends of a nucleic acid molecule. *Contrast with* endonuclease. *See also* nuclease.

exonuclease III \ek-sō-'n(y)ü-klē-ās 'thrē\ A $3' \rightarrow 5'$ exonuclease, isolated from *Escherichia coli*, that starts from each $3'$ end of a DNA duplex, degrading each $3'$ strand and leaves single-stranded $5'$ extensions. It is used in conjunction with S_1 nuclease to create deletions in cloned DNA molecules.

exopolysaccharides \ek-sə-ˌpäl-i-'sak-ə-ˌrīdz\ Water-soluble polysaccharides (gums) produced extracellularly by a range of microorganisms. Exopolysaccharides have novel and unique physical properties and have found a wide range of industrial applications. *See also* alginate, curdlan, dextran, gellan gum, polytran, pullulan, xanthan gum.

exothermic reaction \ˌek-sō-'thər-mik rē-'ak-shən\ A chemical reaction during which heat is released into the surroundings. Thus no energy input is required, and under appropriate conditions the reaction occurs spontaneously and reaches completion. Exothermic reactions have a negative enthalpy change. Microbial growth is often exothermic, and therefore fermentors are normally fitted with cooling facilities to remove excess heat and so maintain constant temperature. *Compare with* endothermic reaction.

explant \ek-'splant\ A piece of tissue used to initiate a tissue culture.

exponential phase \ˌek-spə-'nen-chəl 'fāz\ *See* log phase.

expression vector \ik-'spresh-ən 'vek-tər\ A cloning vector designed such that an inserted gene is expressed (i.e., transcribed into mRNA, which then directs the synthesis of protein at the ribosome) in the host organism. This process involves ensuring that the cloned gene is under the control of a promoter sequence appropriate to the host organism and that the gene is read by the RNA polymerase in the correct reading frame.

extinction coefficient \ik-'stiŋ(k)-shən kō-ə-'fish-ənt\ *See* absorption coefficient.

extraction factor \ik-'strak-shən 'fak-tər\ A factor that determines the degree of separation of a solute in liquid-liquid extraction and is defined as

$$E = \frac{DV_s}{V_f}$$

where D is the distribution coefficient, and V_s and V_f are the volumes of solvent and feed, respectively. In continuous processing these volumes may be replaced by the respective flow rates.

extractive distillation \ik-'strak-tiv ‚dis-tə-'lā-shən\ The addition of a third component to a mixture to decrease the volatility of one of the components and thus allow the separation of mixtures that otherwise would not be separated by distillation. *See also* azeotropic distillation.

extractive fermentation \ik-'strak-tiv ‚fər-mən-'tā-shən\ A fermentation process by which the broth is continuously removed from the reactor; the product is extracted by a solvent or solution that is immiscible with the fermentation broth and with which the product has a partition coefficient greater than one. The extracted broth (raffinate) is then returned to the fermentor. The advantage of this system is that the product can be removed before any feedback limitation on growth rate occurs. However, it is not used frequently because there are few suitable solvent systems with necessary solubility properties that do not themselves interfere with the fermentation process. *See also* adsorption fermentation.

extremophiles \ik-'strēm-ō-‚fīlz\ Organisms that require extreme physicochemical conditions for optimum growth. *See* halophiles, psychrophiles, thermophiles.

F

F_ab \\'ef 'ā 'bē\\ *See* immunoglobulin G.

F_c \\'ef 'sē\\ *See* immunoglobulin G.

facilitated membrane transport \\fə-'sil-ə-tāt-əd 'mem-ˌbrān 'tran(t)s-pō(e)rt\\ The transport of a substance across a membrane with the aid of a carrier molecule. The carrier molecule is confined to the membrane but is able to complex with the transferring substance on the outside of the membrane. The molecule then moves across the membrane to the opposite side, where the complex dissociates and releases the transferred substance. The carrier molecule is then able to return to the feed side and repeat the process. The rate of transfer follows saturation kinetics, with rate leveling at high concentrations of transferred molecules, because there is only a finite concentration of the carrier molecules available. Facilitated transport has been demonstrated in the transport of D-glucose across membranes of vertebrate erythrocytes, with proteins termed permeases or translocases as the carriers.

facultative \\'fak-əl-ˌtāt-iv\\ Able to grow in the presence or absence of an environmental factor, used to describe an organism. *See also* facultative anaerobe.

facultative anaerobe \\'fak-əl-ˌtāt-iv 'an-ə-ˌrōb\\ An organism, usually a bacterium or fungus, that can adapt its metabolism to enable it to survive and grow in either the presence or absence of oxygen. *See also* anaerobe, obligate anaerobe.

FAD \\'ef 'ā 'dē\\ Flavine adenine dinucleotide.

FAGE \\'ef 'ā 'gē ē\\ *See* pulse-field gel electrophoresis.

falling-film evaporator \\'fȯl-iŋ-ˌfilm i-'vap-ə-ˌrāt-ər\\ An evaporator in which the feed solution falls under gravity as a film down the walls of tubular heat exchangers. The vapor–liquid separation usually takes place at the bottom of the heat exchanger. The pressure drop across the tubes is usually very small, and so the boiling point of the liquid is substantially the same as that of the vapor. Hence, the evaporator is used to concentrate heat-sensitive materials and is also useful for processing foaming liquids.

fast-atom bombardment (FAB)   \\'fast-ॻat-əm bäm-'bärd-mənt ('ef 'ā 'bē)\\ A
technique used to produce mass spectra of difficult-to-volatilize substances, such
as peptides and nucleotides, or thermally unstable compounds. The procedure
involves the bombardment of a metal plate coated with the sample by a stream
of fast atoms. The kinetic energy of these atoms is dissipated in a number of
ways, some of which lead to the volatilization and ionization of the sample. By
maintaining an appropriate electric gradient from the plate, either positive or
negative ions can be passed into the mass analyzer. Fast atoms (usually helium,
argon, or xenon) are produced by ionizing the appropriate gas and accelerating
the ions through an electric field. These fast ions are then fed into a gas chamber,
where they collide with atoms of the same gas. Charge exchange occurs as shown
in the equation. It provides a stream of fast atoms, the majority of which retain the
direction and energy of the original fast ions. Any excess fast ions and other ions
produced by this charge exchange can be removed, leaving a beam of fast atoms.

$$Xe^{+\bullet}(fast) + Xe \longrightarrow Xe(fast) + Xe^{+\bullet}$$

By passing a silica tube through the FAB probe it is possible to provide a continuous
flow of substances contained in a stream of solvent, such as the eluant from a liquid
chromatograph, into a mass spectrometer, thus obtaining a continuous FAB mass
spectrum. This process is known as continuous-flow FAB.

fast protein liquid chromatography (FPLC)   \\'fast 'prō-ॻtēn 'lik-wəd ॻkrō-mə-'täg-
rə-fē ('ef 'pē 'el 'sē)\\ A commercial variant of HPLC systems used for protein
separation and recovery.

fed-batch culture   \\'fed-ॻbach 'kəl-chər\\ A system intermediate between batch and
continuous processes. The term describes batch cultures that are fed continuously,
or sequentially, with fresh medium without the removal of culture fluid.

feedback   \\'fēd-ॻbak\\ The use of a reaction variable as a control function to influence
the reaction.

feedback control   \\'fēd-ॻbak kən-'trōl\\ A closed-loop system for controlling a continuous process, in which control is exercised by the generation of an error signal that
operates on the control functions. The error signal is a measure of the discrepancy
between the actual behavior of the system at any time and the preassigned nominal
behavior. This system can react retroactively to departures from nominal behavior
that have already occurred. *See also* anticipatory control, feedforward control.

feedback inhibition   \\'fēd-bak ॻin-(h)ə-'bish-ən\\ A control mechanism in which the
activity of an enzyme associated with an early step of a multistep pathway is inhibited
by a product (usually the end product) produced further along the pathway. For
example, in bacteria, isoleucine is synthesized from threonine in five steps, the
first step being catalyzed by threonine deaminase. When isoleucine concentrations
reach high levels, isoleucine increasingly inhibits this enzyme by binding to the
enzyme at a regulatory site that is distinct from the catalytic site. This allosteric
interaction causes a reversible conformational change in the enzyme and causes
an alteration to the structure of the active site, rendering the enzyme inactive.

When isoleucine levels drop sufficiently, the enzyme–isoleucine complex dissociates, threonine deaminase again becomes active, and more isoleucine is synthesized. Feedback inhibition is an important mechanism in any organism for preventing the overproduction of a particular compound. However, it is an unwanted effect in biotechnological processes in which one aims for maximum yield of the end product. Feedback effects are therefore usually overcome either by the continuous removal of the end product or more frequently by using mutant strains that are insensitive to feedback inhibition, e.g., the strain may have a mutant enzyme with a modified regulatory site that cannot bind the end product. Also known as end product inhibition.

feeder callus \\'fēd-ər 'kal-əs\\ *See* nurse callus.

feeder layer technique \\'fē-dər 'lā-ər tek-'nēk\\ A method for culturing single-plant protoplasts. Protoplasts in low density are spread on filter paper or film mats placed on layers of feeder-plant cells. In this way single cells receive growth factors produced by the feeder cells, as well as growth factors from the nutrient medium. Under these conditions protoplasts synthesize cell walls. Then cell division is initiated; cell division ultimately results in callus formation. The method is also used in animal-cell culture.

feedforward control \\'fēd-ˌfȯr-wərd kən-'trōl\\ A system for controlling a continuous process. It is based on the measurement of disturbing influences that precede any deviation from the controlled condition so that corrective action can be taken before any deviation occurs. The system requires a precise model of the effect of the disturbing influence on the controlled condition and of effects of control-parameter adjustments on the process. *See also* anticipatory control, feedback control.

fermentation \\fər-mən-'tā-shən\\ In biochemistry, a metabolic process whose main purpose is the conservation of free energy via ATP synthesis that uses substrate-level phosphorylation rather than oxidative phosphorylation. In microbiology and biotechnology, any process that produces a useful product by the mass culture of microorganisms. The required product may be microbial cells (biomass), enzymes, metabolites, or the modification of compounds. Because of the different usage of this word by biochemists and microbiologists, many journal editors do not allow the use of this word in research papers.

fermentation beer \\fər-mən-'tā-shən 'b(i)er\\ Fermentation broth after removal of cell debris (i.e., after filtration), used particularly in the United States.

fermenter \\fər-'ment-ər\\ An organism grown in a fermentor.

fermentor \\fər-'mənt-ər\\ A vessel in which organisms can be grown under sterile and controlled conditions that are optimum for product formation.

ferromagnetism \\fer-ō-'mag-nə-ˌtiz-əm\\ A magnetic property of a substance that arises from the cooperative effects of unpaired electrons and is indicated by a very large interaction between the substance and an applied magnetic field. The phenomenon occurs only in the solid state in materials such as metals and metal oxides. Ferromagnetic materials like iron oxides have been used as fillers for ion-exchange resins and in enzyme encapsulation to provide easy and rapid settling on application of a magnetic field. The magnetic properties of such substances can be used as aids to separation in downstream processing.

Ferrovibrio \fer-ō-'vib-rē-ō\ *See Leptospirillum ferrooxidans.*

FET \'ef 'ē 'tē\ *See* field effect transistor.

fetal calf serum \'fēt-ᵊl 'kaf 'sir-əm\ Serum collected from a calf fetus. It is an essential component of many mammalian tissue- and cell-culture media because it contains many important growth factors, attachment and spreading factors, and low-molecular-weight nutrients essential for cell growth. Because of the nature of its source and the great demand for use in tissue-culture laboratories, fetal calf serum is expensive. Because of the expense and the problem of batch variation of different samples, considerable effort is being put into the development and production of chemically defined serum-free growth media.

fetoscopy \fēt-'äs-kə-pē\ The aspiration of fetal blood from vessels at the base of the umbilical cord. Blood samples thus obtained are used for prenatal diagnosis. Like most invasive methods, this method carries a risk of inducing spontaneous abortion. *Contrast with* amniocentesis, chorionic villus sampling.

fibrin \'fī-brən\ A protein formed from fibrinogen that is polymerized to form, together with red blood cells, a blood clot. During blood clotting, the serum protein fibrinogen is cleaved by the enzyme thrombin to give fibrin and two peptides, fibrinopeptides A and B. This soluble fibrin then polymerizes to give the insoluble fibrin clot.

fibrinogen \'fī-brin-ə-jən\ *See* fibrin.

fibrinolysis \fī-brə-'näl-ə-səs\ *See* plasmin.

fibroblast growth factor (FGF) \'fīb-rə-ˌblast 'grōth 'fak-tər ('ef 'jē 'ef)\ Members of a family of single-chain human proteins of 14–18 kDa, which are mitogenic for cells of mesoderm origin such as fibroblast and vascular smooth muscle cells and keratinocytes and as such have potential for use as wound-healing agents.

ficin (EC 3.4.22.3) \'fis-ᵊn\ A thiol protease (MW 25,500) produced by water extraction of the latex of figs. It is the least used of the commercially available thiol proteases (the others are papain and bromelain), but it has been used as a chillproofing enzyme.

Fick's law of diffusion \'fiks 'lȯ 'əv dif-'yü-zhən\ A law that relates the concentration of a substance with distance from a boundary and is applicable for all types of diffusing systems (i.e., liquid, gas, solid, or membrane). The law is normally stated in terms of the rate of diffusion in a given direction (flux, N) as a function of the concentration gradient (dC/dx):

$$N = -D\left(\frac{dC}{dx}\right)$$

where D is the diffusivity of the diffusing substance, and x is the distance in the direction of diffusion.

ficoll \\'fī-kȯl\\ A nonionic synthetic polymer of sucrose, MW ~400,000. It is used particularly as a density-gradient centrifugation medium and in two-phase aqueous partitioning.

field effect transistor (FET) \\'fē(ə)ld i-'fekt tranz-'is-tər ('ef 'ē 'tē)\\ The electronic component of miniaturized electrochemical-type sensors. These sensor chips (~30-μm diameter) are similar to those used in computers, except that the metal gate that controls the transistor current is replaced by an organic or biochemical material to produce modified FET or ChemFET. The response of the sensing material to a change in environment is to drain current off the field effect transistor. This current is usually held at a fixed value while the voltage change from the gate necessary to maintain the current is monitored. *See also* ChemFET.

field-flow fractionation \\'fē(ə)ld-'flō ˌfrak-shə-'nā-shən\\ A separation technique based on the application of a force field across a continuous flat ribbonlike channel containing a fluid sample. The species dissolved or suspended in the fluid tend to become layered by the action of the force field toward one wall of the channel and separated by the velocity profile of the liquid in the channel.

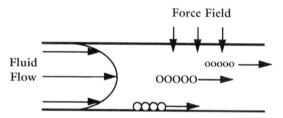

Various force fields have been used:

- *crossflow field-flow fractionation* requires the channel to have semipermeable walls so that a fluid flow is established from the channel, which exerts a pressure gradient perpendicular to the fluid flow. Separation is on the basis of variation in diffusion coefficients.
- *electrical field-flow fractionation* also requires semipermeable walls across which a potential difference is applied. Separation occurs because of variation in electrophoretic mobility and sometimes in pH of the flowing fluid.
- *sedimentation field-flow fractionation* uses the increased gravitational field generated in a centrifuge to effect separation in a channel fixed in the rotating bowl. In this process, separation is based on mass or molecular weight differences.
- *thermal gradient field-flow fractionation* uses a temperature gradient applied across a narrow (100-mm) channel. In this process, the separation is primarily dependent on the variation in diffusion coefficients of the components.

FIGE \\'ef 'ī 'gē 'ē\\ *See* pulse-field gel electrophoresis.

film boiling \\'film 'bȯi(ə)l-iŋ\\ A boiling phenomenon that results in a continuous film of vapor that periodically collapses into the bulk of the liquid.

film coefficients \'film kō-ə-'fish-ənts\ *See* heat transfer.

filter aids \'fil-tər 'ādz\ Substances, often diatomaceous earths, that contain large noncompressible particles. They are added to a fluid to increase the filtration rate. Filter aids are particularly useful with finely divided solids and slimy flocs. The filter aid should be of low bulk density to minimize the settling tendency and should also be porous and chemically inert to the system. By providing a bulky precipitate, the filter aid prevents blinding (clogging) of the filter. *See also* diatomaceous earth, fuller's earth, kieselguhr, montmorillonite.

filter press \'fil-tər 'pres\ A device, of which there are several designs, for pressurized filtration. Pressure greater than atmospheric is provided at the filter surface by a liquid pump or gas pressure. Advantages include increased filtration rate and a large filtration area per unit area of floor space. *See also* plate and frame filter.

filtration \fil-'trā-shən\ A process used in downstream processing for the removal of solids from fluids. As an alternative to centrifugation, filtration generally requires a lower capital investment. However, this advantage is often offset by higher operating costs. A number of different designs and operating regimes are available and are described under separate headings. *See also* crossflow filtration, dead-end filtration, filter press, membrane filtration, microfiltration, plate and frame filter, rotary vacuum filter, ultrafiltration.

fimbriae \'fim-brē-ē\ Short filamentous structures on a bacterial cell that are not involved in motility but have a function in assisting the cell to adhere to surfaces.

fingerprinting \'fiŋ-ger-print-iŋ\ *See* DNA fingerprinting.

flagellum \flə-'jel-əm\ A long, thin, helical structure attached to the bacterial cell at one end and free at the other. The movement of the flagellum gives the cell motility.

flame ionization detector (FID) \'flām ̩ī-ə-nə-'zā-shən di-'tek-tər ('ef 'ī 'dē)\ A common type of gas chromatographic detector that has a wide linear range and high sensitivity. It consists of a hydrogen-air flame polarized in an electrostatic field. The combustible sample components are transported by the carrier gas into the detector, where they are ignited and ionized. The ionized compounds are collected at the electrodes, and the current produced is displayed as a chromatographic peak on a chart. The detector does not respond to permanent gases or water vapor, so it is ideally suited for analysis of aqueous or atmospheric samples. The response of the FID depends on the nature of the organic substance and decreases with substitution of hydrogen atoms by halogen, nitrogen, oxygen, or sulfur. Sample collection is possible only if the column effluent is split into two streams before the detector.

flash chromatography \'flash ̩krō-mə-'täg-rə-fē\ A modification of column chromatography, analytical or preparative, which uses a low applied pressure ($5-10$ lb/in.^{2} or $34.5-69$ kPa) to the top of the column to increase throughput. The packing material for biosystems is often a modified silica to give high porosity with a high loading capacity. The technique may be used in the presence of particulate matter because of the low packing density of the column and, with a suitable choice of packing, can also be applied to ion-exchange, affinity, and gel filtration (size-exclusion) chromatography.

flash evaporator \\'flash i-'vap-ə-ˌrāt-ər\\ A device in which the feed solution is fed into a container held at a temperature above the boiling point of the solvent so that rapid evaporation of the solvent occurs. The device may also operate under reduced pressure to lower the required temperature of the boiler. Flash evaporators are used in particular for solvent recovery from downstream processing raffinates.

flash fermentation \\'flash ˌfər-mən-'tā-shən\\ A method of removing volatile products from a fermentor by circulating the broth from the fermentor to a vacuum chamber, where the volatiles are removed. It is used for ethanol production.

floc \\'fläk\\ An association of suspended solids that provides loose aggregates of larger bulk than the original particles.

flocculation \\ˌfläk-yə-'lā-shən\\ A process of agglomeration of fine particles in a fluid dispersion that facilitates the separation of the solid particles by either sedimentation or filtration. Small particles settle slowly in a fluid. However, if they can be gathered together into larger agglomerates (floc), the rate of sedimentation increases and the resulting sediment is less dense and easier to filter. Some systems readily flocculate with gentle agitation, but others require the addition of chemicals. For example, aluminium sulfate, alum, and ferrous sulfate are used in water treatment; phosphoric acid is used in single-cell protein production. *See also* Stokes law.

flotation \\flō-'tā-shən\\ A technique for separating solids from liquids whereby gas bubbles introduced into the suspension attach to the solid particles. The particles are carried to the liquid surface, from which they can be removed. The gas, often air, may be blown through the liquid as a stream of fine bubbles. In an alternative two-stage process the gas may be dissolved in the liquid under pressure and then, when the pressure is released, the gas nucleates on the suspended solids.

flotation reagent \\flō-'tā-shən rē-'ā-jənt\\ A surface-active substance that, when added in trace amounts to a suspension of solids in a liquid, promotes the attachment of gas bubbles to particular solid particles and thus aids flotation. Other reagents may also be added to increase the stability of the resulting foam and to facilitate the removal of the froth of solids.

flow cytometry \\'flō sī-'täm-ə-trē\\ A technique used to sort cells or chromosomes in a mixed population. For example, a suspension of metaphase chromosomes, released from mitotic cells of an appropriate cell line, is first stained with either one or two fluorescent dyes, depending on whether the machine is single- or dual-beam. (If cells are to be separated, the desired cell type is first labeled with a specific fluorescent dye or substrate.) As the chromosomes pass through a flow chamber they are illuminated by a laser beam and the fluorescence emission, which differs for each chromosome, is measured. The chromosome suspension continues through the flow chamber and exits from the machine as a high-speed liquid jet through a small-diameter orifice (5–10 μm). If the laser beams detect the chromosome of choice (e.g., chromosome 5), a short-voltage pulse applied to the emitting jet electrically charges the droplet containing the chromosomes. This droplet then passes between two charged plates, is deflected away from the waste-collection vessel, and is collected. In this way droplets containing chromosome 5 are collected, whereas other droplets containing unwanted chromosomes (or no chromosomes) go to waste. Two chromosome types can be sorted at once from the same input sample, with each deflected from either side of the waste-collection vessel. Flow cytometric data are usually displayed as a frequency distribution (i.e., a histogram) that shows the number of events detected as a function of either one or two fluorescence intensity values, depending on whether one or two dyes are used. This display is called a flow karyotype, by analogy with the karyotype of cytogenetics.

flow injection analysis \\'flō in-'jek-shən ə-'nal-ə-səs\\ A rapid method of continuous quantitative analysis in which the sample is injected into a continuous carrier system in small-bore tubing. The sample and solution are pumped at constant velocity. Various analytical processes (such as reagent addition, dialysis, and solvent extraction) can be carried out as required before the carrier stream and sample are passed through a flow-through cell in an appropriate detector. With automatic injection, this technique can be used to process large numbers of samples.

flow karyotype \\'flō 'kar-ē-ə-ˌtīp\\ *See* flow cytometry.

fluidization \\ˌflü-əd-ə-'zā-shən\\ A process by which solid particles may be suspended in a vessel by the upflow of a fluid (liquid or gas). The fluid flow results in an expansion of the bed of particles until, at high flow rate, the particles become entrained and are lost from the reactor.

fluidized-bed adsorption \\'flü-ə-ˌdīzd-ˌbed ad-'sorp-shən\\ A technique for the recovery of solutes from an unfiltered feed solution using a selective adsorbent in a fluidized-bed configuration. This technique can eliminate centrifugation or filtration steps in downstream processing. *See* fluidization.

fluidized-bed reactor \\'flü-ə-ˌdīzd-ˌbed rē-'ak-tər\\ A reactor in which the upward passage of a fluid maintains the solid particles (for example, immobilized enzyme particles or microbial cells on an inert support) within the fluid. This design allows good mixing of the reactants and minimizes channeling of the bed.

fluorescein diacetate (FDA) \\ˌflü(-ə)r-'es-ē-ən dī-'as-ə-ˌtāt ('ef 'dē 'ā)\\ A stain commonly used to determine cell viability. FDA is absorbed by cells, and esterase activity within the cells deacetylates the dye, which becomes fluorescent. Cells are viewed under a microscope with an ultraviolet attachment. Viewing of cells on a light microscope with white-light illumination reveals all cells, whereas ultraviolet illumination reveals only "living" cells, which fluoresce brightly. *See also* dye exclusion tests.

fluorescence \\ˌflü(-ə)r-'es-ᵊnts\\ A molecular process whereby radiation is emitted from a substance after stimulation by absorption of radiation. The absorption of radiation promotes electrons within the molecule, which then return to the ground state by a series of processes, one of which produces the emitted fluorescence. The emitted radiation may be in the visible or ultraviolet region, but because of the electronic processes involved, the wavelength emitted is always greater than the wavelength absorbed. The emitted radiation is always measured perpendicular to the incident radiation to eliminate interference. Once the incident radiation is removed, fluorescence ceases (unlike phosphorescence).

$$I = FI_0(1 - e^{-acl})$$

where I is the intensity of fluorescence, F is the quantum fluorescence yield, I_0 is the intensity of initial radiation, e is an exponential function, a is the molar absorption coefficient, c is the concentration of fluorophore, and l is the path length of the cell. For low concentrations of fluorophore over short path lengths, this equation tends toward

$$I = \frac{F}{I_0 acl}$$

fluorimetry   \\flü(-ə)r-'im-ə-trē\ An analytical technique that uses the process of fluorescence. Because of the nature of fluorescence and the limitation placed on the types of molecule that exhibit fluorescence, the technique can be highly selective and very sensitive. Unlike other common spectroscopic techniques, the intensity of fluorescence is related to the intensity of the incident radiation. This relationship provides a useful instrumental variable. An additional advantage is that sterilizable probes are now available so that fermentor profiles can be obtained for compounds like NAD(P)H, which will fluoresce at 460 nm after radiation at 340 nm. The system is responsive to NAD(P)H levels within the cell and thus provides a useful detection system for monitoring certain forms of biomass within a fermentor.

fluorography   \\flü(-ə)r-'äg-rə-fē\ The introduction of a scintillant, or "fluor" into a gel or on the surface of a chromatogram, such that low-energy emitters (such as β-emissions from ^{3}H-labeled materials present in the gel or chromatogram) can be detected by using a photographic film. These β-emissions cannot pass through a gel or even through the protective coating of an X-ray film. By converting this energy to visible light by interaction with a scintillant, the light produced can penetrate the gel and film coating and leave a latent image.

flush end   \\'fləsh 'end\ *See* blunt end.

flux   \\'fləks\ A kinetic parameter that measures the rate of transport across fluids and membranes.

foam   \\'fōm\ An aggregation of gas bubbles and liquids forming a stable raft on the surface of a liquid. Foaming is a problem in fermentor design that requires addition of chemicals and occasionally mechanical devices to prevent excessive foaming, which causes loss of fermentor contents. *See also* antifoam agents.

forced-circulation evaporator \fō(ə)rst-ṣər-kyə-'lā-shən i-'vap-ə-ṛāt-ər\ An evaporator in which the liquor is circulated across the tubular heated surface by means of a circulating pump. This process allows separation of the functions of heat transfer, vapor–liquid separation, and crystallization. Circulation is maintained regardless of the rate of evaporation. This type of evaporator is well suited to crystallization operations because solids do not settle out and plug the heating tubes. The heat exchanger can be placed inside the feed chamber, where scaling of the exchanger tubes is not a problem.

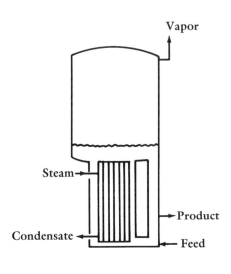

forced convection \'fō(ə)rst kən-'vek-shən\ The relative movement of fluid elements caused by external forces such as mechanical stirring or air sparging. *See also* free convection.

foreign protein \'fȯr-ən 'prō-ṭēn\ *See* heterologous protein.

forward mixing \'fȯr-wərd 'miks-iŋ\ In fluid transport, the phenomenon in which some elements of the fluid are accelerated relative to the average element and thus have a reduced residence time in the reactor. Forward mixing leads to a reduction in reactor efficiency because optimum residence time is not achieved. In forward mixing, elements can short-circuit the normal flow pattern in, for example, packed columns. *See also* axial mixing, back mixing.

fouling \'fau̇(ə)l-iŋ\ The deposition of material onto surfaces that may result in an eventual breakdown of the process. Membranes, filter cloths, and devices containing small holes are particularly vulnerable to fouling. In fermentors, fouling by deposition of biomass, proteins, or polysaccharides can cause problems with heat exchangers, baffles, and interior exposed walls.

frameshift mutation \'frām-ṣhift myü-'tā-shən\ An alteration in the structure of DNA that results in a change of the reading frame (triplet code) being used by the enzyme RNA polymerase. As a result of this mutation, the amino acid sequence of the protein coded for by the region of DNA containing the mutation is completely altered beyond the point of mutation and normally results in the production of a useless protein.

free convection \'frē kən-'vek-shən\ The relative movement of fluid elements under the influence of variations in the natural environment (e.g., temperature and concentration). *See also* forced convection.

free solution capillary electrophoresis \'frē sə-'lü-shən 'kap-ə-ˌler-ē i-ˌlek-tra-fə-'rē-səs\ *See* capillary electrophoresis.

freeze-drying \'frēz-ˌdrī-iŋ\ A process in which a solvent, often water, is removed by sublimation from the frozen state. Because the vapor pressure of the frozen solvent is considerably below atmospheric pressure, freeze-drying requires the use of high vacuum. The technique is used widely to concentrate and dry materials that cannot be dried satisfactorily by other methods. The advantages of freeze-drying include maintenance of sterile conditions, no foaming, and minimal loss of volatile and labile products because of the low temperatures involved. Also known as lyophilization when the solvent is water. *See also* culture maintenance, lyophilization.

freezing and thawing \'frēz-iŋ 'and 'thȯ-iŋ\ A method sometimes used for breaking cells. Ice crystals formed in the frozen cell rupture the cell wall or membrane. The method is rather time-consuming and not particularly efficient.

Freund's adjuvant \'frȯin(d)z 'aj-ə-vənt\ *See* adjuvant.

Freundlich isotherm \'frȯin(d)-lik 'ī-sə-ˌthərm\ An equation relating the quantity of material adsorbed on a surface (q) with that in solution (c) at constant temperature. It applies to adsorption in cases where there is an exponential distribution of heats of adsorption.

$$q = k_F c^{\beta}$$

where k_F is the Freundlich constant. When $\beta > 1$, the isotherm is favorable to adsorption; when $\beta < 1$, the isotherm is unfavorable to adsorption.

Froude number \'früd 'nəm-bər\ A dimensionless group that relates rotational inertial forces with gravitational forces acting on an element of the fluid. It involves the rotational speed of the impeller (N) in revolutions per minute and the impeller diameter D_i:

$$\mathrm{Fr_i} = \frac{N^2 D_i}{g}$$

where $\mathrm{Fr_i}$ is the Froude number and g is the gravitational constant. It is applied to unbaffled tanks and reactors because in baffled systems $\mathrm{Fr_i}$ is insignificant, vortexing is negligible, and the Reynolds number adequately describes the fluid hydrodynamics. *See also* Reynolds number.

fructans \'frək-tanz\ High-molecular-weight polysaccharides of D-fructose linked 1,2- or 1,6- by glycosidic bonds. They are common in plants.

fuller's earth \'fůl-ərz 'ərth\ A type of diatomaceous earth used as a filter aid or adsorbent. *See also* diatomaceous earth, filter aids.

fungal resistance \'fəŋ-gəl re-'zis-tən(t)s\ *See* transgenic plant.

fungi \'fən-jī\ A group of eukaryotic organisms devoid of chlorophyll. Usually they have cell walls of cellulose or chitin and are nonmotile, although they may produce motile reproductive cells. Reproduction is commonly by sexual or asexual spores. They do not possess stems, roots, leaves, or vascular systems. Most fungi consist of septate or nonseptate filaments (individually called hyphae and collectively called mycelium) whose dimensions can vary considerably. There are two main types of mycelium: septate (in which the cells, with one or two nuclei, are separated by cross walls) and coenocytic (lacking septa). Reproductive structures can usually be differentiated from vegetative structures, and the range of distinct forms is used as a basis for classifying fungi into five groups: *Ascomycotina*, *Basidiomycotina*, *Deuteromycotina*, *Mastigomycotina*, and *Zygomycotina*.

Fungi imperfecti \\ˈfən-jī im-pər-ˈfek-tī\\ *See Deuteromycotina.*

fusion protein \\ˈfyü-zhən ˈprō-ṭēn\\ A protein consisting of all or part of the amino acid sequences of two or more proteins, formed by fusing the two protein-encoding genes. This fusion is often done deliberately, either to put the expression of one of the genes under the control of the strong promoter for the first gene or to allow the gene of interest (which is difficult to assay) to be more easily studied by substituting some of the protein with a more easily measured function (e.g., by fusing with the β-galactosidase gene, the product of which can easily be measured using a chromogenic substrate). Fusion proteins have also been designed to facilitate secretion of cloned proteins from cells and to aid purification of the target protein. *See* affinity tag, reporter gene.

fusogen \\ˈfyü-zō-jen\\ *See* chemical fusogen.

fusogenic agent \\ˈfyü-zō-jen-ik ˈā-jent\\ Any agent (e.g., compound or virus) that can be used to fuse cells.

G

G-CSF \'ge-'sē 'es 'ef\ Granulocyte-colony stimulating factor. *See* cytokines.

GAG \'gē 'ā 'gē\ The region of RNA in a retrovirus that codes for a structural protein that associates with the RNA in the core of the viral particle. Other retroviral genes include ENV and POL.

galactans \gə-'lak-tənz\ Polysaccharides, essentially polymers of galactose. *See also* agar, carrageenans.

β-galactosidase \'bāt-ə gə-ˌlak-'tō-sə-ˌdās\ *See* lactase, reporter gene.

galactoside \gə-'lak-tə-ˌsīd\ *See* glycoside.

gamma \'gam-ə\ Compounds beginning with γ- are listed under their roman names.

gas chromatography (GC) \'gas ˌkrō-mə-'täg-rə-fē ('jē 'sē)\ A type of chromatographic separation in which the sample, consisting of a mixture of volatile components, is partitioned between the vapor phase and a nonvolatile liquid adsorbed on the surface of an inert matrix. An inert carrier gas transports the components through the column until they reach the detector. GC may be used to monitor components in a process stream, to purify and separate mixtures, or for quantitative analysis.

gas flooding (channeling) \'gas 'fləd-iŋ ('chan-ᵊl-iŋ)\ The formation of regions in an immobilized cell or enzyme reactor where the provision of nutrients or substrate is prevented by gaseous products that coalesce and are unevenly distributed throughout the reactor.

gas holdup \'gas 'hōl-dəp\ The relative volume of gas that is present in the gas-liquid dispersion in a fermentor. If a fermentor has a broth volume V_b which, when sparged, increases to V_{disp}, then the gas holdup, E_G, is given by

$$E_G = \frac{V_{disp} - V_b}{V_{disp}}$$

121

Only the submerged air bubbles are considered in gas holdup. Gas in the foam on top of a broth does not contribute to gas holdup. Gas holdup is important because it determines the transfer of oxygen from the gas to the liquid phase. If it is too low, the production rate in most aerobic fermentations will fall.

gas lift   \\'gas 'lift\\ A device by which a gas flow is used to lift a liquid and thereby promote circulation, as, for example, in airlift fermentors. *See also* airlift fermentors, deep-shaft airlift fermentor.

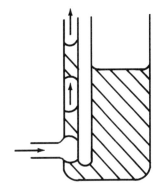

gas-phase protein sequencer   \\'gas-ˌfāz 'prō-ˌtēn 'sē-kwən-sər\\ *See* Edman degradation.

gasohol   \\'gas-ə-ˌhȯl\\ An automobile fuel produced by mixing gasoline (petrol) with ethanol. Gasohol is particularly attractive in developing countries where sucrose-rich plants such as sugar cane and sweet sorghum are readily available as raw materials for cheap ethanol production by fermentation. Brazil, in particular, adopted this approach by establishing and implementing the Brazilian National Alcohol Program. *See also* ethanol.

gateway sensor   \\'gāt-ˌwā 'sen-sȯ(ə)r\\ A type of sensor that can be connected to a computer to yield further information about a process, e.g., a dissolved oxygen sensor that can provide information about oxygen transfer rate and hence fermentor performance.

gel   \\'jel\\ A solid or semisolid mixture of a solid and a liquid, often consisting of a polymeric substance saturated with liquid as, for example, agar, agarose, polyacrylamide, and silica gels.

gel electrophoresis   \\'jel i-ˌlek-trə-fə-'rē-səs\\ A type of electrophoresis in which ions are transported through a gel under the influence of an electric current. Typical matrices are agarose, starch, and polyacrylamide. The separation factors depend on both the charge-mass ratio and interactions between the charged species and the gel matrix. *See also* polyacrylamide gel electrophoresis.

gel exclusion chromatography   \\'jel iks-'klü-zhən ˌkrō-mə-'täg-rə-fē\\ *See* gel filtration.

gel filtration \\'jel fil-'trä-shən\\ A chromatographic purification technique for separating molecules according to size. The column material consists of beads that contain pores of various sizes. As the mixture to be fractionated passes through the column, large molecules are excluded from the pores and enter only the spaces between the beads. These molecules, eluted first from the column, are referred to as the void volume. Smaller molecules that can enter the pores elute in order of decreasing size because the smaller the molecule, the greater the number of pores it can enter and thus the longer it will be held on the column. Below a minimum size, all remaining molecules will elute together because they can all pass through all the pores. This remainder is referred to as the included volume. Different grades of gel filtration media can be purchased, depending on the fractionation range required. Gel filtration media available include cross-linked dextrans (Sephadex and Sephacryl) and cross-linked agarose (Sepharose and Bio-Gel). Also known as gel exclusion chromatography, gel permeation chromatography, size exclusion chromatography.

gel filtration chromatography \\'jel fil-'trä-shən ˌkrō-mə-'täg-rə-fē\\ *See* gel filtration.

gellan gum \\'jel-ən 'gəm\\ A polysaccharide biopolymer excreted by *Pseudomonas elodea* and commercially available for use in microbiological media as a gelling agent to replace agar. The repeating unit consists of a linear tetrasaccharide. The gum is also a potential alternative source of rhamnose (instead of quercetin).

gel permeation chromatography \\'jel ˌpər-mē-'ä-shən ˌkrō-mə-'täg-rə-fē\\ *See* gel filtration.

gel resin \\'jel 'rez-ᵊn\\ An ion-exchange resin in which the polymeric support consists of a polymeric resin (usually cross-linked polystyrene, which has a solid gel-like structure). Because there is not a defined pore structure, the resin has a low affinity for large molecules and very poor kinetics of adsorption. In addition, large molecules that enter the structure are difficult to remove. Thus the resins are susceptible to fouling and subsequent loss of capacity. However, the resins have a high capacity for small ions and are used in water purification and softening processes. *Compare with* macroporous resin.

GEM \\'jem\\ An abbreviation for genetically engineered microorganisms.

geminiviruses \\'jem-ə-nī-'vī-rəs-əz\\ One of the two classes of DNA viruses (the other is caulimoviruses) that infect plants. They have potential as cloning vectors in plants.

GenBank database \\'jēn-ˌbaŋk 'dāt-ə-ˌbās\\ A compilation of nucleotide sequences both genetic (e.g., genes, gene fragments, intrans, cDNA) and synthetic. This database overlaps considerably with the other major nucleic acid sequence database, the EMBL nucleotide sequence database.

gene bank \\'jēn 'baŋk\\ *See* gene library.

gene cloning \\'jēn 'klōn-iŋ\\ The process whereby a gene is inserted into an appropriate vector (e.g., plasmid), which then transports the gene into a host cell (usually a bacterium). Once within the host, the vector produces numerous identical copies of itself and the gene it carries. Division of the host cell produces progeny that all contain vector molecules. The vector molecules can undergo further replication.

This process results in the formation of a colony or clone of identical host cells that each contain one or more copies of the gene-containing vector. The gene is now said to be cloned.

gene gun   \'jēn 'gən\ *See* biolistic process.

gene library (genomic library)   \'jēn 'lī-ˌbrer-ē (ji-'nō-mik 'lī-ˌbrer-ē)\ A collection of recombinant clones, which among them contain all the DNA present in a particular organism.

gene probe   \'jēn 'prōb\ A nucleic acid molecule that can be used to detect, by complementary base-pairing, another nucleic acid molecule that has a complementary or homologous sequence. The probe is invariably labeled (*see also* biotin, nick translation) to allow autoradiographic or enzymatic detection of the hybridization reaction. *See also* colony hybridization, Southern blotting.

gene therapy   \'jēn 'ther-ə-pē\ The treatment of disease by the insertion of a new gene or genes into either embryonic or adult cells. A number of diseases result from a single gene defect (e.g., sickle-cell anemia, thalassemia, phenylketonuria, cystic fibrosis, Duchenne's dystrophy, glycogen storage diseases, hemophilia, emphysema, diabetes mellitus, and Tay-Sachs disease). These diseases should ultimately be amenable to gene therapy. However, many severe technical hurdles must be overcome, and many ethical issues arise, particularly when using embryonic tissues. After obtaining the exact piece of DNA coding for the normal gene (however, the genes involved in many single-gene disorders have yet to be identified), it should be possible to remove tissue involved in the disease from the patient. After inserting the gene into cells from this tissue, the cells must then be replaced in the body. The gene must be inserted into the cell in such a way as to be under the normal control mechanism for gene expression in that cell.

At present, bone marrow cells are the best target for initial gene therapy trials. Bone marrow can be removed easily from the patient, cells can be manipulated in vitro to introduce new DNA, and the cells can be returned to the patient via the bloodstream. Currently, the preferred method for gene transfer uses viral vectors (retrovirus, adenovirus, and herpes simplex virus) that are able to achieve high-efficiency gene transfer, although physical methods such as direct injection, liposome-mediated gene transfer, and receptor-mediated gene transfer are also used. The first human gene therapy was carried out in September 1990, when a four-year-old girl suffering from a rare hereditary disease that results from a deficiency of the catabolic enzyme adenosine deaminase (ADA) was injected with one billion white blood cells engineered to contain a copy of the ADA gene.

genetic code   \jə-'net-ik 'kōd\ *See* codon.

genetic engineering   \jə-'net-ik ˌen-jə-'ni(ə)r-iŋ\ Gene manipulation. Genetic engineering is defined by the 1978 Genetic Manipulation Regulations as "The formation of new combinations of heritable material by the isolation of nucleic acid molecules, produced by whatever means outside the cell, into any virus, bacterial plasmid, or other vector system so as to allow their incorporation into a host organism in which they do not naturally occur, but in which they are capable of continued propagation." Also known as recombinant DNA technology.

genetic mapping \jə-'net-ik 'map-iŋ\ The producti)n of a representation of the relative positioning and genetic distance separating genes, based on the frequency of recombination. *See* centimorgan.

genome \'jē-ˌnōm\ The complete, single-copy set of genetic instructions for an organism.

genomic library \ji-'nō-mik 'lī-ˌbrer-ē\ *See* gene library.

genomics \ji-'nō-miks\ The use of large-scale DNA sequencing programs to identify disease-causing genes, which can then be modulated to affect the cellular function of those genes.

genotoxic agent \jē-nə-'täk-sik 'ā-jənt\ An agent that is toxic as a result of its ability to damage DNA, e.g., alkylating agents used in cancer chemotherapy.

genotype \'jē-nə-ˌtīp\ The genetic constitution of an organism. It is defined by the base sequence of the genome.

germ cell \'jərm 'sel\ The male and female reproductive cells: egg and sperm cells in mammals, ovum and pollen in plants.

germplasm \'jərm-ˌplaz-əm\ An unidentified material transmitted from generation to generation through the gametes. An old term, referring essentially to genes.

germplasm storage \'jərm-ˌplaz-əm 'stōr-ij\ Preservation of plant parts to preserve the genotype (germplasm) of plant species. Plant-breeding programs are generally based on adapting ancient plant varieties and their relatives. As plant-breeding programs proliferate, there is a danger that original genetic information will be lost as these ancient varieties are replaced by modern varieties. Germplasm is most easily stored by cryopreservation of shoot tips, embryos, and meristems.

ghost peaks \'gōst 'pēkz\ Peaks on a chromatogram caused by substances not present in the original sample. These peaks may be caused by septum bleed, decomposition of solutes, or carrier-gas contamination.

Gibberella fujikuroi \jib-ə-'rel-ə ˌf(y)ü-jē-'kü-rói\ A fungus belonging to the *Ascomycotina* that is grown in submerged culture to produce the secondary metabolite, gibberellic acid.

gibberellic acid (GA₃) \jib-ə-'rel-ik 'as-əd ('jē 'ā 'thrē)\ Naturally occurring plant hormones, the gibberellins, which have a complex chemical structure containing a gibbane nucleus with four cyclic carbon rings. Although gibberellins have been isolated from plant materials, the best source is from the fungus *Gibberella fujikuroi*, which is grown in submerged culture. *See also* gibberellins.

gibberellins \jib-ə-'rel-ənz\ Plant-growth substances related to gibberellic acid (GA₃). They are diterpenoids containing four cyclic carbon rings, and they have the ability to cause stem elongation when applied to intact plants in the light. Many endogenous gibberellins have been found in plants. Other synthetic plant-growth substances have been prepared from gibberellic acid. Gibberellic acid, a metabolic product of the fungus *Gibberella fujikuroi*, can be obtained from the fungal growth medium.

Gibbs-Donnan equilibrium \'gibz-'dän-ən ē-kwə-'lib-rē-əm\ *See* Donnan (membrane) equilibrium.

globulins \'gläb-yə-lənz\ One of the four major categories of seed storage proteins. *See also* seed storage proteins.

glucans \'glü-kanz\ Polysaccharides composed of D-glucose units. They can have either straight or branched chains. The glycosidic linkages can be α-1,4, as in amylase and dextran, β-1,4, as in cellulose, β-1,3, as in cellose, or 1,6, as in pustulan. Branched glucans include amylopectin and dextran.

glucoamylases (amyloglucosidases) \glü-kō-'am-ə-ḷās-əz (ˌam-ə-ḷō-glü-'kō-si-ˌdās-ez)\ Exoamylases (i.e., enzymes that cleave off terminal sugar residues from polysaccharides) produced commercially from *Aspergillus* species or *Rhizopus* species, both of which produce the enzymes extracellularly. They have low specificity, hydrolyzing α-1,4-glucosidic bonds in branched oligosaccharides and, at a slower rate, α-1,3- and α-1,6-bonds. They are used principally in the saccharification of starch to produce high-glucose syrups.

glucose isomerase \'glü-ˌkōsī-'säm-ə-ˌrās\ *See* high-fructose syrups.

glucose oxidase (E.C. 1.1.3.4) \'glü-ˌkōs 'äk-sə-ˌdās\ An enzyme that catalyzes the reaction:

$$\beta\text{-}\delta\text{-glucose} + O_2 \longrightarrow \text{D-glucose-}1,5\text{-lactone} + H_2O_2$$

The enzyme is used extensively as an antioxidant in the food industry, usually in conjunction with catalase. The catalase consumes the H_2O_2 produced during the oxidation reaction, which would otherwise denature the enzyme. Commercial sources of the enzyme are *Aspergillus niger*, *Penicillium glaucum*, and *P. notatum*. The enzyme is used to prevent changes in the color and flavor of foods during processing, transportation, and storage (e.g., in beers, wines, sauces, citrus fruit drinks, cake mixes, and instant soup).

glucose syrup \'glü-ˌkōs 'sər-əp\ Purified, concentrated aqueous solutions of glucose and higher saccharides produced by the hydrolysis of corn (maize) or potato starch. Syrups are classified according to their dextrose equivalent (DE), which is a measure of the extent of starch hydrolysis. It is used as a sweetening agent in confectionery. Also known as confectioners' glucose, corn starch hydrolysate, corn syrup, starch syrup.

glucoside \'glü-kə-ˌsīd\ *See* glycoside.

glutamate \'glüt-ə-ˌmāt\ *See* glutamic acid.

glutamic acid (glutamate) \glü-'tam-ik 'as-əd ('glüt-ə-ˌmāt)\ An amino acid of considerable importance to industry as a feed and food supplement, particularly as monosodium glutamate, which is used as a flavor enhancer. It is produced industrially by the growth of *Corynebacterium glutamicum* or *Brevibacterium divaricatum*, using starch hydrolysates, molasses, ethanol, or acetate as substrate. Worldwide production is more than 100,000 tons per year.

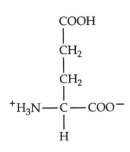

glutelins \'glüd-əl-ənz\ One of the four major categories of seed storage proteins. *See also* seed storage proteins.

glycan \'glī-ˌkan\ Polysaccharide. The term *glycoprotein glycan* thus refers to the carbohydrate portion of the glycoprotein.

glycine betaine \'glī-ˌsēn 'bēt-ə-ˌēn\ A derivative of the amino acid glycine, which accumulates in plant cells undergoing certain environmental stresses. In both drought and salt stress, plant cells must maintain their water content. In a large number of plant species water is maintained by the accumulation

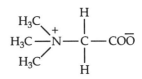

of glycine betaine within the cell, which acts as an osmoprotectant compound. Introduction of the genes for glycine betaine production in other plants may produce stress tolerance.

glycobiology \ˌglī-kō-bī-'äl-ə-jē\ The study of the functions of oligosaccharides in biology. It includes the study of lectins, glycoproteins, and glycolipids.

glycocalyx \ˌglī-kō-'kā-liks\ Polysaccharide components outside the bacterial cell wall. *See also* capsule, slime layer.

glycoform \'glī-kō-ˌfȯ(ə)rm\ Any one of a number of microheterogeneous forms of a glycoprotein, arising from differences in their glycosylation patterns. Any glycoprotein exists in several different glycoforms; the variety is determined by the glycosylation pattern. The amount of microheterogeneity in glycoforms can be quite significant. A typical glycoprotein has several glycosylation sites with a heterogeneous population of oligosaccharides on some or all of the sites, even within a single tissue. In other tissues the same glycoprotein may exist in another range of glycoforms. Variations in the oligosaccharide component of glycoproteins can affect the protein's biological activity, immunogenicity, and half-life. Because a number of cloned glycoproteins are destined for use as therapeutic agents, a detailed description of the glycoform being used is essential to accurately define the product being described and so satisfy the regulatory authorities.

glycolysis \glī-'käl-ə-səs\ The principal route of carbohydrate breakdown and oxidation in cells. It involves a series of anaerobic reactions whereby glucose is metabolized to pyruvic acid with the production of ATP. The series of reactions is known as the glycolytic pathway or the Embden–Meyerhof pathway.

glycoprotein \ˌglī-kō-'prō-ˌtēn\ A protein containing one or more covalently linked carbohydrate moieties. The carbohydrate moiety can be either a single sugar residue or a chain (linear or branched) of the same or dissimilar sugar residues. Linkage to the protein is invariably via the side chain of an asparagine, serine, or threonine residue. The oligosaccharide moiety of a glycoprotein plays an important role in defining several biological properties of the glycoprotein, including clearance rate, immunogenicity, and biological activity. *See also* glycoform.

glycosidases \glī-'kō-sə-ˌdās-əs\ A group of hydrolases that attack glycosidic bonds in carbohydrates and glycoproteins. *See also* α-amylase, β-amylase, cellulase, glucoamylases, invertase.

glycoside \'glī-kə-ˌsīd\ A compound that has a glycosidic bond between the anomeric carbon of a carbohydrate unit (either a mono- or oligosaccharide) and a non-carbohydrate group (aglycon). When the anomeric carbon atom involved is on glucose, the glycoside is a glucoside; when the anomeric carbon is on galactose, the compound is a galactoside. *See*, for example, cardiac glycoside.

glycosidic bond \glī-kə-'sid-ik 'bänd\ The bond between the anomeric carbon of a carbohydrate and some other group. For example, the C—O bond in dissacharides is an *O*-glycosidic bond (or linkage); the C—N bond between the sugar and base in nucleotides is an *N*-glycosidic bond.

glyphosate (*N*-phosphonomethyl glycine) \'glif-ə-ˌsāt 'en-ˌfäs-fō-nō-meth-əl 'glī-ˌsēn\ An extensively used, nonselective herbicide that is effective in controlling most of the world's worst weed species. It is not toxic to animals and is rapidly degraded by soil microorganisms. Glyphosate specifically inhibits the enzyme 5-enolpyruvylshikimate-3-phosphate synthase, which is on the pathway involved in the synthesis of the amino acids tryptophan, tyrosine, and phenylalanine. Tolerance to glyphosate has been engineered into various crops.

GMM \'jē 'em 'em\ An abbreviation for genetically modified microorganisms.

GMO \'jē 'em 'ō\ An abbreviation for genetically modified organisms.

gradient elution \'grād-ē-ənt ē-'lü-shən\ A chromatographic technique in which the composition of the mobile phase (eluant) is continuously varied during elution. *See also* isocratic elution, stepwise elution.

Graetz number (Gz) \'grets 'nəm-bər ('jē 'zē)\ *See* heat transfer.

Gram stain \'gram 'stān\ An important differential stain (introduced by Christian Gram) used to separate bacteria into two groups, Gram-positive and Gram-negative, based on differences in the peptidoglycan content of the cell wall. Cells are first stained with crystal violet, washed, and treated with an iodine solution. Iodine forms a complex with crystal violet and stains the cells a purple-blue. The stained cells are washed with alcohol or acetone. The stain complex will not be removed from the walls of some bacteria (Gram-positive). After decolorization, a red counterstain, such as basic fuchsin or safronin, is used to stain Gram-negative cells red and make them easier to detect under microscopic examination. The Gram stain is thought to become trapped in pores of the peptidoglycan of the cell wall during dehydration with alcohol. In Gram-negative cells the peptidoglycan layer is thinner. The difference in peptidoglycan layer is responsible for the difference in ease of rupturing Gram-positive and Gram-negative bacteria. Gram-negative cells are most easily broken, being lysed by low levels of detergents such as sodium dodecyl sulfate (SDS) or isooctylphenoxypolyethoxyethanol (Triton X-100), whereas Gram-positive cells need lysozyme treatment first.

granulocyte colony stimulating factor (GCSF) \'gran-yə-lō-ˌsīt 'kǎl-ə-nē 'stim-yə-ˌlāt-iŋ 'fak-tər ('jē 'sē 'es 'ef)\ *See* cytokines.

granum \'gra-nəm\ *See* thylakoids.

GRAS organisms \'gras 'ȯr-gə-ˌniz-əmz\ Organisms that are rated "generally recognized as safe" for use in the food industry by the U.S. Food and Drug Administration (FDA). Selected strains of molds, bacteria, and yeasts are currently used as sources of enzymes for food processing. The most commonly used are *Aspergillus oryzae*, *Aspergillus niger*, and *Bacillus subtilis*. They are therefore the preferred organism of choice for a given situation because inadvertent contamination of foodstuffs by these organisms or their products should not present a human health threat. However, the use of several other organisms as sources for specific enzyme preparations has been approved by the FDA (e.g., *Mucor miehei* for the production of rennet).

GRASE \'grās\ "Generally recognized as safe and effective." Under the federal Food, Drug, and Cosmetic Act, before a new drug for human consumption is allowed on the market, experts in the field must find it GRASE for uses recommended by its labeling. This finding involves examination of the safety and effectiveness of the drug based primarily on clinical tests on human subjects.

Grashof number (Gr) \'grās-ˌhȯf 'nəm-bər ('jē 'är)\ A dimensionless group associated with heat and mass transfer by natural convection.
In heat transfer

$$Gr = d^3 \, g \Delta\rho \, \frac{\rho_{av}}{\eta^2}$$

where d is the distance over which heat transfer takes place (m), e.g., tube diameter; g is the gravitational constant (m/s^2); η is the viscosity (kg/ms); $\Delta\rho$ is the density difference (kg/m^3); and ρ_{av} is the average density of the fluid (kg/m^3).
In mass transfer

$$Gr = \frac{L^3 g \, \Delta\rho_A}{\rho\nu^2}$$

where L is the length of the system; $\Delta\rho_A$ is the density difference of A between two points in the transfer field; and ν is the kinematic viscosity of the fluid. *See also* Peclet number, Schmidt number, Sherwood number.

griseofulvin \ˌgriz-ē-ō-'fúl-vən\ A nontoxic systemic antifungal antibiotic originally detected in cultures of *Penicillium griseofulvum*, but commercially produced using *P. patulum*. Griseofulvin, with a fungistatic mode of action against fungi other than yeasts and the mastigomycotina, is used in treatment and control of fungal diseases of the hair, skin, and nails. It is thought to interfere with chitin synthesis in growing hyphae.

growth hormone \'grōth 'hȯr-ˌmōn\ *See* human growth hormone.

Grunstein–Hogness method \'grün-stīn–'hȯg-nes 'meth-əd\ *See* colony hybridization.

5'-guanylate (guanosine monophosphate, GMP) \'fīv-ˌprīm 'gwän-ə-ˌlāt ('gwän-
ə-sīn ˌmän-ō-'fäs-ˌfāt, 'jē 'em 'pē)\ A flavor enhancer produced industrially in
several ways: (1) by the action of nuclease (5'-phosphodiesterase) from *Penicillium
citrinum* or *Aspergillus* species on yeast RNA, (2) by chemical phosphorylation of guanosine, (3) by the enzymatic (microbial) conversion of 5'-xanthosine
monophosphate to GMP, and (4) by chemical synthesis. More than 1000 tons of
GMP is produced annually in Japan.

$$H$$

habituation \hə-ˌbich-ə-'wā-shən\ The acquired ability of plant cells to grow and divide independently of exogenous plant-growth substances (hormones).

Hagen–Poiseuille law \häg-ən–pwä-'zói 'ló\ *See* Darcy law.

halophiles \'hal-ə-ˌfīlz\ Organisms that require the presence of minimum concentration (>~3%) of sodium chloride in the environment to achieve optimum growth. Such concentrations are toxic to freshwater microorganisms.

hammer mill \'ham-ər 'mil\ A mechanical device used to shred bulk solid materials (e.g., sugar cane, beet). The device consists of a rotating shaft on which hinged flails or hammers are fixed. The shaft is contained in a shell or cylinder through which the feed material is passed to emerge at the other end in a shredded form.

Hanes–Wilkinson plot \'hānz-'wil-kin-sən 'plät\ A mathematical treatment of enzyme kinetic data used for the determination of the maximum reaction rate and the Michaelis constant. Rearrangement of the Michaelis–Menton equation gives

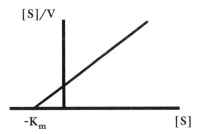

$$\frac{[S]}{V} = \frac{[S]}{V_{max}} + \frac{K_m}{V_{max}}$$

where a plot of $[S]/V$ versus $[S]$ gives a straight line of gradient of $1/V_{max}$, with a y-intercept of K_m/V_{max} and an x-intercept of $-K_m$. *See also* Eadie–Hofstee plot, Lineweaver–Burke plot, Michaelis–Menton equation.

haploid cells \'hap-lóid 'selz\ Cells that have a single set of chromosomes.

HART \'härt\ *See* hybrid arrest translation.

harvesting \'här-vəst-iŋ\ The recovery of microorganisms from a liquid culture, usually by filtration or centrifugation, to give a pellet of microorganisms.

HAT medium \\'āch 'ā 'tē 'mēd-ē-əm\\ Tissue culture medium (containing hypoxan-thine, aminopterin, and thymidine) that is used as a selection medium. For example, nonfused myeloma cells and plasma cells (lymphocytes) cannot survive in this medium, but fused hybridoma cells can (*See* monoclonal antibody). Similarly, the medium can be used to select transfected mammalian cells by using the thymidine kinase (Tk) gene as a selectable marker for transfected cells. Tk^- cells are used, and transfection is achieved by using the Tk gene with the gene or genes of interest linked to it. Tk^- cells die in HAT medium, whereas Tk^+ cells (transfected cells) survive.

headspace sampling \\'hed-ṣpās 'sam-pliŋ\\ Sampling of a volatile solute by with-drawing a gas-phase sample from an enclosed space above a solid or liquid sample for subsequent analysis. In particular, the technique is used in gas chromatography.

heart cut \\'härt 'kət\\ A chromatographic technique whereby two or more partially resolved peaks are eluted from one column and then directed onto another column of different polarity or operating under different conditions to improve resolution.

heat exchanger \\'hēt iks-'chānj-ər\\ A device in which heat is transferred from one fluid stream to another without the streams coming into physical contact. One fluid is heated and the other is cooled. Common types of heat exchanger include double pipe, shell and tube, and plate and frame. The choice depends on the duty required and the types of fluid being carried.

heat transfer \\'hēt 'tran(t)s-fər\\ A term used to describe the movement of heat energy from one body (e.g., a fluid) to another. The transfer of heat to and from a reactor is controlled by a series of resistances that involve film coefficients in fluids on either side of the jacket or coil wall as well as the thermal conductivity of the material of the jacket or coil. The rate of heat transfer (Q) is given by

$$Q = UA\Delta T$$

where U is the overall heat transfer coefficient (W/m^{2}K), A is area (m^2), and ΔT is temperature difference (K). U is composed of the film heat transfer coefficients on either side of the jacket or coil wall (h_i and h_o in W/m^{2}K), the wall thickness x (m), and its thermal conductivity k (W/mK):

$$\frac{1}{U} = \frac{1}{h_i} + \frac{x}{k} + \frac{1}{h_o}$$

These film coefficients (h) can be evaluated by semiempirical correlations, which depend on the nature of the fluid. For example, for Newtonian fluids:

$$\text{Nu} = k(\text{Re})^a(\text{Pr})^b$$

where k, a, and b are constants and Nu is the Nusselt number, Pr is the Prandtl number, and Re is the Reynolds number. Problems arise with non-Newtonian fluids, for example in airlift reactors under laminar flow conditions and with non-Newtonian fluids

$$Nu = 1.75\,\delta^{0.33}Gz^{0.33}$$

where δ is the correction for non-Newtonian behavior $(3n + 1)/4n$, n is the power-law fluid flow behavior index, and Gz is the Graetz number, which is MC_p/kL where M is the mass flow rate of fluid through a tube (kg/s), L is the length of the tube (m), C_p is the specific heat of the fluid (J/kgK), and k is the thermal conductivity of fluid (W/mK). *See* non-Newtonian fluid.

heavy chain \hev-ē 'chān\ *See* immunoglobulin G.

height equivalent to a theoretical plate (HETP) \'hīt i-'kwiv(-ə)-lənt 'tü 'ā ˌthē-ə-'ret-i-kəl 'plāt ('āch 'ē 'tē 'pē)\ A measure of the performance of a column contactor or chromatographic column, defined as the height of the column that, under the process conditions, is equivalent to a theoretical plate. The theoretical plate is defined as a device at which perfect contact occurs so that the streams leaving it are in equilibrium. HETP is a step function dividing the column into equal portions, unlike the height of a transfer unit (HTU), which is a continuous function.

height of a transfer unit (HTU) \'hīt 'əv 'ā 'tran(t)s-fər 'yü-nət ('āch 'tē 'yü)\ A measure of the performance of a column contactor as used in distillation or adsorption. A transfer unit (N_T) is a dimensionless parameter that relates the column height (h), overall mass-transfer coefficient (k_o), interfacial area of any sorbent species (a), volumetric flow rate of the fluid (F), and column cross section (s) according to the equation:

$$N_T = \frac{k_o ahs}{F}$$

from which the height of a transfer unit (H) can be calculated:

$$H_T = \frac{h}{N_T} = \frac{F}{k_o as}$$

The advantage of HTU is that it includes the ratio of mass transfer coefficient and flow rate; this inclusion reduces the problems associated with the variation of the mass transfer coefficient with flow rate.

helicase \'hel-ə-ˌkās\ An enzyme preparation rich in β-1,3-glucanase activity, prepared from snail internal organs, used to degrade yeast cell walls to produce spheroplasts.

hemicelluloses \ˌhem-i-'sel-yə-ˌlōs-ez\ High-molecular-weight polysaccharides found in the cell walls of higher plants, together with cellulose and lignin. In many plants the major hemicellulose is a polymer comprising about 80% pentoses (of which about 80% are D-xylose) and containing side chains of other sugars and sugar acids.

hemolysis \hi-'mäl-ə-səs\ The lysis of red blood cells.

heparin \'hep-ə-rən\ An acidic mucopolysaccharide from animal tissues capable of interacting with basic residues on the surface of a protein. It consists of equal amounts of D-glucosamine and D-glucuronic acid, α-1,4-glycosidically linked, and also contains O- and N-sulfate links. Heparin inhibits many enzymes involved with nucleic acid metabolism and has been used as an affinity ligand for the purification of many such enzymes (e.g., DNA and RNA polymerase, ribonuclease, restriction endonucleases, and DNA topoisomerase). Heparin also inhibits blood clotting by inhibiting the conversion of prothrombin to thrombin and fibrinogen to fibrin. It is therefore used in the collection of whole blood and in other chemical situations, such as in heart-lung machines. The fractionation of heparin to produce a group of new oligosaccharides with reduced risk of hemorrhage while maintaining antithrombotic efficacy has been investigated with *Cytophaga heparina*. Some fractions have been found to possess additional fibrinolytic and tumor-growth-inhibiting activities.

hepatitis B virus (HBV) \hep-ə-'tīt-əs 'bē 'vī-rəs ('āch 'bē 'vē)\ A small DNA virus (genome size about 3 kilobase pairs) of the hepadenovirus family. In humans, HBV is responsible for infectious hepatitis. The primary infection results in liver damage, which usually resolves within a matter of months. However, in 5–10% of cases, the disease persists for many years. Epidemiological data also show a link between HBV infection and liver cancer, although the exact role of HBV has yet to be determined. The genome encodes four polypeptides (using overlapping sequences in different reading frames). Two of these polypeptides represent the viral surface antigen (HBsAg or "Australia" antigen) and core antigen (HBcAg), which are used to diagnose HBV infections.

heptose \'hep-ṭōs\ *See* carbohydrates.

herbicide resistance (plants) \'(h)ər-bə-ṣīd ri-'zis-tən(t)s ('plants)\ *See* transgenic plants.

Herschel–Buckley equation \'her-shəl 'bək-lē i-'kwā-zhən\ A rheological equation applicable to plastic flow, which relates the applied stress (τ) to the shear rate (γ) in the form:

$$\tau = k\gamma^n + \tau_0$$

where τ_0 is yield stress and k is a constant.

heterodimer \'het-ə-rō-ˌdī-mər\ A dimer in which the two subunits are different.

heteroduplex \het-ə-rō-'d(y)ü-pleks\ A DNA molecule formed by base-pairing between two DNA strands that are not completely complementary. *See also* cross hybridization.

heterofermentative \het-ə-rō-fər-'ment-ət-iv\ Of or relating to a bacterial culture that ferments glucose or another sugar to lactic acid and other products, usually carbon dioxide and ethanol. *See also* lactic acid bacteria. *Compare with* homofermentative.

heterokaryon \het-ə-rō-'kar-ē-än\ A cell in which two or more nuclei, originating from different cell types, are present in a single cytoplasm. If the two nuclei fuse, the heterokaryon develops into a hybrid cell.

heterologous \het-ə-'räl-ə-gəs\ Derived from or associated with a species different from that being referred to.

heterologous protein \het-ə-'räl-ə-gəs 'prō-ˌtēn\ A protein not native to the organism in which it is expressed, for example, a human gene expressed in *Escherichia coli*. Also known as foreign protein.

heterotroph \'het-ə-rə-ˌtrōf\ An organism that requires preformed organic molecules as its source of energy or substrate for biosynthesis. Heterotrophs include animals, fungi, and most (but not all) bacteria. *Compare with* autotroph.

heterozygous \het-ə-rə-'zī-gəs\ *See* allele.

HETP \'ach 'e 'te 'pe\ *See* height equivalent to a theoretical plate.

Heurty process \'hyür-te 'präs-es\ *See* whey.

Hevea brasiliensis \'he-ve-ə ˌbra-sil-e-'en-səs\ A cultivated plant that is the source of natural rubber. Latex obtained from this plant contains hydrocarbons in the molecular weight range $1–2 \times 10^6$.

hexose \'hek-ˌsōs\ *See* carbohydrates.

hierarchical control \hi-(ə-)'rär-ki-kəl kən-'trōl\ The use of a computer to oversee the control functions of other computers, which are themselves controlling individual parts of a process.

high-fructose syrups (HFS) \hī-'frək-tōs 'sər-əps ('ach 'ef 'es)\ Solutions containing high fructose proportions, prepared by the isomerization of glucose. The enzyme glucose isomerase isomerizes glucose (which has only 65% of the sweetness of sucrose) into a mixture of fructose (>50%), glucose, and other sugars (high-fructose syrup or isoglucose). Fructose is about 1.5 times sweeter than sucrose. Because glucose can be prepared readily from starch, HFS has been replacing sucrose in many food applications. The process is particularly attractive in countries where starch crops are in abundance but locally produced sugar is in short supply or absent, or where sucrose costs are high (often unrealistically high as a result of farm subsidies). Glucose produced by starch hydrolysis is used as the starting material for HFS production. Microorganisms most commonly used for the production of glucose isomerase are *Bacillus coagulans*, *Actinoplanes missouriensis*, *Arthrobacter* species, and *Streptomyces* species. Either the extracted enzyme is used in an immobilized form, or immobilized whole cells that contain the enzyme are used. Further enrichment using industrial-scale chromatography can give syrups containing up to 70% fructose. More than 5 million tons of HFS is currently produced annually worldwide.

high-performance capillary electrophoresis (HPCE) \hī-pə(r)-'fȯr-mən(t)s 'kap-ə-ˌlər-ē i-ˌlek-trə-fə-'rē-səs ('ach 'pē 'sē 'ē)\ *See* capillary electrophoresis.

high-performance liquid chromatography (HPLC) \hī-pə(r)-'fȯr-mən(t)s 'lik-wid ˌkrō-mə-'täg-rə-fē ('ach 'pē 'el 'sē) \ Generally, high-resolution column chromatographic techniques. Improvements in the nature of column-packing materials for a range of chromatographic methods (e.g., gel filtration, ion exchange, reversed phase) has yielded smaller rigid beads with greater uniformity in size and shape. This improvement allows packing in columns with minimum spaces between beads and thus minimizes peak broadening of eluted molecules caused by diffusion within these spaces. Minimum peak broadening results in considerably increased resolution over the more conventional soft gels. Because of the close packing of spheres, high pressure is needed to pump solvents through the column, so columns are packed in stainless steel tubes. Because of this need for high pressure, HPLC is often inadvertently referred to as high-*pressure* liquid chromatography.

high-performance thin-layer chromatography (HPTLC) \hī-pə(r)-'fȯr-mən(t)s 'thin-ˌlā-ər ˌkrō-mə-'täg-rə-fē ('ach 'pē 'tē 'el 'sē)\ *See* thin-layer chromatography.

high-rate filter \'hī-ˌrāt 'fil-tər\ *See* percolating filter.

hn RNA \'ach 'en 'är 'en 'ā\ Heterogeneous nuclear RNA. *See also* intron.

hirudin \hir-'üd-ᵊn\ A highly potent polypeptide thrombin inhibitor (anticoagulant) produced by the medicinal leech. Several closely related hirudin variants have been described, each containing 65–66 amino acids. Hirudin has potential for the treatment or prophylaxis of several thrombotic diseases.

Hofmeister series \'hȯf-mīs-tər 'si(ə)r-ēz\ The arrangement of anions in order of their efficiency in salting out of hydrophilic colloids. The anions in this series possess strong dehydrating properties and so can precipitate hydrophilic colloids. Also known as lyotropic series.

hollow-fiber reactor \'häl-ō-ˌfī-bər rē-'ak-tər\ A reactor design used particularly for immobilized enzymes and the growth of mammalian cells. Hollow fibers are packed into bundles within a cylindrical vessel. Nutrient is delivered through the lumen of each fiber while the cells grow in the extracapillary space. Fibers can be made from a range of porous semipermeable membranes. The pore size is small enough to retain cells but large enough to allow the passage of nutrients and waste materials.

holoenzyme \ˌhō-lō-'en-ˌzīm\ *See* cofactor.

homofermentative \ˌhō-mə-fər-mən-'tā-tiv\ Of or relating to a bacterial culture that ferments glucose or another sugar to lactic acid as a single product. *See also* lactic acid bacteria. *Compare with* heterofermentative.

homology \hō-'mäl-ə-jē\ The degree of identity between two nucleotide sequences. For example, 95% homology indicates that 95 nucleotide positions out of 100 are identical in the two sequences.

homopolymer tailing \ˌhō-mə-'päl-ə-mər\ A method for producing sticky ends in a DNA molecule by adding a polymer of identical nucleotides (e.g., polydeoxycytosine) to the 3′-OH termini by using the enzyme terminal transferase in the presence of a single deoxynucleotide. This enzyme allows the DNA fragment to be linked to an appropriate vector by complementary base-pairing where the appropriate homopolymer (in this case polydeoxyguanosine) has been added to the vector by the same technique.

homozygous \ˌhō-mə-'zī-gəs\ *See* allele.

horseradish peroxidase (HRP) \\'hȯrs-ˌrad-ish pə-'räk-sə-ˌdās ('ach 'är 'pē)\\ *See* peroxidase.

host range (hr) mutants \\'hōst 'ranj ('āch 'är) 'myüt-ᵊnts\\ Pathogens that have lost virulence as a result of serial passage in cultured cells. This loss of virulence allows the mutant to be used as a live, *attenuated* vaccine. For example, Sabin's oral vaccine for polio uses poliovirus type 1 strain, which contains 57 separate base substitutions compared to the parent strain. Because there is always the possibility that these naturally occurring mutants may revert to virulent forms, such vaccines must be regularly checked. The genetic engineering of deletion mutants will probably be used in the future to produce more controlled and hence safer attenuated organisms. *See also* BCG.

hotstart PCR \\'hät-ˌstärt 'pē 'sē 'är\\ *See* polymerase chain reaction.

HPCE (high-performance capillary electrophoresis) \\'āch 'pē 'sē 'ē\\ *See* capillary electrophoresis.

HPLC \\'āch 'pē 'el 'sē\\ *See* high-performance liquid chromatography.

HPTLC (high-performance thin-layer chromatography) \\'āch 'pē 'tē 'el 'sē\\ *See* thin-layer chromatography.

HRP \\'ach 'är 'pē\\ An abbreviation for the enzyme horseradish peroxidase.

HTU \\'ach 'tē 'yü\\ *See* height of a transfer unit.

Hughes press \\'hyüz 'pres\\ A device used to disrupt cells before downstream processing. It operates by forcing a frozen cell paste through a small orifice under high pressure. This method of cell disruption is usually confined to laboratory-scale work. *See also* liquid shear.

human gene therapy \\'hyü-mən 'jēn 'ther-ə-pē\\ *See* gene therapy.

Human Genome Project (HGP) \\'hyü-mən 'jē-ˌnōm 'präj-ekt ('āch 'jē 'pē)\\ A large-scale, multinational effort to map and eventually sequence the three billion nucleotides within the human genome. The main objectives of the HGP are the generation of a dense (2-centimorgan resolution) genetic linkage map, of a physical map of the human genome, and of the nucleotide sequence of the human genome. The mapping of the nucleotide sequence relies on the development of new, faster DNA sequencing technology. Because of the low information content and technical difficulties, most tandemly repeated DNA sequences in the human genome will not be sequenced. However, multiple sequence determinations will be made at some loci in the human genome to assess the prevalence and importance of mutations and polymorphisms. Started in 1990, the HGP is expected to take at least 15 years.

human growth hormone (somatotropin) \\'hyü-mən 'grōth 'hȯr-ˌmōn (sō-ˌmat-ə-'trō-pən)\\ A 191-residue protein (MW 22,000) produced in the anterior pituitary gland and needed for longitudinal growth of the skeleton. Growth hormone deficiency results in dwarfism. Originally in very short supply, because it had to be purified from cadaver pituitaries, it is now being produced from *Escherichia coli* by recombinant DNA technology. Unlike other clinically useful products produced by recombinant technology (e.g., insulin and interferon), the world market for the material is not great. However, the hormone may also be of value in treating burns, wounds, and fractured bones. These applications would considerably enhance the market for the genetically engineered product.

humanized antibody \\'hyü-mə-ˌnīzd 'ant-i-ˌbäd-ē\\ *See* chimeric antibody.

humectants \\hyü-'mek-tən(t)s\\ Substances that absorb moisture. They are used to maintain the water content of materials such as baked products, tobacco, and glue. Examples are glucose syrup, invert sugar, and honey.

humic acid \\'hyü-mik 'as-əd\\ A group of naturally derived organic acids of high molecular weight and varying structure present in soils and peat and forming the coloring matter in surface waters. It is often the cause of fouling of ion-exchange resins and other adsorbents used in raw-water treatment.

hyaluronic acid (HA) \\hīl-yü-'rän-ik 'as-əd ('āch 'ā)\\ A glycosaminoglycan forming part of the connective-tissue matrices of all vertebrates and also found as part of the capsular material that surrounds *Streptococcus* species. It is a viscoelastic biopolymer that has great importance in ophthalmic surgery, where it is used to replace the aqueous humor of the anterior chamber during cataract surgery and intraocular lens implantation. HA is also a major component of synovial fluid; it can be used to replace fluid lost during joint surgery and to treat joint inflammation. HA has considerable potential in the health care and cosmetic industry. The main source of HA at present is extraction from rooster combs, but a number of processes that produce HA by the fermentation of streptococci are also in use.

hybrid arrest translation \\'hī-brəd ə-'rest tran(t)s-'lā-shən\\ A method for identifying the protein coded for by a cloned gene. The denatured cloned gene is mixed with an mRNA population and base-pairs with its corresponding mRNA. The mRNA population is then translated in an in vitro protein synthesis system, and the protein products are identified by gel electrophoresis and autoradiography. This experiment is repeated with the mRNA population *without* the added gene. It will give the same protein pattern, but with an extra protein band. This extra protein is the protein encoded for by the cloned gene.

hybrid release translation \\'hī-brəd ri-'lēs tran(t)s-'lā-shən\\ A method for identifying the protein coded for by a cloned gene. The denatured gene is immobilized on nitrocellulose, and an mRNA population is passed through the filter. The DNA base-pairs with its corresponding mRNA, whereas other mRNA species do not bind. The DNA–RNA hybrid is then dissociated by heating, and the mRNA is translated in an in vitro protein synthesis system. The protein product, identified by immunoprecipitation or gel electrophoresis and autoradiography, represents the protein encoded for by the cloned gene.

hybridization \\hī-brəd-ə-'zā-shən\\

(1) Any mechanism that allows the exchange of nuclear material between one cell and another similar but genetically different individual and results in the formation of offspring with genotypes different from either of the parents. In general, hybridization is expressed as sexual hybridization in eukaryotic

organisms and as parasexual hybridization in prokaryotes, most eukaryotic tissue cultures, and certain eukaryotes that do not exhibit true sexuality.

(2) The formation of a double-stranded molecule by complementary base-pairing between two single-stranded DNA molecules, or a single-stranded DNA molecule and an RNA molecule. *See also* gene probe.

hybridization probe \hī-brəd-ə-'zā-shən 'prōb\ *See* gene probe.

hybridoma cell \hī-brə-'dō-mə 'sel\ An immortalized, antibody-secreting cell line created by fusing an antibody-secreting plasma cell (lymphocyte), which produces the desired antibody, with a myeloma cell line (an immortal, antibody-secreting tumor cell). The myeloma cell line used is usually a mutant that has lost the ability to produce its own antibodies, so that the resultant hybridoma cell line secretes only the desired antibody.

hybrids \'hī-brədz\ *See* cell hybrids.

hydrocolloids \hī-drə-'käl-öidz\ A class of food additives used to enhance the physical properties of food by their ability to thicken and form gels. Compounds used include bacterial polysaccharides from *Pseudomonas* species and xanthan gum.

hydrocyclone \hī-drə-'sī-klōn\ A device for the separation and concentration of particles in suspension in a fluid. It is operated by pumping the fluid suspension tangentially into the top of the cyclone to produce a centrifugal action and vortexing. The cover of the cyclone has a downward-extending tube that extends into the vortex and removes part of the stream as an overflow product. The remainder of the flow, containing most of the solids, travels down the walls of the cone and is removed in a partially concentrated form at the cone apex. Various sizes are available, constructed of materials that include plastics, ceramics, and metals. Advantages are low capital cost and the ability to make separations based on small differences in particle size.

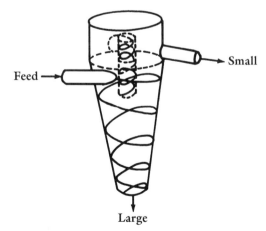

hydrolases \hī-'dräl-ə-sēz\ An enzyme classification covering all enzymes that catalyze the hydrolytic cleavage of C—O, C—N, C—C, and some other bonds. The systematic name always includes hydrolase, and the recommended name is often formed by adding the suffix "-ase" to the name of the substrate. Hydrolases are one of the six main classes (E.C. 1) used in enzyme classification, and they play an important role in biotechnology. *See,* for example, glycosidases, proteases. *See also* Enzyme Commission number.

hydrophilic \hī-drə-'fil-ik\ *See* lipophobic.

hydrophilic–lipophilic balance (HLB) \hī-drə-'fil-ik–lī-pə-'fil-ik 'bal-ən(t)s ('āch 'el 'bē)\ A measure of the emulsifying and solubilizing properties of a surface-active agent. Surface-active compounds contain both hydrophilic and lipophilic groups, and the ratio of the contribution from these groupings influences the surface activity. Thus, a water-in-oil (W/O) emulsifier has a low HLB (3.5–6); an oil-in-water (O/W) emulsifier has an intermediate value (8–18); and a solubilizing agent has a high value (15–18). Ranges of HLB values for other surface-active agents include those for wetting agents (7–9) and detergents (13–15).

Various experimental methods are detailed in texts on colloid and surfactant chemistry for the experimental determination of HLB values. These values can also be obtained using empirically determined group numbers. The values of common hydrophilic and lipophilic groups are available and can be substituted into the following formula:.

$$HLB = \Sigma(\text{hydrophilic group numbers}) = n(\text{lipophilic group numbers}) + 7$$

Good agreement between calculated and experimentally determined values are found over a wide range of compounds. *See also* surface-active agent.

hydrophobic \hī-drə-'fō-bik\ *See* lipophilic.

hydrophobic chromatography \hī-drə-'fō-bik ˌkrō-mə-'täg-rə-fē\ A chromatographic separation technique for proteins based on the hydrophobic interaction between the protein and the chromatographic support. The support is usually an agarose derivatized with long-chain alkyl hydrocarbon chains to provide a hydrophobic surface. Interaction between these hydrophobic areas and hydrophobic "patches" or "pockets" on the protein provides a means of discrimination, although electrostatic interactions and hydrogen bonding may also be involved. Proteins are bound to the column matrix in high salt concentration (e.g., 2 M ammonium sulfate) and eluted by decreasing the ionic strength, thereby decreasing the hydrophobic interactions between the column matrix and protein. Proteins are therefore eluted in order of increasing hydrophobicity.

hydroponics \hī-drə-'pän-iks\ A plant-culturing system in which plant roots are developed not in soil but in a mass of inert particles (usually sand or gravel) that is moistened with a prepared nutrient solution by capillary action. Although this method provides abundant water and nutrients to the plants, aeration of root systems is poor. *Compare with* nutrient film technique.

hydroxyapatite \hī-ˌdrak-sē-'ap-ə-tīt\ A naturally occurring hydrated calcium phosphate that can be used as a chromatographic support for the chromatographic separation of proteins.

hyoscyamine \hī-ə-'sī-ə-ˌmēn\ An anticholinergic compound isolated from the plant *Hyoscyamus niger.*

hyperchromic effect \ˌhī-pər-'krō-mik i-'fekt\ The observation that the UV absorbance at 260 nm of a solution of single-stranded DNA molecules is approximately 30% more than would be displayed by the same concentration (nucleotides per cubic decimeter) of double-stranded DNA. The heterocyclic rings of DNA absorb at 260 nm, but this absorbance is reduced when the nucleotides are involved in hydrogen bonding and base stacking. The denaturation (melting) of DNA can therefore be monitored by the increase in absorbance at 260 nm, referred to as hyperchromicity. *Compare with* hypochromic effect.

hyperchromicity \ˌhī-pər-krō-'mis-ət-ē\ *See* hyperchromic effect.

hyperfiltration \ˌhī-pər-fil-'trā-shən\ *See* reverse osmosis.

hypervariable regions \ˌhī-pər-'ver-ē-ə-bəl 'rē-jənz\ *See* DNA fingerprinting.

hypha \'hī-fə\ The basic unit of a fungus. A fungal filament (pl. hyphae).

hypochromic effect \ˌhī-pə-'krō-mik i-'fekt\ The decrease in absorbance at 260 nm observed when single-stranded DNA molecules renature to give double-stranded DNA. *Compare with* hyperchromic effect.

hypophase \ˌhī-pə-'fāz\ The more dense phase in a two-phase system, especially where the two phases may be similar in properties (e.g., two-phase aqueous systems). *Compare with* epiphase.

hypotonic solution \ˌhī-pə-'tän-ik sə-'lü-shən\ The solution with the lower osmotic pressure of two compared solutions. *Compare with* isotonic solution.

hypsochromic shift \ˌhip-sō-'krō-mik 'shift\ The movement of a characteristic spectral band to shorter wavelength, higher frequency, or higher energy following the insertion of a substituent into the molecular structure. The converse of bathochromic shift. Also known as blue shift.

I

IAA \ˈī ˈā ˈā\ Indoleacetic acid. *See also* auxins.

ice genes \ˈīs ˈjēnz\ *See* ice-nucleating bacteria.

ice-nucleating (ice plus) bacteria \ˈīs-ˈn(y)ü-klē-ˌāt-iŋ (ˈīs ˈpləs) bak-ˈtir-e-ə\ Bacteria possessing proteins that enable the bacteria to nucleate crystallization in supercooled water. The proteins appear to provide a template that mimics the crystal structure. Such nucleation is detrimental in agriculture because these bacteria initiate frost formation in plants that would otherwise supercool. The formation of ice crystals in the cells, rather than the low temperature, causes damage.

Attempts have been made to control the population of ice-nucleating bacteria in plants by treating plants with genetically engineered ice minus strains in which the genes for the ice-nucleating protein have been deleted. Released in high enough concentrations, these engineered organisms should compete with and outgrow the ice plus bacteria. Early field trials were encouraging, but the deliberate release of genetically engineered microorganisms into the environment caused considerable debate. Ice-nucleating bacteria include certain species of *Erwinia, Pseudomonas,* and *Xanthomonas.*

idiophase \ˈid-ē-ə-ˌfāz\ The phase of a culture during which products other than primary metabolites are synthesized. Unlike primary metabolites, metabolites produced during the idiophase have no obvious role in cell metabolism. The metabolites produced during the idiophase are referred to as secondary metabolites.

idiotype \ˈid-ē-ə-ˌtīp\ The characteristics of an antibody that makes it unique. The idiotype of an antibody is determined by the molecular structure of its variable region. *See* immunoglobulin G.

IEF \ˈī ˈē ˈef\ *See* isoelectric focusing.

143

IgG \ī 'jē 'jē\ *See* immunoglobulin G.

IMAC \ī 'em 'ā 'sē\ *See* immobilized metal-affinity chromatography.

immiscibility \im-(m)is-ə-'bil-ət-ē\ The property whereby fluids form two distinct phases under all relative proportions. *See also* miscibility, partially miscible substances.

immobilization \im-ō-bə-lə-'zā-shən\ The conversion of enzymes or cells from the free mobile state to the immobilized state, either by attachment to an appropriate support or by entrapment in an appropriate matrix. *See also* cell immobilization, enzyme immobilization.

immobilized biocatalyst \im-'ō-bə-‚līzd ‚bī-ō-'kad-ᵊl-əst\ Enzymes or cells immobilized on a solid support and used to catalyze a biochemical reaction. *See also* cell immobilization, enzyme immobilization.

immobilized metal-affinity chromatography (IMAC) \im-'ō-bə-‚līzd ‚met-ᵊl-ə-'fin-ət-ē ‚krō-mə-'täg-rə-fē ('ī 'em 'ā 'sē)\ A chromatographic protein purification technique involving the adsorption of the protein to a support matrix as a result of coordination between an immobilized metal ion and an electron donor. Usually, the metal ion is immobilized (usually chelated) on a support, and the protein binds to the metal ions via electron-donor groups on the protein surface (e.g., cysteine, arginine, and histidine have electron-donor atoms in their side chains). In an alternative but less commonly used approach, the metal ion, immobilized on the protein surface, forms a coordination bond with an electron-donor group attached to the support matrix. Protein is eluted either by changing the pH, by ligand exchange (i.e., eluting the column with competing electron-donor solutes), or by the addition of chelating agents.

immortalized cells \im-'ort-ᵊl-‚līzd 'selz\ Cells that show continuous proliferation and can be subcultured indefinitely. Such cells are necessarily malignant in nature. The culture of normal cell lines is generally very difficult. Even for those that can be cultured, indefinite subculturing is impossible because of built-in senescence. Malignant cells, or hybrids of normal and malignant cells (e.g., hybridomas), overcome this built-in senescence and are said to be immortal.

immunoadsorbent \im-yə-nō-ad-'sor-bənt\ An affinity matrix formed by linking an antibody preparation to an insoluble support.

immunoaffinity chromatography \im-yə-nō-ə-'fin-ət-ē ‚krō-mə-'täg-rə-fē\ Purification of a compound (ligate) by using an affinity column in which the ligand is an antibody.

immunoassay   \im-yə-nō-'as-ā\ Any assay method that uses antibodies for detecting and quantifying biological molecules (e.g., hormones or proteins) or microorganisms. The most commonly used methods are radioimmunoassay (RIA) and enzyme-linked immunosorbent assay (ELISA). Many commercial immunoassay kits are available. Most of the methods are based on the use of monoclonal antibodies, which gives much greater specificity than the use of polyvalent antisera.

immunoblotting   \im-yə-nō-'blät-iŋ\ The use of antibodies to detect a specific protein on a nitrocellulose sheet following protein (western) blotting (*see* protein blotting). The nitrocellulose sheet is incubated in a solution containing an antibody (primary antibody) to the protein of interest. Following appropriate washing steps, the antigen–antibody reaction that occurs is then visualized by incubating with a second antibody (anti-IgG), which binds to the primary antibody. The second antibody is used linked to a marker enzyme, usually horseradish peroxidase or alkaline phosphatase. After washing, the sheet is incubated in a chromogenic substrate solution appropriate for the enzyme; a colored band is produced. The position of this colored band indicates the position of the protein of interest.

immunogen   \i-'myü-nə-jən\ A substance used to induce an immune response.

immunoglobulin   \im-yə-nō-'gläb-yə-lən\ A class of globular protein that shows antibody activity. The five classes of human immunoglobulin are referred to as IgA, IgD, IgE, IgG, and IgM. *See also* immunoglobulin G.

immunoglobulin G (IgG)   \im-yə-nō-'gläb-yə-lən 'jē ('ī 'jē 'jē)\ The major immunoglobulin component of serum. It makes up 75% of the total immunoglobulin content. (The other immunoglobulins are IgA, IgD, IgE, and IgM). When an antigen (e.g., a protein) is injected into an animal, a large number of IgG molecules are synthesized, each one capable of binding to a different epitope on the antigen. The IgG molecule has a molecular weight of about 150,000 and comprises two light (L) chains of MW ~25,000 and two heavy (H) chains of MW ~50,000 that are held together by disulfide bridges. When the amino acid sequences of different IgG molecules are compared, it is found that each molecule has a variable region at the end of both the light and heavy chains (V_L and V_H, respectively) as well as a common constant region (C_L and C_H). The variable regions on each chain combine to form two identical antigen binding (ab) sites on each IgG molecule. The constant regions are involved in the mediation of common effector functions of IgG molecules, such as the binding of complement reactivity with rheumatoid factors or the transfer of antibodies across the placental membrane. The IgG molecule can be cleaved with the proteolytic enzyme papain (P) to give two identical F_{ab} fragments that contain the antigen binding site and a crystallizable fragment (F_c) that comprises part of the constant region. *See also* antiserum, B-lymphocytes, bispecific antibodies, chimeric antibody, epitope, phage display antibodies, single-chain antibody.

immunoglobulin G

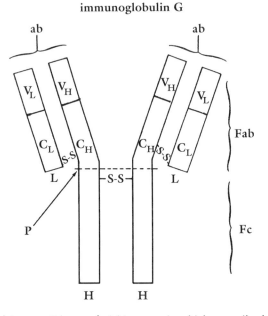

immunosensor \\ˌim-yə-nō-'sen-sər\\ A biosensor in which an antibody is used to detect the molecule of interest. *See also* biosensor.

immunotherapy \\ˌim-yə-nō-'ther-ə-pē\\

(1) A method that enhances a patient's immunological response to a disease (e.g., enhancement of the immune response by injection with BCG).

(2) The use of exogenous antibodies (such as monoclonal antibodies) or drug-linked antibodies to treat disease. *See also* bispecific antibodies, chimeric antibodies, immunotoxin.

immunotoxin \\ˌim-yə-nō-'täk-sən\\ A conjugate formed by linking an antibody to a toxin so that both retain their biological activity. Immunotoxins are being developed as a means of directing toxins (drugs) to target cells. For example, monoclonal antibodies against tumor-cell-surface antigens can be linked to toxic molecules. On injection into the patient the toxic molecule (drug) is directed to the tumor cell, where it exerts its toxic effect while leaving other tissues unaffected. *See also* bispecific antibodies, chimeric antibody.

impeller \\im-'pel-ər\\ *See* agitator.

in-line analysis \\'in-'līn ə-'nal-ə-səs\\ The analysis of a process by the use of direct-reading probes inserted into the process stream. This technique is preferred when possible because it eliminates any errors caused by removing a sample from the process either by a continuous sample line or by a discrete sample, and it provides the facility for continuous monitoring of the process. Suitable probes are available for determinants such as temperature, pH, dissolved oxygen, pressure, impeller speed, and liquid and gas flow rates. *See also* off-line analysis, on-line analysis.

in situ hybridization \\'in 'si-tü ˌhī-brəd-ə-'zā-shən\\ A technique for identifying the position of a gene on a chromosome. Cell preparations on glass slides are treated to denature their DNA (without losing the identity of the overall chromosome structures) and then washed with an appropriate radiolabeled probe. Hybridization occurs between the probe and the corresponding gene on the chromosome, and this hybridization can be detected by autoradiography.

in vitro \\'in 'vē-trō\\ Literally, *in glass,* pertaining to a biological reaction taking place in an artificial apparatus, as distinct from observations made in living organisms (in vivo). *Compare with* in vivo.

in vitro mutagenesis \\'in 'vē-trō ˌmyüt-ə-'jen-ə-səs\\ Any in vitro method that produces a mutation at a specific position in a DNA molecule. *See also* site-directed mutagenesis.

in vitro packaging \\'in 'vē-trō 'pak-ij-iŋ\\ The construction of infective λ-bacteriophage particles that contain a DNA sequence to be cloned within the genome. Transfection of bacteria with λ-DNA (containing the insert to be cloned) is highly inefficient. By packaging this recombinant molecule within an infective phage particle, the DNA can be introduced efficiently into bacterial cells. When the assembled virus is added to the bacterial culture, the normal λ-phage infective process can take place.

in vitro protein synthesis \\'in 'vē-trō 'prō-ˌtēn 'sin(t)-thə-səs\\ The addition of an mRNA sample to a cell-free translation system, usually in the presence of one or more radiolabeled amino acids, resulting in the translation of the mRNA molecule. The newly synthesized radiolabeled protein can be observed in a number of ways (e.g., gel electrophoresis followed by fluorography) to identify the protein coded for by the mRNA.

in vivo \\'in 'vē-vō\\ Literally, *in life,* pertaining to a biological reaction that takes place in a living cell or organism. *Compare with* in vitro.

inclusion bodies \\in-'klü-zhən 'bäd-ēz\\ *See* protein inclusion bodies.

incompatibility groups \\in-kəm-ˌpat-ə-'bil-ət-ē 'grüps\\ *See* compatibility.

incubator \\'iŋ-kyə-ˌbāt-ər\\ A controlled environmental system, room, or cabinet used to maintain microbial and cell cultures. It may be used to control atmosphere, gas composition, humidity, light, and temperature.

individual-specific autoantibodies (IS antibodies) \\ˌin-də-'vij-(ə)-wəl-spi-ˌsif-ik ˌôt-ō-'ant-i-ˌbäd-ēz ('ī 'es 'ant-i-ˌbäd-ēz)\\ Antibodies produced by humans against their own cellular components. These human autoantibodies increase in number from birth until approximately two years after birth, and then remain constant for decades or possibly for the individual's life span. The complement of such autoantibodies present in an individual is unique, and when physically separated they comprise an "antibody fingerprint", which can be used to identify people, animals, or their body fluids. IS antibodies are not to be confused with autoantibodies produced against normal cellular components in some disease states such as rheumatoid arthritis.

indoleacetic acid \\'in-ˌdōl-ə-ˌsēt-ik 'as-əd\\ *See* auxins.

induced mutation \in-'d(y)üst myü-'tā-shən\ *See* mutation.

inducer molecule \in-'d(y)ü-sər 'mäl-i-ˌkyü(ə)l\ *See* induction of a gene.

induction of a gene \in-'dək-shən 'əv 'ə 'jēn\ Switching on of the expression of a gene or group of genes (*operon*) by an inducer molecule (e.g., a chemical such as a hormone or a substrate) or other stimulus (e.g., heat).

induction of λ-phage \in-'dək-shən 'əv 'lam-də-ˌfāj\ Excision of integrated λ-DNA from the bacterial genome, leading to a change to the lytic mode of infection, caused by a chemical or other stimulus.

inert support \in-'ərt sə-'pō(ə)rt\ Any nonreactive matrix. Inert supports are used for a range of purposes, including the immobilization of cells and enzymes, the growth of anchorage-dependent cells, immobilizing affinity groups for affinity chromatography, the attachment of a charged group to give an ion-exchange resin, and the solid-phase synthesis of peptides or oligonucleotides. Many materials have been used, including glass, polymers, silica, and cellulose.

inferential mass flow meter \ˌin-fə-'ren-chəl 'mas 'flō 'mē-tər\ *See* mass flow sensor.

infrared spectrophotometer \ˌin-frə-'red ˌspek-trō-fə-'täm-ət-ər\ An instrument for measuring the adsorption of infrared radiation by a sample. Three major types are available:

(1) *dispersive,* in which a prism or grating causes the radiation to be split into individual wavelength bands that are then scanned across the sample;

(2) *Fourier transform infrared (FTIR),* in which the complete wavelength range is incident on the sample simultaneously, to give an overall spectrum of the sample; and

(3) *nondispersive,* in which filters are used to select the appropriate wavelength required for analysis.

The nondispersive type is used exclusively for quantitative measurement (e.g., on-line gas analyzer for carbon dioxide in the exhaust gas from a fermentor), whereas the other types may be used both qualitatively and quantitatively. The speed of FTIR, together with its powerful data analysis, allows it to be used as an on-line chromatographic detector.

infrequent cutters \in-'frē-kwənt 'kət-ərz\ *See* low-frequency restriction enzymes.

inherent kinetics \in-'hir-ənt kə-'net-iks\ *See* intrinsic kinetics.

initiation codon \in-ˌish-ē-'ā-shən kō-ˌdän\ A specific sequence of three bases (GUG or AUG) in mRNA to which a molecule of transfer RNA (tRNA) carrying a methionine residue (eukaryote) or a formyl methionine residue (prokaryote) binds in the first step of protein synthesis. Also known as a start codon.

inoculum \in-'äk-yə-ləm\ The initial culture of an organism or mixed culture, added to a suitable medium to promote the propagation of a larger number of cells (pl. inocula).

5′-inosate (inosine monophosphate) \'fīv-ˌprīm 'i-nō-ˌsāt (i-nō-sīn ˌmän-ō-'fäs-ˌfāt)\ A flavor enhancer produced commercially by the action of nuclease (5′-phosphodiesterase) from *Penicillium citrinum* and *Aspergillus* species on yeast

RNA to produce the constituent ribonucleotides. In a second step, adenosine monophosphate is converted to inosine monophosphate by a 5′-AMP deaminase from *Aspergillus oryzae*.

insect resistance in plants \'in-sekt ri-'zis-tən(t)s 'in 'plantz\ *See* transgenic plant.

λ-insertion vector \'lam-də-in-ˌsər-shən 'vek-tər\ A λ-phage vector produced by deleting a nonessential segment of DNA. This deleted region contains most of the genes involved in the integration and excision of the λ-prophage from the bacterial genome. Such deleted vectors are consequently nonlysogenic and can exhibit only the lytic infection cycle. If a DNA molecule is inserted into the normal λ-DNA molecule, only about 3 kilobase pairs of DNA can be inserted before the DNA is too large to be packaged into the λ-phage head. However, using a λ-insertion vector, fragments of DNA ranging in size from 3 to 9 kilobase pairs can be cloned. *See also* cosmid, λ-phage, λ-replacement vector.

insertional inactivation \in-'sər-shən-ᵊl in-ˌak-tə-'vā-shən\ A cloning strategy in which insertion of a piece of DNA into a vector inactivates a gene carried by the vector. The loss of this gene product can be used to detect transformants (e.g., the loss of resistance to a particular antibiotic).

insulin \'in(t)-s(ə-)lən\ A polypeptide hormone that controls the level of blood sugar. About 20% of sufferers of *diabetes mellitus* are dependent on insulin injections to control their condition, and for more than 50 years the hormone has been isolated from pig or cow pancreas. However, both sources of the hormone differ slightly in their amino acid sequence from human insulin. Consequently, almost all patients develop anti-insulin antibodies, which result in side effects and a need for increased dosage. The human insulin gene has now been cloned in *Escherichia coli* and was the first genetically engineered protein to be tested in humans. The use of this cloned material is now routine, and, although it is too early to tell, it is anticipated that many of the side effects of animal-derived insulin will not occur with human insulin.

integration \ˌint-ə-grā-shən\ Insertion of viral DNA within the host chromosome, e.g., the integration of phage DNA in the bacterial chromosome.

integrative plasmids \'int-ə-ˌgrāt-iv 'plaz-mədz\ *See* episomes.

intensifying screen \in-'ten(t)-sə-fī-iŋ 'skrēn\ A solid layer of material (e.g., calcium tungstate) that fluoresces (emits visible light) when struck by β-particles or X-rays. These screens are used to detect high-energy emissions, such as those produced by ^{32}P or ^{125}I. These high-energy emissions pass straight through a photographic emulsion without being absorbed and therefore do not leave a latent image. When an intensifying screen is placed on the side of the film opposite to the source, the emitted energy is converted to light that forms a latent image on the film. It is used in autoradiography.

intercalating agent \in-'tər-kə-ˌlāt-iŋ 'ā-jənt\ Any planar molecule that can be inserted between two adjacent stacked base pairs in the DNA double helix. *See,* for example, acridine orange, ethidium bromide.

interferons   \int-ə(r)-'fi(ə)r-änz\ A family of proteins, originally identified as compounds produced by virally infected cells, that produce an antiviral state in other cells. They have been classified into three subtypes (α, β, and γ) with a number of proteins in each group. In addition to their antiviral effect, interferons inhibit cellular proliferation (hence, they are a potential anticancer drug) and modulate the immune system. These properties have made interferons sought-after compounds for clinical testing. Originally only prepared in small (and not very pure) quantities from human white blood cells, they are now being produced in quantity by at least three pharmaceutical companies that use recombinant DNA technology. This process provides sufficient material to carry out detailed clinical trials. *See also* cytokines.

interleukins   \in-tər-'lü-kənz\ *See* cytokines.

intermittent-discharge centrifuge   \int-ər-'mit-ᵊnt-dis(h)-'chärj 'sen-trə-ˌfyüj\ A centrifuge in which openings along the periphery of the bowl are closed by valves that can be opened intermittently. An alternative design allows the centrifuge bowl to form a horizontal slit through which the solids are discharged. In these designs the solids can be accumulated as a relatively dry cake, so little fluid is lost.

international unit (IU) of enzyme activity   \int-ər-'nash-ən-ᵊl 'yü-nət ('ī 'yü) 'əv 'en-ˌzīm ak-'tiv-ət-ē\ The amount of enzyme capable of producing 1 μmol of reaction product in 1 min under optimal (or defined) reaction conditions.

intervening sequences   \int-ər-'vēn-iŋ 'sē-kwən(t)-səs\ *See* intron.

intrinsic kinetics   \in-'trin-zik kə-'net-iks\ Kinetics of an enzyme in free solution. For an immobilized enzyme, partitioning effects affect the kinetic parameters obtained (still obeying Michaelis–Menton-type kinetics), and the kinetics are referred to as *inherent* kinetics. If diffusional limitations occur, the kinetics obtained are *effective* kinetics.

intron   \'in-trän\ A DNA sequence, found within the coding region of most eukaryotic genes, that interrupts the code for the gene product. The full gene sequence is initially transcribed into hn RNA (heterogeneous nuclear RNA), and then the intron sequences are removed (by cutting and splicing) to give the final mRNA molecule, which is then translated at the ribosome to give the protein product. The protein-encoding parts of the gene are referred to as *exons*. Bacterial genes do not contain introns and therefore do not possess the appropriate enzymes to remove these sequences. For this reason it is usually impossible to express a cloned eukaryotic gene in bacteria. A cDNA copy of the mammalian mRNA must be made first and then cloned into the bacterium. Also known as intervening sequences.

inverse PCR   \'in-ˌvərs 'pē 'sē 'är\ *See* polymerase chain reaction.

invert sugar   \'in-ˌvərt 'shüg-ər\ *See* invertase.

invertase (sucrose) (E.C. 3.2.1.26)   \in-'vərt-ˌās ('sü-ˌkrōs)\ An enzyme that hydrolyzes sucrose into an equal mixture of glucose and fructose (invert sugar). The enzyme is isolated from *Saccharomyces cerevisiae* or *S. carlsbergensis*. Invert sugar, somewhat sweeter than sucrose, is used mostly in food and confectionery products as a humectant to hold moisture and prevent drying. Invert sugar is used in the preparation of artificial honey, in the brewing industry, in jam manufacture, and in making soft-centered chocolates.

ion chromatography \'ī-än ͵krō-mə-'täg-rə-fē\ A variant of ion-exchange chromatography that uses a highly efficient column composed of small ion-exchange resin beads. The ion-exchange functionality of these beads is often restricted to a thin surface coating (*pellicular resins*), thereby providing rapid exchange kinetics and thus sharp chromatographic peaks. Ion chromatography has developed into a versatile technique because of the commercial availability of a number of different columns for separating inorganic and organic anions and cations. Detection of the eluting solutes is usually achieved in commercial units by conductometric or spectrophotometric sensors, although other suitable sensors may be used for particular applications. *See also* ion-exchange chromatography.

ion exchange \'ī-än iks-'chānj\ A process in which adsorption of one or several ions on a surface is accompanied by a simultaneous desorption or displacement of an equivalent amount of another ionic species of the same charge. This exchange often occurs on a surface of a polymeric molecular network that possesses functional groups capable of carrying an ionic charge. The counterions associated with these fixed charges are partially free and may exchange for other ions of the same charge in solution. The extent of this exchange depends on the nature and concentration of the exchanging ions. The material may be naturally occurring minerals (e.g., zeolites and aluminosilicates) or derivatized polymers of biological (cellulose, dextran) or synthetic (cross-linked polystyrene) origin.

ion-exchange chromatography \'ī-än-iks-͵chānj ͵krō-mə-'täg-rə-fē\ A type of chromatography in which the solid support is an ion-exchange material used to separate mixtures of charged molecules or ions. This exchange can be carried out on a preparative scale or as a modification of HPLC. *See also* ion-exchange resin.

ion-exchange fermentation \'ī-än-iks-͵chānj ͵fər-mən-'tā-shən\ A process by which products are removed from the culture broth by adsorption onto ion-exchange media. *See also* adsorption fermentation, of which this is a modification.

ion-exchange resin \'ī-än-iks-͵chānj 'rez-ᵊn\ A synthetic organic polymer, often based on cross-linked polystyrene, that has been derivatized by the addition of charged groups to produce materials that exchange counterions when suspended in aqueous solutions. Cationic exchangers have fixed acidic substituents based on, for example, sulfonic acid (SO_3H) (strong acid exchangers) or carboxylic acid (COOH) (weak acid exchangers). Anionic exchangers have fixed substituents based on, for example, quaternary ammonium (R_4N^+, type I) or ethoxyamine ($-NH_2^+CH_2OH$, type II) groups, or amines (R_3N), which are weak base exchangers. Other functional groups may also be attached to the resin skeleton to provide more selective behavior. These functional groups may be similar to those used in affinity chromatography. The degree of derivatization and the extent of cross-linking of the resin determines the overall capacity for ion exchange. *See also* gel resin, macroporous resin.

ion-exclusion chromatography \'ī-än-iks-'klü-zhən ͵krō-mə-'täg-rə-fē\ The separation of electrolytes from nonelectrolytes by the use of an ion-exchange resin using the principle of Donnan membrane equilibrium. This principle requires that ions have a lower concentration within the pores of the resin or membrane than in the external solution. Nonelectrolytes are not affected by this principle, so the concentration within the pores will equal that of the external solution. Thus, a partial separation of nonelectrolytes from electrolytes is possible. Under favorable conditions in a column, complete separation may be achieved. *See also* Donnan membrane equilibrium.

ion pair \'ī-än 'pa(ə)r\ Ionic species of opposite charge, which may associate to form pairs that carry an effective zero charge under certain conditions of high concentration or low dielectric constant of the environment..

$$A^+ + B^- = A^+B^-$$

ion-pair partitioning \'ī-än-'pa(ə)r pär-'tish-ən-iŋ\ The separation and concentration of a charged species into an immiscible second phase by the formation of ion pairs. This partitioning can be achieved by selection of appropriate ionic species dissolved in the aqueous phase to form ion pairs, which can then partition between the two phases. More commonly, a compound will already exist as an ion pair in the organic phase (e.g., quaternary ammonium halides), and this solution is put in contact with the aqueous phase that contains the anion to be extracted. This anion will exchange with the halide ions from the quaternary ammonium compound and thus be extracted into the organic phase. Because ion pairs carry an effective zero charge, they can be present in a nonpolar organic solvent. Also known as liquid ion exchange, phase-transfer catalysis.

ion-selective electrode \'ī-än-sə-'lək-tiv i-'lek-ˌtrōd\ An electrochemical device that responds selectively to variations in activity of one or more ionic species in the presence of other ions. Contact between the reference electrode or solution and the solution under test is through a membrane across which a potential difference develops that is proportional to the activity of the ion. The pH electrode for hydrogen ion activity measurements is the most common example of such electrodes, in which a glass membrane separates the reference electrode and test solution. Other types of membranes include the following:

(1) homogeneous solid-state single-crystal membranes, e.g., LaF_3/F^- and AgX/X^- (where X is a halide), or compacted disk membranes of an insoluble salt, e.g., $Ag_2S \bullet CuS/Cu^{2+}$;

(2) heterogeneous membranes (AgX/X^-) in which the active material, an insoluble metal salt, is dispersed in an inert matrix such as silicon rubber;

(3) the incorporation of an organic liquid ion-exchanger into an inert membrane.

In this third type of membrane, the organic compound forms a complex of low aqueous solubility with the ion under test. Examples of such electrodes include analysis of cations such as calcium, ammonium, and potassium and anions such as nitrate.

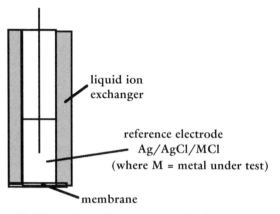

liquid ion
exchanger

reference electrode
Ag/AgCl/MCl
(where M = metal under test)

membrane

Because all of these electrodes are not specific for the ionic species under test, selectivity factors supplied by the manufacturer must be used where interferences may occur. Also, these electrodes depend on the Nernst equation, so the potential change obtained per decade change of activity depends on the charge (valency) on the ion. For example, this value is 59 mV per decade change for the monovalent cation H^+ whereas the value is 29.5 mV per decade change for the divalent cation Ca^{2+}; therefore, the precision of measurements decrease with the higher valency ions.

ion-selective field effect transistor \ī-än-sə-'lək-tiv 'fē(ə)ld i-'fekt tranz-'is-tər\ *See* ChemFET.

ionic strength \ī-'än-ik 'streŋ(k)th\ A function that allows the comparison of properties in solutions of different electrolytes (i.e., differing in ionic concentration or ionic charge). Ionic strength (I) is given by half of the summation of the molar concentration (C_i) of each of the ionic species in solution, multiplied by the square of their valences (z_i).

$$I = \frac{1}{2} \sum_i C_i z_i{}^2$$

ISCOMS \'iz-kämz\ Immune-stimulating complexes, complex structures (micelles) formed from antigenic molecules. Such complexes can be highly immunogenic.

ISFET \'iz-fet\ Ion-selective field effect transistor, used as a sensor for monitoring a particular ion. *See also* ChemFET, field effect transistor.

iso-pH \'ī-sō-,pē-,āch\ *See* isoelectric point.

isoamylase \ī-sō-'am-ə-lās\ *See* debranching enzyme.

isocratic elution \ī-sō-'kra-tik ē-'lü-shən\ A chromatographic elution technique in which the composition of the eluate remains constant. *See also* gradient elution, stepwise elution.

isoelectric focusing \ī-sō-i-'lek-trik 'fō-kəs-iŋ\ An electrophoretic method for separating proteins according to their isoelectric points (pI). Electrophoresis of the protein or protein mixture to be analyzed is carried out in a support medium such as acrylamide or agarose in the presence of ampholytes. Ampholytes are a synthetic mixture of polyamino, polycarboxylic acids, which under the presence of an electric field form a pH gradient. As the proteins being analyzed electrophorese through the gel, the charge on the protein is continuously changing because of changing pH of the environment until eventually the protein arrives at a pH value equal to its isoelectric point, at which point it stops or "focuses". The method is capable of resolving proteins that differ in their pI values by as little as 0.001 of a pH unit. The method is therefore particularly useful for analyzing purified proteins because minor postsynthetic modifications, such as the presence of one or more extra phosphate groups, can be detected by this method. Although essentially an analytical method, it can also be used preparatively.

isoelectric point (pI) \ˌī-sō-i-'lek-trik 'pȯint ('pē 'ī)\ The pH value at which a molecule that carries charged groups of opposite polarity (known as zwitterions or dipolar ions) has zero charge. This point occurs when the total positive and negative charges on the molecule balance each other out. Proteins, for example, carry both a number of positively ($-NH_3^+$) and negatively ($-COO^-$) charged groups. The relative number of these groups differs from protein to protein; therefore, different proteins tend to have different isoelectric points. When placed in a solution at a pH level below its isoelectric point, a protein has an overall positive charge, and when above its isoelectric point, it has an overall negative charge. However, at a pH value equal to its isoelectric point the number of positive and negative charges balance each other out, resulting in a zero overall charge. *See also* isoelectric focusing.

isogenic \ˌī-sō-'jen-ik\ Genetically identical.

isoglucose \ˌī-sō-'glü-kōs\ *See* high-fructose syrups.

isograft (isogenic graft) \ˈī-sō-ˌgraft (ī-sō-'jen-ik 'graft)\ A graft between genetically identical individuals, e.g., identical twins.

isomer(s) \ˈī-sə-mərz\ Compounds displaying isomerism.

isomerases \ī-'säm-ə-ˌrās-əz\ An enzyme class covering all enzymes that catalyze geometric or structural changes within a molecule. Isomerases are one of the six main classes (E.C. 5) used in enzyme classification. *See also* Enzyme Commission number.

isomerism \ī-'säm-ə-ˌriz-əm\ The ability of compounds with the same formula and molecular weight to possess different structural forms. Different types of isomerism are found; e.g., structural (atoms are joined together in different arrangements) and stereoisomeric (different arrangements in space(s) of the same grouping). *See also* chirality, cis isomers, dextrorotatory, E-isomers, levorotatory, optical isomerism, racemate, trans configuration, trans isomers, Z-isomers.

isopentenyladenine \ˌī-sō-'pent-ə-nil-ˌad-ᵊn-ēn\ *See* cytokinins.

isosbestic point \ī-säs-'bes-tik 'pȯint\ The defined wavelengths in the spectra of compounds in equilibrium at which the absorbance is constant and does not change as the relative concentrations of the compounds in equilibrium are varied. The presence of such isosbestic points may be taken to confirm the presence of such equilibria (for example, one isosbestic point infers the presence of two compounds in equilibria; two points, three compounds; etc.) The absence of an isosbestic point implies that the compounds in solution are not in equilibrium and that a process such as the one shown may be occurring:

$$A \rightleftarrows B \rightarrow C$$

Isosbestic points are useful for quantitative analysis because systems in equilibrium obey the Beer–Lambert law only at these wavelengths. *See also* Beer–Lambert law.

isoschisomere (isoschizomer) \ī-sō-'siz-ə-mər\ Two restriction enzymes that have the same target sequence are described as a pair of isoschizomers. For example, HpaII and MspI both recognize and cleave the following sequence:.

'5 CCGG 3'

3' GGCC 5'

isotonic solution \ī-sō-'tän-ik sə-'lü-shən\ A solution that has the same osmotic pressure as another solution with which it is being compared, but not necessarily the same chemical composition. *Compare with* hypotonic solution.

isotropic membrane \ī-sō-'träp-ik 'mem-ˌbrān\ A synthetic membrane that has a regular (symmetric) pore size throughout the film thickness. Isotropic membranes are generally used for microfiltration. *Compare with* anisotropic membranes.

itaconic acid \it-ə-'kän-ik 'as-əd\ Methylenesuccinic acid. It is produced industrially by the growth of *Aspergillus terreus* on either pure sugars or crude carbohydrates. Its major use is in the formation of polymers and in the synthesis of N-substituted pyrrolidones for use in detergents, shampoos, pharmaceuticals, and herbicides.

jasmine oil \\'jaz-mən 'ȯi(ə)l\ A compound isolated from plants of the *Jasminum* genus. It is used in the cosmetics industry as a perfume.

jet reactor \\'jet rē-'ak-tər\ A type of fermentor in which the gas and feed solution enter at the base of the tank in such a way as to produce a jet of material that provides the energy required to agitate the contents of the tank. It is used on the industrial scale for the production of yeast from whey. *See also* airlift fermentors.

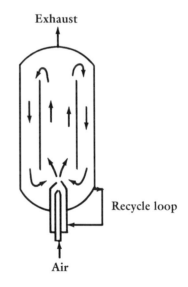

Exhaust

Recycle loop

Air

K

K_a \\'kā 'ā\\ *See* association constant.

K_d \\'kā 'dē\\ *See* dissociation constant.

K_L \\'kā 'el\\ In aeration studies of a submerged liquid culture, the mass-transfer coefficient (cm/h). K_L may be considered the reciprocal of the resistances to the transfer of oxygen from the gas to the liquid phase. It is usually considered in a combined term, $K_L a$.

$K_L a$ \\'kā 'el 'ā\\ In aeration studies, the volumetric transfer coefficient. It is a measure of the aeration capacity of a fermentor, where a is defined as the gas–liquid interface area per liquid volume. It is extremely difficult to measure K_L and a separately, but $K_L a$ can be determined by a sulfite oxidation technique, static gassing out, dynamic gassing out, or an oxygen balance technique.

karyotype \\'kar-ē-ə-ˌtīp\\ The chromosomal content of a cell. It can be abnormal in both overall quantity and the form of individual chromosomes. It is determined by light microscopy.

kbp (kb) \\'kā 'bē 'pē ('kā 'bē)\\ Abbreviation for 1000 base pairs of DNA. Because of the considerable size of genomic DNA and fragments derived from such DNA, the size of DNA is not referred to by molecular weight but more conveniently in terms of the number of kilobase pairs. Many workers abbreviate kbp to kb.

keratin \\'ker-ət-ᵊn\\ A structural protein with a high sulfur content found in hair, feathers, horn, and shells of vertebrates. It is sometimes used as an adsorbent material for effluent treatment.

kieselguhr \\'kē-zəl-gu̇(ə)r\\ A form of diatomaceous earth. It is commonly used as a filter aid for filtering bacteria or gelatinous suspensions and has a voidage of approximately 0.85. *See also* diatomaceous earth, filter aids.

kilobase pair \\'kil-ə-ˌbās 'pa(ə)r\\ *See* kbp.

kimal \\'kim-ᵊl\\ A type of porous particulate aluminium oxide (alumina) used as a chromatographic support.

kinases \\'kī-ˌnās-əz\\ Enzymes that catalyze the transfer of a phosphate group from ATP to another substrate. *See also* polynucleotide kinase.

kinematic viscosity \\kin-ə-'mat-ik vis-'käs-ət-ē\\ The ratio of the dynamic viscosity to density of a fluid measured at a constant temperature. *See also* viscosity.

kinetic viscosity \\kə-'net-ik vis-'käs-ət-ē\\ *See* kinematic viscosity.

kinetin (6-furfurylaminopurine) \\'kī-nə-tən ('siks-'fər-fə-ˌril-a-ˌmīn-ō-ˌpyú-rēn)\\ Probably the best-known cytokinin used in plant cell and tissue culture. It does not occur naturally in plants.

kinins \\'kī-nənz\\ In animals, peptides that are derived from physiological protein precursors (kininogens). They are found in the bloodstream and cause contraction of smooth muscle and dilation of blood vessels. Plant kinins are more commonly referred to as cytokinins. *See* cytokinins.

Kjeldahl nitrogen determination \\'kel-däl 'nī-trə-jən di-ˌtər-mə-'nā-shən\\ A method of determining total nitrogen content of a substance by digesting with concentrated sulfuric acid in the presence of a catalyst and thereby converting the nitrogen to ammonia. The amount can be determined by titration after the solution is made alkaline and the ammonia is distilled from the reaction mixture into an acid solution. Once determined, the value can be used (multiplied by 6.25) as an approximate indication of protein content (milligrams of protein per gram of dry weight).

Klenow fragment \\'klen-ō 'frag-mənt\\ Part of the *Escherichia coli* DNA polymerase I molecule, produced by the treatment of *E. coli* DNA polymerase I with subtilisin. The fragment still has the polymerase and $3' \rightarrow 5'$ exonuclease activity, but lacks the $5' \rightarrow 3'$ exonuclease activity of the original enzyme. It is used to "fill out" $5'$ or $3'$ overhangs at the ends of DNA molecules produced by restriction nucleases. In particular, this method is often used to radiolabel DNA molecules by using radiolabeled nucleotides. *See also* nick translation.

Koji \\'kō-jē\\ A fungal proteolytic enzyme preparation produced by growing *Aspergillus oryzae* on a solid substrate such as rice or soybean. The enzyme is produced commercially for treating flour proteins and is used as a digestive aid called taka-diastase.

Koji process \\'kō-jē 'präs-es\\ A traditional solid-substrate fermentation process for the fermentation of grain and soybeans by *Aspergillus* species. The cooked substrate is inoculated with a pure culture and grown in shallow layers, where amylases and proteases are produced to break down the substrate and form Koji. This process forms the basis of a number of fermentations, including the production of organic acids (such as citric acid), rice wine (sake), and other sweetened rice fermentations.

L

label \\'lā-bəl\\ A distinguishing feature or tag that can enable a particular molecule or group to be recognized. Labels include isotopes, radioactive or heavy atoms, immune labels, antibodies or antigens, and colored or fluorescent dyes. *See also* biotinylation, ELISA.

β-lactam antibiotics \\'bāt-ə 'lak-təm ˌant-i-bī-'ät-iks\\ Antibiotics in which the main structural feature is a *β*-lactam, a four-membered ring in which a carbonyl group and a nitrogen are joined in an amide linkage. These antibiotics inhibit enzymes involved in the synthesis of the bacterial cell wall. *See*, for example, cephalosporins, penicillin G.

β-lactamases \\'bāt-ə 'lak-tə-ˌmās-əz\\ *See* penicillinases.

lactase (β-galactosidase) (E.C. 3.2.1.23) \\'lak-ˌtās ('bāt-ə-gə-ˌlak-'tō-sə-ˌdās)\\ An enzyme (usually isolated from *Aspergillus niger*) that hydrolyzes lactose to glucose and galactose. It is used industrially to convert waste whey (essentially a lactose solution) from the cheese industry to a sweeter solution of glucose and galactose that is of use to the confectionery and baking industry. It is also used to hydrolyze lactose in skimmed milk to give a product suitable for consumption by people deficient in intestinal lactase, who would otherwise experience unpleasant intestinal effects. It also helps to prevent the crystallization of lactose in ice cream.

lactic acid \\'lak-tik 'as-əd\\ An organic acid with a range of industrial uses, especially as an acidulant and preservative in the food industry, where it is used as an additive to soft drinks, jams, syrups, and fruit juices. About half of the world's supply is produced industrially by the growth of *Lactobacillus* species on hydrolyzed starch or sugar solutions. It is also produced on a small scale by the fermentation of lactose in whey by organisms such as *L. bulgaricus*.

lactic acid bacteria \\'lak-tik 'as-əd bak-'tir-ē-ə\\ Gram-positive, nonsporulating bacteria (rods or cocci) that produce lactic acid as a major or sole product of fermentative metabolism. The major genera are *Enterococcus, Lactobacillus, Lactococcus, Leuconostoc, Pediococcus,* and *Streptococcus*. *See also* heterofermentative, homofermentative.

lactifers \'lak-ti-fərz\ *See* latex.

Lactobacillus \ˌlak-tō-bə-'sil-əs\ A genus of Gram-positive, nonsporulating, rod-shaped bacteria. Most species are homofermentative, but some are heterofermentative. They are commonly found in decaying plant materials, milk, or milk products. *L. bulgaricus* is used for the preparation of yogurt, *L. acidophilus* for acidophilus milk, and other species in the production of cheeses, pickles, sauerkraut, and silage.

lag phase \'lag 'fāz\ An initial period of time when growth does not occur after the introduction of a microorganism into a nutrient medium that supports its growth. The lag phase may be considered as a period of adaptation to the new environment.

lambda \'lam-də\ Compounds beginning with λ- are listed under their roman names.

lambda cloning vector \'lam-də 'klōn-iŋ 'vek-tər\ A lambda phage that has been genetically engineered for use as a cloning vector. *See* λ-insertion vector, λ-replacement vector.

lamellar settlers \lə-'mel-ər 'set-lərs\ Tanks (settlers) with plates or tubes (*lamellae*) inserted to increase settling efficiency by decreasing the distance a particle must settle and increasing the specific surface area of plates. These lamellae are fixed at inclined angles so that cells or solids slide down and are collected. Particles that adhere to the lamellae can be removed by occasional vigorous agitation. Several types of settler are available.

laminar flow \'lam-ə-nər 'flō\ Movement of a fluid parallel to the walls of the container. The characteristic of this type of flow is a low Reynolds number, the precise range of which depends on the vessel or channel geometry. Also known as streamline flow or viscous flow. The converse of turbulent flow. *See also* Reynolds number, turbulent flow.

Langmuir adsorption isotherm \'laŋ-myür ad-'sȯrp-shən 'īs-ə-ˌthərm\ An isotherm that relates the amount of a species adsorbed (q) to that remaining in solution (c) at constant temperature. The Langmuir isotherm is applicable to a system in which the adsorbing surface is completely homogeneous and there is negligible interaction between the adsorbed species, such that the adsorbing species forms a monomolecular layer on the adsorbent. Thus

$$q = \frac{Qk_L c}{1 + k_L c}$$

where Q is the maximum adsorbed concentration and k_L is the equilibrium constant.

laser Doppler anemometer \'lā-sər 'däp-lər ˌan-ə-'mäm-ət-ər\ A noncontact optical device for the measurement of fluid velocity based on the scattering of laser light by small particles in the fluid. The shift of frequency that occurs as the result of the Doppler principle allows the velocity of the fluid to be calculated.

latex \\'lā-teks\\ A milky emulsion of plant secondary metabolites (including terpenes, terpenoids, and alkaloids) that, on vulcanization, forms rubber. It is derived from a number of plant sources and produced in specific cells known as lactifers. Natural rubber is prepared from latex obtained from the plant *Hevea brasiliensis.*

leachate \\'lē-ˌchāt\\ The solution that leaves a leaching operation. It consists of solvent that contains the leached material as solute.

leaching \\'lēch-iŋ\\ Removal of soluble components from an inert matrix by the percolation of a solvent to produce a soluble portion (leachate) and a residue, as, for example, in the extraction of sugar from sugar beet or oil from rapeseed. *See also* microbial leaching.

lectins (phytohemagglutinins) \\'lek-tinz ('fīt-ō-ˌhē-mə-'glüt-ᵊn-ənz)\\ A class of proteins that bind to specific carbohydrates. They are classified according to their carbohydrate binding specificity. For example, the plant lectin concanavalin A is a D-mannose binding lectin. A whole range of lectins with different carbohydrate binding specificities have been identified. Originally thought to exist only in plants (from which most commonly used lectins are derived), lectins are now known to be widespread and to be involved in cell–cell attachment. Many, but not all, lectins, are glycoproteins. When immobilized on a suitable support material, they can be used for the separation and purification of glycoproteins, polysaccharides, and glycolipids and have been used for many years to agglutinate cells by linking cell surface carbohydrates. When linked to an appropriate marker (e.g., horseradish peroxidase), they are also used to detect glycoproteins on protein blots. Certain lectins are used to induce mitogenic activity in tissue culture. The term *lectin* is derived from the Latin *legere*, meaning to pick out or choose.

legumes \\'leg-yümz\\ *See Leguminosae.*

Leguminosae \\lə-ˌgyüm-ə-'nō-sē\\ One of the largest and most economically important families of flowering plants. One of the main reasons for their importance is their ability to fix atmospheric nitrogen in root nodules and to use the nitrogen compounds produced for growth (*see* nitrogen fixation). Such crops therefore need little or no nitrogen fertilizer. Crops include soybeans, lentils, groundnuts, and clovers (as animal fodder).

Lentinus edodes \\'lent-ə-nəs ē-'dō-ˌdēz\\ The edible shiitake mushroom. In Japan it is grown on logs of the shii tree (*Pasania cuspidata*) or oak trees (*Quercus* species). Infected wooden plugs are hammered into holes in the logs. After a year the logs are dampened. Mushrooms may be harvested from the rotting logs for about 5 years.

Leptospirillum ferrooxidans (Ferrovibrio) \\lep-tə-spī-'ril-əm ˌfer-ə-'äk-sə-denz (ˌfer-ō-'vib-rē-ō)\\ One of the organisms involved in bacterial leaching of minerals. It oxidizes iron or the iron moiety of pyrite, but not sulfur. *See* microbial leaching.

Leuconostoc \\'lü-kə-ˌnäs-täk\\ A genus of Gram-positive, nonsporulating bacteria that produce cocci in chains. They grow on fermentable carbohydrates with the production of lactic acid, CO_2, and ethanol. In environments in which sucrose is present, *L. mesenteroides* produces dextran slimes, which can lead to blockage of pipes in sugar refineries. This dextran has also been used medically as a blood plasma extender.

levorotatory \ˌlē-və-'rōt-ə-ˌtȯr-ē\ Able to rotate the plane of polarized light to the left, said of a chiral compound. Denoted by the prefix L-, as in L-dextrose. The converse of dextrorotatory. *See also* chirality.

library \'lī-brer-ē\ *See* gene library.

ligand \'lig-ənd\ A chemical species possessing lone pairs of electrons that can be donated to electron-deficient molecules such as metal ions, thus forming a metal–ligand complex (coordination complex). Ligands can possess one electron pair as in the case of ammonia, H_3N: monodentate ligand, or several as in the glycinate ion [$:NH_2COO:$]⁻ bi- or polydentate ligand. The term can also be used to describe a substance that interacts specifically and reversibly with a biological substance that is to be purified by affinity chromatography. The ligand (e.g., an enzyme cofactor, a triazine dye, or a monoclonal antibody) is usually immobilized to a solid support.

ligand-specific chromatography \ˌlig-ənd-spi-'sif-ik ˌkrō-mə-'täg-rə-fē\ *See* affinity chromatography.

ligases \'lī-ˌgās-əz\ An enzyme class covering all enzymes that catalyze the joining together of two molecules coupled with the hydrolysis of a pyrophosphate bond in ATP or a similar triphosphate. The systematic name is based on *X : Y ligase.* Use of the term *synthetase* is no longer recommended due to confusion with the term *synthase. See also* DNA ligase.

ligate \'lī-ˌgāt\ The biological substance that is specifically bound by an immobilized affinity ligand.

ligation \lī-'gā-shən\ *See* DNA ligase.

light chain \'līt 'chān\ *See* immunoglobulin G.

light scattering \'līt 'skat-ər-iŋ\ The scattering of electromagnetic radiation from polymers and their solutions, which provides information on the size and shape of molecules as determined by the weight-average molecular weight and radius of gyration.

lignin \'lig-nən\ An abundant aromatic polymer found as a major component of lignocellulose. It is formed by the polymerization of monomers such as *p*-hydroxycinnamyl alcohols, in which formation of a number of different bonds gives an irregular structure. These bonds are difficult to hydrolyze; thus lignin is highly resistant to microbial attack. Because of this resistance, lignin prevents access of microorganisms to the cellulose and hemicellulose components of lignocellulose. Straw, for example, is not extensively used as animal feed because very little of the cellulose polymer is accessible to the gut flora of ruminants. The development of efficient delignification processes, both to increase availability of the plant biomass for microbial degradation and to degrade lignin to usable aromatic compounds, is therefore receiving considerable effort. Methods for lignin breakdown will offer new routes to many important aromatic intermediates central to the chemical industry. Unfortunately, few microorganisms are known to degrade lignin. Both white-rot and soft-rot fungi can extensively degrade lignin, and brown-rot fungi and some bacteria can modify lignin. Of these, the white-rot fungus *Phanerochaete chrysosporium* shows considerable potential for industrial application to lignocellulose degradation.

lignocellulose \lig-nō-'sel-yə-ˌlōs\ Cellulose, hemicellulose (a collection of polysac-charides), and lignin. All species of woods and woody plants (including straw and bagasse) contain these three major components. For example, softwood contains 40% cellulose, 28% hemicellulose, and 28% lignin. Lignocellulose, derived from the agriculture and forestry industries, is the most abundant source of biomass available to the biotechnological industries for the production of fermentable sugars and carbohydrates. However, because of its crystalline-ordered structure, the insoluble nature of the cellulose fibers, and the protective nature of the lignin sheath surrounding the cellulose fibers, lignocellulose is almost resistant to direct enzymatic attack. Expensive and energy-demanding pretreatments (e.g., grinding, acid treatment, or steaming) of lignocellulose are therefore required to dissociate lignin from the polysaccharides and thus provide a substrate from lignocellulose that is suitable for microbial degradation. The development of methods for the direct microbial degradation of the lignin component of lignocellulose should help to overcome this problem in the future. *See also* lignin.

Lineweaver–Burk plot \'līn-wēv-er-'bərk 'plät\ A mathematical analysis of en-zyme reaction data that allows straight-line graphs to be obtained from re-action rate data as a function of sub-strate concentration thereby enabling determination of the maximum rate of reaction and the Michaelis constant. A reciprocal of the Michaelis–Menten equation is used:

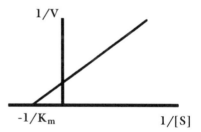

$$\frac{1}{V} = \frac{K_m + [S]}{V_{max}[S]}$$

where V is the rate of reaction, $[S]$ is the substrate concentration, V_{max} is the maximum rate of reaction, and K_m is the Michaelis constant. A plot of $1/V$ versus $1/[S]$ gives as gradient K_m/V_{max}, as y-intercept $1/V_{max}$, and as x-intercept $-1/K_m$. *See also* Eadie–Hofstee plot, Hanes–Wilkinson plot, Michaelis–Menten equation.

linker \'liŋ-kər\

(1) A short chain (usually a C_5—C_8 alkyl chain) that attaches either a ligand or an antibody to the support medium in an affinity column matrix. Such linkers are necessary to hold the ligand or antibody away from the support matrix thus avoiding any steric interference which may prevent the compound being purified from binding to the ligand or antibody.

(2) A short, synthetic, double-stranded polynucleotide containing one or more restriction endonuclease recognition sequences, used to attach sticky ends to a blunt-ended molecule. The linker is attached to blunt-end DNA by using T4 ligase and then cleaved by using the appropriate endonuclease to generate sticky ends. This DNA molecule can now be linked, by complementary base-pairing of sticky ends, to any other DNA fragment generated using the same restriction enzyme.

γ-linoleic acid (GLA) \'gam-ə ¸lin-ə-'lā-ik 'as-əd ('jē 'el 'ā)\ 6,9,12-Octadecatrienoic acid, a compound of importance to the pharmaceutical industry as a precursor in prostaglandin synthesis. It is currently produced from plant seeds (*Oenothera* species or *Boraginaceae*), but productivity is low because a long growth period and a huge area for harvesting seed are required. GLA is also produced by some fungi (e.g., *Mortierella* species), and production from this source is currently being investigated.

lipases \'lī-¸pās-əs\ Enzymes that hydrolyze triglycerides to free fatty acid, partial glycerides, and glycerol. Their natural substrates are triglycerides of long-chain fatty acids that are insoluble in water. Lipases, which are water soluble, hydrolyze the ester bonds at the interface between the aqueous phase and the insoluble substrate phase. Lipases are produced commercially from porcine and bovine pancreas and from microorganisms such as the yeast *Candida* and from *Aspergillus*, *Rhizopus*, and *Mucor* species. Commercial applications of lipases include the hydrolysis of oils for soap manufacture, use as a digestive aid, promotion of interesterification of oils and fats, formation of fatty acyl esters, enhanced flavor formation in certain cheeses, and use in the manufacture of cheese and butter flavors and in detergents.

lipofectin \¸lī-pə-'fek-tin\ *See* cytofectins.

lipophilic \¸lī-pə-'fil-ik\ Having an affinity for nonpolar species. The converse of hydrophilic. Also known as hydrophobic.

lipophobic \¸lī-pə-'fō-bik\ Having no affinity for nonpolar species. The converse of lipophilic. Also known as hydrophilic.

liposomes \'lī-pə-¸sōmz\ Small spheres whose walls are layers of phospholipid molecules. They are formed by mixing dry phospholipids (e.g., egg yolk or soybean lecithins) with water. As they form, liposomes entrap water and any water-soluble solutes that are present. Because of this entrapping ability, they are useful as drug-delivery systems. *See also* cytofectins.

liquefaction \¸lik-wə-'fak-shən\ A biological or chemical process whereby an insoluble solid organic medium is converted into a liquid or soluble form. An example is the process whereby starch granules are dispersed or gelatinized in solution and then partially hydrolyzed by thermostable α-amylase. *See also* starch.

liquid chromatography \'lik-wəd ¸krō-mə-'täg-rə-fē\ The separation and concentration of solutes by partition between a solid adsorbent and a mobile solvent phase. The adsorbent is usually in the form of a uniformly packed column to minimize diffusion effects. It produces a series of discrete bands that contain the separated components. Various adsorbents are used for particular types of separation. *See also* high-performance liquid chromatography.

liquid ion exchange \\'lik-wəd 'ī-än iks-'chānj\\ Liquid–liquid extraction in which the species being extracted from the aqueous phase undergoes an ion-exchange reaction transfer with ionic groups present in the organic phase as part of the organic extractant. Also known as ion-pair partitioning, phase-transfer catalysis.

liquid–liquid extraction \\'lik-wəd–'lik-wəd ik-'strak-shən\\ The partition of solutes between immiscible or partially miscible solvent phases. The partition may be assisted by the incorporation of molecules in either phase. Thus, in an aqueous–organic system, ionic salts in the aqueous phase may increase the partition of an organic compound (salting out). Similarly, compounds in the organic phase may combine with an inorganic solute, increasing its partition into the organic phase by forming a nonpolar complex. *See also* ion-pair partitioning, liquid ion exchange, two-phase aqueous partitioning.

liquid shear \\'lik-wəd 'shi(ə)r\\ The relative parallel movement of elements in a fluid under an applied force. The application of liquid shear forms the basis of the principal mechanical method for cell breakage used in the large-scale disruption of microorganisms. A suspension of bacteria is forced through a small orifice under very high pressure. The sudden drop in pressure on passing through the orifice causes the cells to expand rapidly and rupture. *See also* Hughes press, X-press.

liquid surfactant membrane extraction \\'lik-wəd sər-'fak-tənt 'mem-ˌbrān ik-'strak-shən\\ *See* emulsion liquid membrane extraction.

lithotroph \\'li-thō-ˌtrōf\\ *See* chemolithotroph.

load cell \\'lōd 'sel\\ A device for measuring the weight of a fermentor or for monitoring the feed rate from a small reservoir. The load cell is usually a solid or tubular steel cylinder, the compressive strain of which may be measured under axial load. Compressive strains over the appropriate range of loading are measured by means of electrical resistance strain gauges, which are cemented to the surface of the cylinder. Changes in resistance with strain are proportional to load. Load cells are available for a range of a few grams to thousands of tons. *See also* strain gauge.

loading \\'lōd-iŋ\\ The level or concentration of a solute in a defined amount of solvent or support material in separation processes, such as liquid extraction or chromatography.

log phase (exponential phase) \\'läg 'fāz (ˌek-spə-'nen-chəl 'fāz)\\ The period of microbial growth when cells grow at a constant maximum rate. After the introduction of a microorganism into a nutrient medium that will support its growth, there is an initial lag phase followed by an interval during which the growth rate of the cells gradually increases (acceleration phase) until the cells grow at a constant maximum rate. This period is referred to as the log, or exponential, phase.

long terminal repeat (LTR) \\'lóŋ 'tərm-ᵊn-əl 'rē-pēt ('el 'tē 'är)\\ An identical sequence of several hundred base pairs, introduced at each end of an integrated provirus during the synthesis and integration of viral DNA into a host cell genome during retrovirus infection.

loop (recycle) reactor \\'lüp ('rē-'sī-kəl) re-'ak-tər\\ A reactor that incorporates a recycle of the contents (*loop*),
together with additional makeup with the feed solution.
One design incorporates an airlift to effect the recycle.
This design has the advantages of no moving mechanical
parts, a high rate of oxygen transfer, and a high volume of
production. However, the disadvantages include problems
with uniform mass transfer and an energy-intensive system
with application limited to microbial systems. *See also*
airlift fermentors.

low-frequency restriction enzymes (LFREs) \\'lō-'frē-kwən-sē ri-'strik-shən 'en-
ˌzīmz ('el 'ef 'är 'ēz)\\ Certain restriction enzymes that produce longer restriction
fragments than the majority of restriction enzymes. Such LFREs are necessary
for mapping and sequencing whole genomes and producing genomic fragments
suitable for analysis by pulse-field gel electrophoresis. Most common restriction
enzymes recognize DNA sequences either four or six kilobase pairs long. For a
genome that is 50% G + C, these enzymes will produce fragments averaging 256
and 4096 kilobase pairs long, respectively. Three restriction enzymes recognize
sequences larger than six base pairs.

(1) *Rsr* II has partial specificity for a seventh base pair (recognition sequence
CGG CCG).

(2) *Not* I (recognition sequence GCGGCCGC) recognizes eight base-pair sequences.

(3) *Sfi* I (recognition sequence GGCCNNNNNGGCC, when N = G, C, T, or
A) also recognizes eight base-pair sequences.

However, for genome sequences of different base composition, other restriction
enzymes may be LFREs for that particular genome sequence. Also known as
infrequent cutters, rare cutters. *See also* restriction endonucleases.

LTR \\'el 'tē 'är\\ *See* long terminal repeat.

luciferases \\lü-'sif-ə-ˌrās-əz\\ *See* reporter gene.

luciferin \\lü-'sif-(ə-)rən\\ *See* bioluminescence.

lumen \\'lü-mən\\ The central cavity of a tube, especially used in the context of capillaries
and hollow fibers.

lux genes \\'ləks 'jēnz\\ Genes involved in coding for the enzyme luciferase.

lyases \\'lī-ˌās-əz\\ An enzyme class covering all enzymes cleaving C—C, C—O,
C—N, and other bonds by eliminating (leaving double bonds or rings) or
conversely adding groups to double bonds. The systematic name is based on the

substrate group lyase, and recommended names include decarboxylase, aldolase, and dehydratase (for the elimination of CO_2, aldehyde, and water, respectively). In the case where the reverse reaction dominates, the term *synthase* is used in the name. Lyases are one of the six main classes (E.C. 4) used in enzyme classification. *See also* Enzyme Commission number.

lymphocytes \'lim(p)-fə-ˌsīts\ *See* B-lymphocytes, T-lymphocytes.

lymphokines \'lim(p)-fə-ˌkīnz\ *See* cytokines.

lyophilization \lī-ˌäf-ə-lə-'zā-shən\ The removal of water from a frozen sample by the application of vacuum; also a drying method for the long-term preservation of microbial cultures. A few drops of culture solution are placed in an ampule and frozen in a dry-ice–ethanol mixture (-80 °C); then the plugged ampule is attached to a vacuum source. Moisture evaporates under vacuum by sublimation. The evaporation maintains the frozen state of the sample until moisture evaporation is complete; the result is a preserved, dry pellet. The dried ampule is sealed under vacuum and stored under refrigeration. Cultures are revived when required by adding a small volume of sterile water and streaking the suspension onto an agar medium. The method is also used to store protein samples in the solid state by lyophilization of aqueous protein solutions.

lyotrophic series \lī-ō-'trō-fik 'si(ə)r-ēz\ *See* Hofmeister series.

lysine \'lī-ˌsēn\ An amino acid, of considerable importance to industry as a feed and food supplement because it is an essential amino acid, found only in low levels in many cereals. It is produced industrially by the growth of *Brevibacterium lactofermentum* on glucose or *Corynebacterium glutamicum* on cane molasses. World production is 40,000 tons per year.

lysis \'lī-səs\ The rupturing of membranes or cell walls (if present) to release the cell contents. The bacterial cell wall consists of a cytoplasmic membrane surrounded by a rigid cell wall. Some species (e.g., *Escherichia coli*) have the cell wall surrounded by a second outer membrane. Lysis can be achieved by physical means such as mechanical force (*see* liquid shear, X-press) or by chemical means such as lysozyme (which digests the cell wall), detergents such as sodium dodecyl sulfate (SDS) or isooctylphenoxypolyethoxyethanol (Triton X-100) (which remove lipid molecules causing disruption of the cell membranes), or EDTA, which removes calcium ions that preserve the overall structure of the cell membrane. Often two or three of these are used together as a mixture.

lysogen \'lī-sə-jən\ *See* lysogenic infection.

lysogeny \lī-'säj-ə-nē\ *See* lysogenic infection.

lysogenic infection \lī-sə-'jen-ik in-'fek-shən\ A bacteriophage infection characterized by retention of the phage DNA molecule in the host bacterium for many thousands of cell divisions. For many lysogenic phages, the phage DNA is integrated into the host genomic DNA (*compare with* lytic infection cycle). This integrated form of viral DNA is referred to as a *prophage*. A bacterium carrying a prophage is physiologically indistinguishable from an uninfected cell. In this stable state the bacterium is referred to as a *lysogen* and is said to exhibit *lysogeny*. The integrated

DNA is carried through many cell divisions but eventually reverts to the lytic mode and lyses the cell. Both λ-phage and M13 exhibit lysogenic infection and have been used as cloning vectors. Infection with λ-phage involves integration into the genomic DNA, whereas M13 does not. *See also* M13.

lysozyme (E.C. 3.2.1.17) \\'lī-sə-ˌzīm\\ An enzyme produced commercially from chicken egg white. The enzyme catalyzes the hydrolysis of β-1,4-glycosidic bonds in the mucopeptide of bacterial cell walls and is therefore used as a method for lysing bacterial cells before product recovery. However, because of the cost, the enzyme is rarely used for large-scale extractions.

lytic infection cycle \\'lit-ik in-'fek-shən 'sī-kəl\\ The series of events displayed by a bacteriophage that replicates and lyses the host cell immediately after the initial infection. The phage attaches to the bacterium and injects its DNA into the cell. The DNA both replicates in the cell and directs the synthesis of capsid proteins; the result is the formation of new phage particles that are released by cell lysis. At no stage is the viral DNA incorporated into the genomic DNA. *Compare with* lysogenic infection.

M

M13 \\'em thər(t)-'ēn\\ A filamentous, lysogenic bacteriophage that contains single-stranded DNA. M13 is used as a cloning vector, in particular to produce single-stranded DNA for sequence determination. Infection of *Escherichia coli* results in the continuous synthesis and release of new phage particles without cell lysis. Bacterial growth continues uninterrupted, although at a slower rate than usual. The genome of M13 is 6407 nucleotides long.

mabs \\'mabz\\ An abbreviation for monoclonal antibodies.

mabzymes \\'mab-ˌzīmz\\ *See* catalytic antibody.

macerozyme \\'mas-ər-ō-ˌzīm\\ An enzyme preparation from *Rhizopus* species, rich in pectinase, that is used to degrade plant cell walls and produce protoplasts.

macrolide \\'mak-rə-ˌlīd\\ Any high-molecular-weight molecule that is not a polymer of smaller subunits of similar chemical type. The term is used to describe a group of antibiotics (macrolide antibiotics) derived from various strains of *Streptomyces*. They all consist of a macrocyclic lactone ring (12-, 14-, or 16-membered ring) to which novel amino or neutral sugars (e.g., erythromycin) are attached.

macromolecule \\ˌmak-rō-'mäl-i-kyü(ə)l\\ A high-molecular-weight polymeric compound (e.g., protein, carbohydrate, or nucleic acid).

macroporous resin \\ˌmak-rə-'por-əs 'rez-ən\\ An ion-exchange resin with an open porous structure capable of adsorbing large ions. These resins have better kinetics for large molecules than do gel resins, but they have a lower overall capacity. Because of the open pore structure, macroporous resins also have better resistance to osmotic shock than do gel resins. Also known as macroreticular resins. *Compare with* gel resin.

macroreticular resin \\ˌmak-rō-rə-'tik-yə-lər 'rez-ən\\ *See* macroporous resin.

magnetic separation techniques \mag-'net-ik ˌsep-ə-'rā-shən tek-'nēks\ The use of magnetic solids to enhance the separation of substances by the application of an external magnetic field. Enzymes immobilized on a magnetic solid can readily be separated from other particulate matter by the application of a magnetic field (i.e., a magnetic filter). The opposite effect can be generated, so that an enzyme attached in this way can be fluidized or stirred magnetically. Similar systems can be designed by attaching species to micromagnetic substances for affinity chromatography, ion exchange, or immunoassay. *See also* diamagnetism, ferromagnetism, paramagnetism.

magnetotactic bacteria (magnetic bacteria) \mag-'nēt-ō-'tak-tik bak-tir-ē-ə (mag-'net-ik bak-tir-ē-ə)\ Bacteria that migrate toward the magnetic north or south even in weak geomagnetic fields. These bacteria synthesize small (50–100 nm) magnetite (Fe_3O_4) particles that are surrounded by a membrane. These magnetite crystals align in rows of 10–20 particles, producing a biomagnet. The presence of this membrane around the particle allows other molecules to be attached to the particle. Potential uses of these particles include the formation of biosensors by immobilizing enzymes or antibodies on the membrane wall, or by producing magnetized conjugates or cells that can be directed within the body using a magnetic field.

MALDI \'mäl-dē\ *See* matrix-assisted laser desorption ionization.

malic acid   \'mal-ik 'as-əd\ A compound used as an acidifying agent in the food industry. It is produced from fumaric acid, either by fermentation with *Paracolobactrum* species or with immobilized *Brevibacterium flavum* as the source of fumarase.

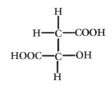

malt \'mȯlt\ Barley grains that have been steeped in water, drained, allowed to germinate, and dried at defined temperatures in a kiln. During this process, the starch and proteins in the grain are partially degraded. The kilning ensures that about 70% of the original dry weight will be available as soluble components such as dextrins, sugars, and proteins when mashed during wort preparation in a brewery.

mammalian cell culture \mə-'mā-lē-ən 'sel 'kəl-chər\ *See* animal cell culture.

manometry \mə-'näm-ə-trē\ A technique for measuring changes in gas pressure that result from the uptake or release of gases as a consequence of chemical or biological action, as in the Gilson respirometer or Warburg manometer.

Manton–Gaulin–APV homogenizer \'mant–en 'gȯl-in–'ā 'pē 'vē hō-'mäj-ə-ˌnīz-ər\ A device for the disruption of cells. The cells are forced through a small orifice onto an impact ring.

Manton-Gaulin
homogenizer head
(schematic)

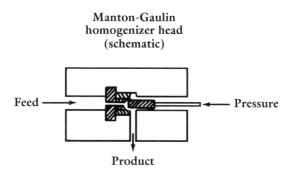

Feed →   ← Pressure

Product

map \\'map\\ A representation of the relative positions of genes or restriction sites and the distance between them. *See* genetic mapping, restriction map.

mass balance \\'mas 'bal-ən(t)s\\ The analysis of a process in which the mass of reactants is correlated with the mass of products according to the law of conservation of mass. That is,

$$[A] + [B] = [C] + [D]$$

mass flow sensor \\'mas 'flō 'sen(t)-sər\\ A type of sensor used to measure fluid flows. Three commercially available types are designed on two main principles; two are flow meters, and the other is an ultrasonic sensor.

(1) A true mass flow meter (e.g.,the axial flow–transverse momentum mass flow meter) uses the axial flow through a driven impeller and turbine in series. The impeller imparts angular momentum to the fluid. The fluid momentum causes a torque to be imparted to the turbine, which is restrained from turning by a spring. The measured torque is proportional to the rotational speed of the impeller and the mass flow rate of the fluid.

(2) In an inferential mass flow meter, a head meter (e.g., orifice or venturi) is used in conjunction with a densitometer. The signal from the head meter is proportional to ρV^2, where V is fluid velocity. Compensation for the density, ρ, gives V proportional to the mass flow.

(3) The latest device uses the scatter of ultrasonic radiation by the fluid and thus provides a sensor that has no contact with the fluid being measured.

mass spectrometer \\'mas spek-'träm-ət-ər\\ An instrument in which a molecule is fragmented into charged species of varying mass-to-charge (m/e) ratio by the application of high-energy electrons in a vacuum. Alternative methods for molecular fragmentation exist (e.g., chemical ionization (CI) and fast-atom bombardment (FAB)). The resulting charged species are separated by a combination of electric and magnetic fields according to their m/e ratios, and are then detected and displayed as a plot of abundance versus m/e. This device aids in clarification of structures of complex molecules. It also serves as a powerful and versatile detector for chromatography and as an on-line monitor for evolved gases and volatile products.

The requirement of biotechnologists to analyze high-molecular-weight compounds has led to the introduction of multisector instruments (MS–MS) that perform both functions of component separation and analysis. The multisector instrument incorporates repeated analyzers, which may include electrostatic analyzers and quadrupole analyzers in addition to double-focus mass spectrometers in various combinations and arrangements. *See also* chemical ionization, electron-impact ionization, electrospray mass spectrometry, fast-atom bombardment, matrix-assisted laser desorption ionization, quadrupole mass spectrometer, time-of-flight mass spectrometer.

mass transfer \'mas 'tran(t)s-fər\ The transfer of a given component in a system across space under the influence of driving forces such as concentration differences and temperature variation. This process may occur within a single phase (e.g., diffusion) or across boundaries (e.g., membranes). The term is used to describe the exchange of materials between the surface of an organism and its external environment (i.e., the uptake of nutrients and the elimination of waste metabolites).

mass-transfer coefficient \'mas-'tran(t)s-fər kō-ə-'fish-ənt\ A quantity that characterizes the extent of mass transfer within a system or across a boundary, defined as the ratio of mass flux to the difference between mass fractions on either side of the boundary. The mass-transfer coefficient can be correlated with the Reynolds (Re) and Schmidt (Sc) numbers in forced convection systems as:

$$Sh = a \, Re^b \, Sc^c$$

where Sh is the Sherwood number, and the coefficients *a*, *b*, and *c* are usually determined experimentally. However, predictions can be made in a number of flow regimes, such as laminar flow:

$$Sh = 0.664 \, Re^{1/2} \, Sc^{1/3}$$

For oxygen gas transfer

$$Sh = 0.13 \, Re^{1/4} \, Sc^{1/3}$$

For natural convective systems, the Grashof number, Gr, replaces the Reynolds number. Thus

$$Sh = a' \, Gr^{b'} \, Sc^{c'}$$

See also Grashof number, Reynolds number, Schmidt number.

Mastigomycotina \mas-ti-gō-,mī-kō-'tin-ə\ A subdivision of the fungi whose members have hyphae without cross walls. The sexual organs are the *antheridium* (male) and the *oogonium* (female). Fusion leads to the formation of *oospores* within *oogonia*. Many *Mastigomycotina* also asexually produce sporangia that contain *zoospores* (which are motile). Many fungi in this group are plant pathogens that cause blights (e.g., potato blight) and damping off, but none are used industrially.

matrix-assisted laser desorption ionization (MALDI) \'mā-triks ə-'sist-əd 'lā-zər dē-'zorp-shən ī-ə-nə-'zā-shən\ A technique for ionizing large biomolecules, particularly proteins, for subsequent mass spectral analysis. The technique involves the laser irradiation of a sample of the analyte crystallized in an organic matrix. The matrix molecules absorb the laser radiation and thus undergo electronic excitation and transfer considerable energy into the matrix–analyte mixture. This energy causes the ejection of molecules and ions from the surface. Because the laser operates in a pulsed mode, these ejected ions also occur in pulses so that a time-of-flight mass spectrometer is the method of choice for mass analysis. The majority of the ions produced carry only single charge, and their mass to charge spectrum thus gives a direct readout of the protein molecular mass.

Maxam–Gilbert method \'maks-əm 'gil-bərt 'meth-əd\ A chemical method for sequencing DNA. The radiolabeled DNA sample to be sequenced is divided into four aliquots. Each aliquot is chemically treated to cause random cleavage at one of the four nucleotides. The fragments generated from each reaction are separated in adjacent lanes on a polyacrylamide gel, and the DNA sequence is read directly by examination of the gel pattern obtained by autoradiography.

maxicells \'mak-sē-selz\ *Escherichia coli* cells that contain mainly plasmids, with very little chromosomal DNA. When certain *E. coli* cells are irradiated with UV light, chromosomal DNA is extensively degraded, but multicopy plasmids within the cell are relatively undamaged. Such plasmids continue to replicate after irradiation. This replication produces cells that contain mostly plasmid DNA and synthesize plasmid gene products almost exclusively. Such systems are ideal for studying plasmid-encoded products. *See also* minicells.

maximum oxygen-transfer rate \'mak-s(ə-)məm ˌäk-si-jən-'tran(t)s-fər 'rāt\ The product of volumetric mass-transfer coefficient and oxygen solubility in the medium used to evaluate oxygen-transfer rate capacity of a fermentor. It may be determined by the sulfite method, whereby sulfite ions are oxidized to sulfate by oxygen in the presence of copper ions as a catalyst. The residual sulfite concentration is determined by titration with iodine. Alternative methods of determination include (1) the use of dissolved oxygen probes to measure the dynamic oxygen balance in the fermentor and (2) the monitoring of the oxygen balance between the inlet and exhaust gas streams.

maximum specific growth rate (m_{max}) \'mak-s(ə-)məm spi-'sif-ik 'grōth 'rāt ('em 'maks)\ *See* critical dilution rate, Monod kinetics.

mediated membrane transport \'mēd-ē-ˌāt-əd 'mem-ˌbrān 'tran(t)s-pō(ə)rt\ *See* facilitated membrane transport.

medium \'mēd-ē-əm\ A solid or liquid substrate that can support the growth of an organism. All organisms require water, sources of energy, carbon, nitrogen, mineral elements, possibly vitamins, and oxygen if aerobic. The medium may consist of pure compounds (defined medium) or crude animal or plant extracts (complex medium). Media may be specially devised for inoculation, isolation, product formation, or sporulation.

medium-pressure liquid chromatography \\'mēd-ē-əm-,presh-ər 'lik-wəd ,krō-mə-'täg-rə-fē\\ A form of partition chromatography involving a solid matrix and a liquid mobile phase, operating under lower pressures (~40 bar) than usually experienced with high-performance liquid chromatography. This pressure level allows the use of glass columns. It may be used on a large scale with liquid flow rates of 3–160 cm^3/min for operations such as separation of proteins.

meiosis \\mī-'ō-səs\\ A form of cell division in which half the normal complement of chromosomes is generated in each of the daughter cells. These daughter cells are haploid, whereas the original cell is diploid.

melting of DNA \\'mel-tiŋ 'əv 'dē 'en 'ā\\ The separation of a DNA molecule into its component chains (denaturation) by heating a solution of the DNA. Hydrogen bonds are disrupted by the heating process. Most DNA samples begin to melt at about 70–80 °C; they are usually completely melted at 95 °C. The midpoint of the temperature range over which the strands of DNA separate is called the melting temperature, denoted T_m.

melting temperature of DNA (T_m) \\'mel-tiŋ 'tem-pə(r)-chu(ə)r 'əv 'dē 'en 'ā ('tē 'em)\\ *See* melting of DNA.

membrane \\'mem-,brān\\ An insoluble porous layer or film of material forming a division between two fluids, across which molecules or ions may be transported. Membrane pores of a defined size can be used in processing, for example, microfiltration, ultrafiltration, reverse osmosis, or dialysis. Various configurations of membrane equipment are possible, including flat sheets; leaf or plate, in which the membrane is cast directly onto the support; spiral wound, consisting of a sandwich of porous sheets and spacers wound into tubes; wide-bore tubes, 10–25-mm i.d.; narrow-bore tubes, 2–10-mm i.d.; and hollow fibers, 1–2-mm i.d.

Two types of synthetic membrane are available, anisotropic (asymmetric) and homogeneous (amorphous). *Anisotropic* membranes have a thin, dense layer (0.1–10-μm thick) supported by a much thicker (2–5-mm), spongy porous substrate. The thin, dense layer achieves the required separation with little resistance to permeation because of its thinness; the backing support provides strength. The openness of the structure allows rapid transport to the permeate, again with little resistance. *Homogeneous* membranes, as their name suggests, have the same structure throughout their thickness.

membrane distillation \\'mem-,brān ,dis-tə-'lā-shən\\ A technique whereby a substance is distilled through a membrane as a result of a temperature difference across the membrane. The pores of the membrane are filled with vapor of the substance being transferred, and the membrane is bounded by two liquid phases. *Compare with* perstraction, pervaporation.

membrane filtration \\'mem-,brān fil-'trā-shən\\ A number of processes in which a polymeric membrane acts as a barrier through which a fluid permeates. The pore size and construction of the membrane allow different-sized particles to be retained and thereby act as a separation process. *See also* dialysis, microfiltration, reverse osmosis, ultrafiltration.

membrane reactor \\'mem-ˌbrān rē-'ak-tər\\ Either a type of fermentor in which a membrane separates the feed and product streams or one in which a biocatalyst is immobilized on the membrane. Two main designs are used:

(1) The membrane separation unit is on-line with a reactor, and the reactor contents are pumped continuously through the separator. The permeate, containing the products, is removed while the retentate, containing the biocatalyst and nutrients, is returned to the reactor. The membrane separation processes that may be involved include micro- or ultrafiltration and pervaporation.

(2) The biocatalyst is immobilized within the membrane. Either a pressure difference drives nutrients and products across the membrane, or the module is diffusion-controlled so that nutrients and products occur on both sides of the membrane. *See also* dialysis fermentor.

membrane separation processes \\'mem-ˌbrān ˌsep-ə-'rā-shən 'präs-es-əz\\ The use of a membrane to assist in separation of reactants and products. This technique can separate mixtures with similar chemical and physical properties, as well as isomers and thermally unstable compounds. The processes can be used by themselves or in conjunction with conventional separation processes. *See also* dialysis, electrodialysis, microfiltration, pervaporation, ultrafiltration.

meristem \\'mer-ə-ˌstem\\ Concentrated regions of active cell division in plants, where most coordinated cell division takes place and from which permanent tissue cultures can be derived (e.g., tips of stems and roots).

Merrifield synthesis \\'mer-i-ˌfē(ə)ld 'sin(t)-thə-səs\\ Generally, the solid-phase synthesis of peptides, a procedure originally pioneered by Bruce Merrifield and his co-workers.

mesophiles \\'mez-ə-ˌfīlz\\ Any microorganisms that grow in the temperature range 20–45 °C. *See also* psychrophiles, thermophiles.

messenger RNA (mRNA) \\'mes-ᵊn-jər 'är 'en 'ā ('em 'är 'en 'ā)\\ An RNA molecule synthesized from a DNA template by the enzyme RNA polymerase. The mRNA sequence is complementary to (i.e., base-pairs) the DNA sequence. *See also* RNA polymerase, transcription.

metabolism \\mə-'tab-ə-ˌliz-əm\\ The physicochemical processes in which nutrients are converted into energy and metabolites. Metabolites may be simpler molecules (produced by catabolism), which in turn may be converted by anabolism to form more complex molecules. *See* anabolism, catabolism.

Metafilters \met-ə-'fil-tərz\ Trade name for pressure leaf filters.

metal oxide silicon field effect transistor (MOSFET) \'met-ᵊl 'äk-ˌsīd 'sil-i-kən 'fē(ə)ld i-'fekt tranz-'is-tər ('mäs-fet)\ Also known as modified field effect transistor. *See* field effect transistor.

metalloproteases \mə-'tal-ə-ˌprōt-ē-ās-ez\ One of the four categories of classification of proteases; the others are acid (or carboxyl), serine, and thiol proteases. All metalloproteases have a metal ion involved in the catalytic mechanism and are therefore inhibited by chelating agents such as EDTA.

methanogen \mə-'than-ə-jən\ A methane-producing strain of obligate anaerobic bacteria. This group includes the following genera: *Methanobacterium, Methanococcus, Methanosarcina.*

methylotroph \'meth-ə-lə-ˌtrōf\ An organism that can grow on organic compounds that contain no carbon-carbon bonds. The genera include *Methylomonas* and *Methylococcus,* which use methane, methanol, or formaldehyde as their sole carbon and energy sources.

MIC \'em 'ī 'sē\ *See* minimum inhibitory concentration.

micellar electrokinetic capillary chromatography (MECC) \mī-'sel-ər i-ˌlek-trō-kə-'net-ik 'kap-ə-ˌler-ē ˌkrō-mə-'täg-rə-fē ('em 'ē 'sē 'sē)\ A capillary electrophoretic technique that allows the resolution of anions, cations, and neutral species by combining electrophoretic mobility with partitioning between the buffer and surfactant micelles. Under the applied potential, the surfactant micelles move along the capillary according to their micellar charge, which in the case of sodium dodecylsulfonate (an anionic micelle) is opposed to the electroendosmotic flow. Thus, the sample migration rates are between those of electroendosmosis and micellar migration, depending on the relative interaction of the components with the micelles.

This technique has an advantage over capillary zone electrophoresis in that both charged and uncharged species can be detected simultaneously. However, resolution is limited if all components are similar in properties such as polarity or hydrophobicity.

micellar enhanced ultrafiltration \mī-'sel-ər in-'han(t)st ˌəl-trə-fil-'trā-shən\ A technique used mainly in the purification of wastewaters for the separation of substances usually too small to be removed by ultrafiltration. The process involves the adsorption of the impurities into micelles formed by surfactants. This adsorption increases the particulate size and enables ultrafiltration to be used successfully. Cationic surfactants are used to form micelles for nonionic organic materials, and anionic surfactants are used for metallic impurities.

micellar extraction \mī-'sel-ər ik-'strak-shən\ An extraction and concentration technique for biomolecules and other species whereby the solute is contained in the aqueous core of a reverse micelle formed by a suitable surfactant molecule. This process allows the extraction of biomolecules from an aqueous phase into an oil phase without altering or denaturing the molecule. The technique consists of mixing the aqueous feed solution with a suitable surfactant and an immiscible organic phase. After extraction of the solute, the two bulk phases are separated and the organic phase is mixed with another aqueous phase to recover the biomolecule. Also known as reverse micellar extraction. *See also* micelle. (Figure: P represents the protein and —O the surfactant molecule.)

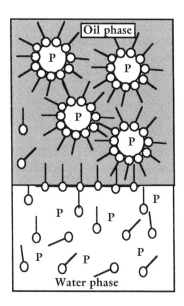

micelle \mī-'sel\ A colloid formed by the reversible aggregation of surface-active molecules above a critical micelle concentration (cmc) in a solvent. In aqueous solution the hydrophobic parts of the molecule aggregate, leaving the polar groups on the surface of the micelle, where they interact (hydrogen bond) with the water molecules (normal micelle). When placed in a nonpolar medium the reverse situation applies, and the polar groups aggregate together, usually in the presence of a core of water molecules leaving the hydrophobic tails to interact with the oil phase (reversed micelle). *See also* critical micelle concentration, micellar extraction.

normal micelle

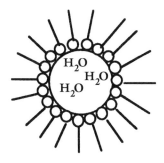

reverse micelle

Michaelis–Menten equation   \mī-'kā-ləs 'men-tən i-'kwā-zhən\ A description of enzyme kinetics that assumes formation of an intermediate enzyme–substrate complex which dissociates to form product and unchanged enzyme:.

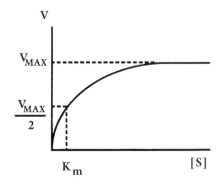

$$E + S \longrightarrow E-S \longrightarrow E + P$$

The rate of reaction (V), is given by

$$V = \frac{V_{\max}[S]}{[S] + K_m}$$

where $V_{\max}$ is the maximum reaction rate, $[S]$ is the substrate concentration, and K_m is the Michaelis constant, which is equal to the substrate concentration at half the maximum rate of reaction. *See also* Eadie–Hofstee plot, Hanes–Wilkinson plot, Lineweaver–Burk plot.

Mickle tissue disintegrator \'mik-ᵊl 'tish-ü dis-'int-ə-ˌgrāt-ər\ *See* bead mill.

microbial leaching \mī-'krō-bē-əl 'lē-chiŋ\ The use of microorganisms to solubilize metals in industrial extraction processes. The extraction of metals from insoluble minerals (mainly mineral sulfides) is achieved through leaching by acidophilic iron-oxidizing and sulfur-oxidizing bacteria. This solubilization of metals from their ores produces concentrated solutions of metals that can be recovered with hydrometallurgical processes. Although a number of microorganisms are involved in these processes, the principal acid-generating microorganism affiliated with the microbial degradation of sulfide mineral is *Thiobacillus ferrooxidans.* This organism oxidizes ferrous iron and reduced-oxidation-state sulfur compounds. At present, the process is largely confined to the copper and uranium industries.

microcarrier \'mī-krō-ˌkar-ē-ər\ A small, beaded matrix ~(100–200-μm diameter) used as a support for the culture of anchorage-dependent cells. Beads can be made of such materials as silica, glass, collagen, DEAE-Sephadex, and polyacrylamide. Cells grown in this way can be treated as a suspension-cell culture, and fermentation technology processes for suspension culture can be used.

***Micrococcus luteus* polymerase** \mī-krō-'käk-əs 'lüt-ē-əs 'päl-ə-mə-ˌrās\ A DNA polymerase that also has $5' \rightarrow 3'$ exonuclease activity. It is used to create small, single-stranded regions in cloned DNA molecules starting from a nick.

microemulsions \mī-krō-i-'məl-shənz\ Transparent, thermodynamically stable dispersions of very small droplets, typically in the range of 10–100 nm. Two types of dispersion, oil-in-water (micelle) and water-in-oil (reversed micelle), may be formed spontaneously by interactions between surfactants and possibly cosurfactants, such as long-chain alcohols. The ability of these microemulsions to solubilize hydrophilic macromolecules in a water-in-oil emulsion provides a mechanism for the extraction of enzymes, separation of biopolymers, and as a novel reaction medium for biochemical reactions. *See also* micellar extraction, micelle, Winsor-type microemulsions.

microencapsulation   \mī-krō-in-ˌkap-sə-'lā-shən\ The encapsulation of enzymes within spherical semipermeable polymeric membranes. This method is commonly used as an enzyme immobilization method but is also used as a drug-delivery system. The membrane material is chosen such that large molecules (e.g., enzymes) are retained within the sphere (artificial cell), although substrate and product molecules can freely diffuse across the membrane. Membrane materials that have been used successfully include cellulose nitrate, epoxy resins, gelatin, nylon, polyamides, polyurethanes, and silicone rubber. *See also* liposomes.

microfiltration   \mī-krō-fil-'trā-shən\ A process for the membrane-filtration removal of particles in the 0.1–10 μm range. The technique can be used in either a crossflow (tangential-flow) or static dead-end configuration. The separation is characterized by the pore size of the symmetrical microporous polymer membrane and the particle size of the impurities. Two levels of particulate removal are generally considered: nominal rating (90–93% removal) and absolute rating (100% removal). The driving force for the process is a pressure difference across the membrane, usually of the order $1-5 \times 10^5$ Pa. The technique may be used for sterile filtration, clarification, cell harvesting, and bacteria or virus separation. *See also* crossflow filtration, dead-end filtration.

microfluorimeter   \mī-krō-flú(-ə)r-'im-ət-ər\ A device for automatically counting cells by means of the fluorescence that results from passing a suitably stained cell through a laser beam. The device can be modified to determine components in the biomass by selective staining (e.g., proteins or nucleic acids). In addition, the use of multiple lasers and optical processing of the fluorescence signal will allow more than one stained species to be counted simultaneously. *See also* flow cytometry.

microheterogeneity   \mī-krō-ˌhet-ə-rō-jə-'nā-ət-ē\ The state of a particular purified preparation of a macromolecule (usually a protein) in which subpopulations of molecules exist that differ slightly in their structure. A whole range of postsynthetic modifications can be responsible for microheterogeneity in a protein. For example, some molecules may be phosphorylated at two positions, some at only one, others not at all. Other causes of microheterogeneity in a protein include acetylation or methylation of lysine side chains and the covalent linkage of polysaccharide side chains. *See also* glycoform, post-translational modification.

microinjection   \mī-krō-in-'jek-shən\ A method for introducing DNA into a cell by injecting it directly into the cell nucleus with an extremely fine needle, usually a glass micropipet that has been drawn out to a diameter of 0.1–0.5 μm.

microplasts   \'mī-krō-ˌplasts\ Small vesicles produced by the fragmentation and subdivision of protoplasts. They contain small amounts of the original genetic material from the parent protoplast and can be fused with a complete protoplast. This fusion introduces new genetic information into the protoplast. It is the equivalent of the use of minicells to transform bacterial cells.

micropropagation   \mī-krō-ˌpräp-ə-'gā-shən\ The clonal propagation of plants in vitro.

microtiter plate   \'mī-krō-ˌtīt-ər 'plāt\ Plastic plates (polystyrene, polypropylene, or polyvinyl) consisting of eight rows of 12 flat-bottomed wells, each with a volume of about 400 μL. Microtiter plates are used in particular for ELISA assays, in which antigen or antibody is adsorbed onto the surface of each well and all antigen–antibody reactions take place in solution inside the well.

mimetics \mə-'met-iks\ Small organic molecules or peptides that resemble the structure of the antigen-binding region (complementarity-determining region) or an antibody molecule. Such compounds mimic the binding and biological activity of that antibody and have potential uses as pharmacological agents. Also known as CDR mimetics.

mineral leaching \'min(-ə)-rəl 'lēch-iŋ\ *See* microbial leaching.

minicells \'min-ē-selz\ Small, nongrowing bodies produced by aberrant cell division at the polar ends of bacteria. They do not contain chromosomal DNA. The introduction of a plasmid, cloned DNA, or phage genome into a minicell results in RNA and protein synthesis, which can be analyzed in the absence of host cell products because genomic DNA is absent. Minicells are particularly useful for analyzing plasmid-encoded gene products. *See also* maxicells.

minimal medium \'min-ə-məl 'mēd-ē-əm\ A defined medium that provides only the minimum number of nutrients needed for the growth of a particular microorganism.

minimum inhibitory concentration (MIC) \'min-ə-məm in-'hib-ə-ͺtōr-ē ͺkän(t)-sen-'trā-shən ('em 'ī 'sē)\ The lowest concentration of an antibiotic or similar compound in a dilution series of broth tubes that shows no growth of the test organism after incubation for a standard time and temperature. Standard-sized drops of 24-h cultures are used as inocula.

minisatellites \ͺmin-ē-'sat-ᵊl-ͺīts\ *See* DNA fingerprinting.

miscibility \ͺmis-ə-'bil-ət-ē\ The ability of two substances to form a single phase at all relative concentrations. *See also* immiscibility, partially miscible substances.

mismatching \mis-'mach-iŋ\ Regions in a double-stranded DNA molecule or DNA–RNA hybrid, where the bases on the respective strands are noncomplementary and therefore do not hydrogen-bond.

mitDNA \'em 'ī 'tē 'dē 'en 'ā\ *See* mtDNA.

mitogen \'mīt-ə-jən\ Any compound that stimulates a cell to divide. Mitogens may be normal physiological effectors such as hormones or growth factors (*see*, for example, epidermal growth factor) present in normal serum, or they may be compounds such as plant lectins or bacterial lipopolysaccharides, which stimulate cultured cells to divide but whose normal in vivo role is not to act as a mitogen.

mitosis \mī-'tō-səs\ The common form of cellular division, in which the normal complement of chromosomes is generated in each of the daughter cells.

mitotic cycle \mī-'tät-ik 'sī-kəl\ *See* cell cycle.

mitotic index \mī-'tät-ik 'in-deks\ The percentage of a total cell population of a culture that, at a given time, exhibits some stage of mitosis.

mobile phase   \\'mō-bəl 'fāz\\ In chromatography, the carrier phase (liquid or gas) that moves relative to the stationary phase (support) to bring about partition and separation.

mobilization (*mob*) genes   \\mō-bə-lə-'zā-shən ('mäb) 'jēnz\\ Genes involved in the transfer of a plasmid from one cell to another. If a plasmid that has had its transfer(*tra*) genes deleted is present in a cell with a plasmid that has *tra* genes, then this plasmid is capable of being transferred to another cell (*see* transfer genes) as long as it contains the mobilization (*mob*) genes. The plasmid is said to be mobilized. Laboratory-used plasmids therefore have their *mob* genes (as well as their *tra* genes) deleted to prevent the possibility of mobilization of recombinant plasmids if they should escape from the laboratory environment.

modified field effect transistor (MOSFET)   \\'mäd-ə-ˌfīd 'fē(ə)ld i-'fekt tranz-'is-tər ('mäs-ˌfet)\\ Also known as metal oxide silicon field effect transistor. *See* field effect transistor.

moiety   \\'mȯi-ət-ē\\ One of two parts. For example, one talks about the carbohydrate moiety of a glycoprotein.

molasses   \\mə-'las-əz\\ The residue remaining after the repeated crystallization of sugar extracted from sugar cane or sugar beet. It contains sucrose, glucose, fructose, and, for molasses from beets, raffinose and some dextrans. It is a byproduct from the manufacture of sugar and comprises slightly purified sucrose. Beet sugar molasses is 48% sucrose, 2% other sugars, and 21% organic nonsugar substances. Cane sugar molasses is 33% sucrose, 21% invert sugar, and 20% organic nonsugar substances. Molasses, together with a nitrogen supplement, is used almost exclusively for the manufacture of baker's yeast.

molecular chaperones   \\mə-'lek-yə-lər 'shap-ə-ˌrōnz\\ *See* chaperones.

molecular diffusivity   \\mə-'lek-yə-lər dif-yü-'siv-ət-ē\\ The rate of diffusion of a molecule. Because diffusivity depends on the physical state of the surrounding fluid, diffusivities in gases are about 10^4 times greater than in liquids. Diffusivities can be estimated on the basis of the kinetic theory, which is much better developed for gases than for liquids.

molecular distillation   \\mə-'lek-yə-lər ˌdis-tə-'lā-shən\\ A distillation process carried out under very low pressures (1.3 N/m^2 or less) in equipment with a short distance between the surface of the liquid and the condensing surface. Because the mean free path of the molecules in the vapor is the same order of magnitude as the distance to the condensing surface, rapid distillation is possible and thermal decomposition is minimized. The technique can be used for the final concentration of vitamins and natural products as an alternative to freeze-drying.

molecular exclusion chromatography   \\mə-'lek-yə-lər iks-'klü-zhən ˌkrō-mə-'täg-rə-fē\\ A chromatographic purification technique for separating molecules according to size. Also known as gel filtration.

molecular sieve   \\mə-'lek-yə-lər 'siv\\ A material, often a zeolite, that contains channels or pores of a defined size which can retain solutes by molecular size interaction.

Molecular sieves can be used to remove solutes from solutions or can act as a chromatographic support. *See also* zeolites.

molecular weight \mə-'lek-yə-lər 'wāt\ The sum of the relative atomic masses of all of the atoms in a given molecule, abbreviated MW, mol wt, and M. Units are Daltons.

molecular weight of polymers \mə-'lek-yə-lər 'wāt 'ev 'päl-ə-mərz\ The molecular weight of polymeric molecules is difficult to define unambiguously because of the statistical nature of the polymeric process and because polymer chains have different lengths. Therefore, the experimentally determined molecular weights of polymers are average values and are defined in terms of the molecular weight of individual molecular species (M_i) and either the number of such molecular species (n_i) as in number average molecular weight (M_n), or the weight of such individual species (w_i) in the weight average molecular weight (M_w).

$$M_n = \frac{\Sigma n_i M_i}{\Sigma n_i}$$

$$M_w = \frac{\Sigma w_i M_i}{\Sigma w_i}$$

Number average molecular weights can be obtained from the analysis of the quantity of end-groups on the polymer chains, and colligative properties such as lowering of vapor pressure, cryoscopy, vapor-phase osmometry, which depend on the number of particles in solution.

Weight average molecular weights are obtained from light scattering or ultracentrifugation measurements. Molecular weight distribution of polymers can be obtained from fractionation of the polymer or from gel permeation chromatography of the polymeric solution.

molar absorption coefficient \'mō-lər əb-'zȯrp-shən ˌkō-ə-'fish-ənt\ *See* absorption coefficient, Beer–Lambert law.

monoclonal antibody \ˌmän-ə-'klōn-ᵊl 'ant-i-ˌbäd-ē\ An immunoglobulin produced by a single clone of lymphocytes. A monoclonal antibody recognizes only a single epitope on an antigen and therefore cannot precipitate antigens by forming a cross-linked three-dimensional precipitation matrix, as is the case with a polyclonal antiserum. In the laboratory, monoclonal antibodies are produced from hybridoma cells. *See also* bispecific antibodies, chimeric antibody, hybridoma cell, phage display antibody.

Monod kinetics \'mō-näd kə-'net-iks\ An expression for the growth rate of microbial species:.

$$\mu = \frac{\mu_{max}[S]}{K_s + [S]}$$

where μ is specific growth rate, μ_{max} is maximum specific growth rate, K_s is saturation constant (Monod coefficient), and $[S]$ is substrate concentration.

monolayer \'män-ō-ˌlā-ər\ A single layer of adsorbed molecules on a support or at a fluid interface; a one-molecule-thick film.

monolayer culture \\'män-ō-ˌlā-ər 'kəl-chər\ A single layer of cells on the surface of a culture vessel. Most cultures of mammalian cells grow by attachment to a glass or plastic surface (showing anchorage dependence). This attachment forms a monolayer.

monomer \\'män-ə-mər\ The repeat unit in a polymeric structure; the simplest molecule capable of producing that polymer.

monosaccharides \\ˌmän-ə-'sak-ə-ˌrīdz\ *See* carbohydrates.

monosodium glutamate (MSG) \\ˌmän-ə-'sōd-ē-əm 'glüt-ə-ˌmāt ('em 'es 'jē)\ A compound much used as a flavor enhancer in the food industry. It is produced by fermentation of *Corynebacterium* or *Brevibacterium* species. World production exceeds 100,000 tons per year. *See also* glutamic acid.

montmorillonite \\ˌmänt-mə-'ril-ə-ˌnīt\ A clay mineral (aluminosilicate) with variable composition and pronounced adsorptive properties, which may be used as a filter aid or adsorbent.

morgan \\'mȯr-gə(n)\ *See* centimorgan.

MOSFET \\'mäs-ˌfet\ Abbreviation for metal oxide silicon (modified) field effect transistor. *See* field effect transistor.

mRNA \\'em 'är 'en 'ā\ *See* messenger RNA.

mtDNA \\'em 'tē 'dē 'en 'ā\ DNA found in mitochondria (mt). Also known as mitDNA.

Mucor \\'myü-kȯ(e)r\ A fungal genus classified in the *Zygomycotina*, which is widespread in soil and often causes storage rots and other spoilage problems in materials contaminated with soil. *M. lacemosus* can be grown in submerged culture to produce ethanol. *M. miehei* and *M. pusillus* both produce enzymes used as rennin substitutes.

multichamber centrifuge \\ˌməl-tī-'chäm-bər 'sen-trə-ˌfyüj\ A modification of a tubular bowl centrifuge in which the bowl consists of a series of concentric tubular sections of increasing diameter that form a continuously enlarging passage for the liquid flow. The feed is introduced into the center of the bowl and hence into the smallest-diameter tube, where it experiences the least centrifugal force. From the center it passes by channels into chambers of increasing diameter, and hence increasing centrifugal force. The heavier particles are deposited in the smaller chambers and the lighter particles in chambers with a larger diameter. The total holding volume for solids is not improved by sectioning the bowl into chambers, but separation of the particles by size improves the clarification of the emerging liquid. The largest devices are used to clarify fruit juices and beer.

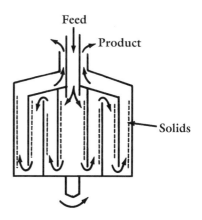

Feed

Product

Solids

multidentate ligand \məl-tē-'den-ˌtāt 'li-gənd\ *See* polydentate ligand.

murine \'myu̇(ə)r-īn\ Pertaining to animals belonging to the *Muridae* family, which includes mice and rats. Many monoclonal antibodies are produced in mice; hence, they are murine monoclonal antibodies.

mutagenic \ˌmyüt-ə-'jen-ik\ *See* mutation.

mutant \'myü-tənt\ *See* mutation.

mutation \myü-'tā-shən\ A chemical change in the DNA of an organism leading to a change in the genetic character. The DNA change is inherited unless the mutation is lethal. A strain exhibiting such a changed characteristic is termed a *mutant*. Each time a microbial cell divides, there is a small probability of an inheritable change occurring (spontaneous mutation). However, mutations are often deliberately introduced (induced) into organisms in an attempt to produce an improved strain. Mutations can be achieved by the use of UV radiation, ionizing radiation, or chemicals (e.g., nitrous acid). Any such compound or energy source that introduces a mutation is known as a *mutagen* and is said to be *mutagenic*. *See also* frameshift mutation, point mutation.

mutualism \'myüch-(ə-)wə-ˌliz-əm\ A system in a culture containing two microbial species in which both species grow faster together than they would grow under the same conditions in pure culture. *See also* amensalism, commensalism, neutralism.

mycelium \mī-'sē-lē-əm\ The collective term for a network of hyphae (pl. mycelia).

mycoplasmas \ˌmī-kō-'plaz-məz\ The smallest living cells known. They are usually spherical, with a diameter varying from 0.3 to 0.8 μm. However, filamentous forms also exist, with diameters of 0.1–0.3 μm and lengths of up to 150 μm. A solution that has been "sterilized" by filtration through a 0.2-μm filter can therefore still contain mycoplasmas. Up to 15% of all cell cultures in the world are contaminated by mycoplasmas, many of which exist symbiotically with growing cells. Bovine mycoplasmas (45%), human oral species (33%), and porcine mycoplasmas (21%) are the main contaminating species. The elimination of mycoplasmal contamination from biotechnological products like monoclonal antibodies and viral vaccines is a major problem, but antibiotics such as BM cycline are proving successful in eliminating the problem.

mycoprotein \ˌmī-kō-'prō-ˌtēn\ A food product (single-cell protein) consisting of fungal mycelia, produced by Rank–Hovis–MacDougall for human consumption. The organism used is a strain of *Fusarium graminearum*. The mycoprotein production process involves continuous fermentation with glucose as substrate and ammonia and ammonium salts as nitrogen sources. Following fermentation, the culture is heat-treated to reduce the RNA content. The mycelium is separated by vacuum filtration.

mycorrhiza \ˌmī-kə-'rī-zə\ An association of a nonpathogenic fungus with the root of a plant. In *ectomycorrhiza*, caused by basidiomycete and ascomycete fungi, the fungi proliferate to form a sheath around the fine roots. In *endomycorrhiza*, the fungi invade root cortical tissues. A form of endomycorrhiza known as vesicular–arbuscular (VA) mycorrhiza has particular agricultural importance. *See also* vesicular–arbuscular mycorrhiza.

mycotoxin   \mī-kə-'täk-sən\ Any low-molecular-weight fungal secondary metabolite capable of producing toxic effects in humans or animals. They are produced most commonly on grains and nuts stored in humid conditions. *Aspergillus flavus* produces a series of aflatoxins that are extremely toxic. These toxins are heat-stable and normally withstand food-preparation temperatures.

myeloma cell   \mī-ə-'lō-mə 'sel\ An immortalized (malignant) antibody-secreting tumor cell derived from a B-lymphocyte. *See also* hybridoma cell.

N

N-terminal sequencing \\'en 'tərm-ən-ᵊl 'sē-kwən-siŋ\\ *See* Edman degradation.

N-terminus \\'en 'tər-mə-nəs\\ *See* amino terminal.

nabla factor \\'nab-lə 'fak-tər\\ *See* del factor.

NAD⁺ \\'en 'ā 'dē 'pləs\\ Nicotinamide adenine dinucleotide.

NADP⁺ \\'en 'ā 'dē 'pē 'pləs\\ Nicotinamide adenine dinucleotide phosphate.

nanofiltration \\nan-ō-fil-'trā-shən\\ A pressure-driven membrane filtration process similar to ultrafiltration but with membrane pore sizes of <0.001 nm and operating at pressures above 50 kPa.

National Collection of Type Cultures (NCTC) \\'nash-ən-ᵊl kə-'lek-shən 'əv 'tīp 'kəl-chərz ('en 'sē 'tē 'sē)\\ A British organization that collects, preserves, and supplies cultures of microorganisms.

natural-circulation evaporator \\nach-(ə-)rəl-ˌsər-kyə-'lā-shən i-'vap-ə-ˌrāt-ər\\ An evaporator in which the circulation is provided by natural convection that is induced by vapor formation in the heat-exchanger tubes contained in the feed vessel. Circulation usually comes through a central downcomer through the heat exchanger.

necrotrophic fungi \\ne-krō-'trō-fik 'fən-jī\\ Fungi that obtain nutrients from dead host cells.

neokink \\'nē-ō-ˌkiŋk\\ A localized distortion in the normal structure of DNA caused by an alteration in the orientation of the phosphodiester bonds. For example, when restriction nucleases bind to DNA, this bond causes a perturbation and effectively unwinds the DNA by 25° (producing a neokink) which causes a widening of the major groove and improves access to the appropriate bases. Base-pairing is not disrupted.

nephelometry \\nef-ə-'läm-ə-trē\\ The quantitative assessment of a suspension of particles by the measurement of scattered light at right angles to the incident

189

beam. The technique, more sensitive than turbidity measurements, can be used at lower particle concentrations and with smaller particles, but it requires more precise optics. The angular dependence of the amount of scattered light is a characteristic of a microorganism and can be used to identify species in favorable conditions.

Nernst equation \'nernst i-'kwā-zhən\ A thermodynamic equation that relates the observed electrode or redox potential to the conditions in a system. The form of the equation is

$$E = E° + \frac{RT}{nF} \ln \frac{[\text{oxidized form}]}{[\text{reduced form}]}$$

where E is observed potential; $E°$ is standard potential; R is the gas constant; T is absolute temperature; n is number of electrons involved in the system; F is the Faraday constant; and square brackets denote concentration or activity of the species involved.

Nernst layer (Nernst–Planck layer) \'nernst 'lā-ər ('nernst–'plaŋk 'lā-ər)\ A thin layer of unstirred solvent that surrounds a particle in a stirred solution. The thickness of this layer is affected to a large part by the rate at which the solution is stirred. This layer causes diffusional limitations for immobilized enzymes by restricting the rate of substrate diffusion to the enzyme.

neuraminic acid \n(y)ür-ə-'min-ik 'as-əd\ A C_9 carbohydrate widely distributed in the animal kingdom. The N- and O-acetyl derivatives are called *sialic acids*. Sialic acids often occur as the terminal sugar on the polysaccharide chains of glycoproteins. The enzyme neuraminidase (E.C. 3.2.1.8) cleaves N-acetyl neuraminic acid from the ends of polysaccharide chains.

neuraminidase \n(y)ür-ə-'min-ə-ˌdās\ *See* neuraminic acid.

neutralism \'n(y)ü-trə-ˌliz-əm\ A system in a culture containing two microbial species in which there is no change in growth rate of either species when compared with their growth rates in pure culture. *See also* amensalism, commensalism, mutualism.

neutrase \'n(y)ü-ˌtrās\ A protease from *Bacillus subtilis* added to cheeses prior to the ripening stage to reduce maturation time.

neutrophilic organism \n(y)ü-trə-'fil-ik 'ȯr-gə-ˌniz-əm\ A microorganism that prefers a neutral medium for growth.

Newtonian fluid \n(y)ü-'tōn-ē-ən 'flü-əd\ A fluid with a constant viscosity that is not influenced by the shear rate. The viscosity of a Newtonian fermentation broth does not vary with changes in the agitation rate.

nick \'nik\ A region in a double-stranded DNA molecule in which the phosphodiester bond is broken on one of the polynucleotide chains.

nick translation \'nik tran(t)s-'lā-shən\ An enzymatic method of radiolabeling or biotin labeling a DNA molecule. Free $3'$-hydroxyl groups are first formed randomly within the unlabeled DNA (*nicks*) by using a nuclease such as pancreatic deoxyribonuclease. DNA polymerase I is then added with one or more radiolabeled or biotinylated nucleotides. As well as being a polymerase (it adds a nucleotide to a free $3'$-OH group), this enzyme also possesses $5' \rightarrow 3'$ exonuclease activity. As the enzyme removes nucleotides (starting at the nick) in the $5' \rightarrow 3'$ direction, it replaces a labeled nucleotide by adding it to the free $3'$-hydroxyl group at the nick. In this way, the nick is shifted along (translated), one nucleotide at a time. The removed nucleotide is replaced by a labeled one. By means of this "old for new" swap of nucleotides, about 50% of the residues in the DNA can be labeled.

nif genes \'nif 'jēnz\ Those genes that code for the enzymes and other proteins involved in the fixation of atmospheric nitrogen. *See also* nitrogen fixation, nitrogenase.

nitrification \ˌnī-trə-fə-'kā-shən\ The microbial conversion of ammonia to nitrate ions. Two major genera are involved in this overall conversion process: *Nitrosomonas,* which convert ammonia directly to nitrate, and *Nitrobacter,* which convert nitrite to nitrate.

nitrilotriacetic acid (NTA) \ˌnī-trə-lō-ˌtrī-ə-'sēd-ik 'as-əd ('en 'tē 'ā)\ A synthetic, metal-complexing organic chemical extensively used to replace polyphosphates in washing powders. Both NTA and polyphosphates are used to sequester calcium ions and thus enhance frothing, but NTA has the advantage that it is biodegradable.

nitrogen fixation \'nī-trə-jən fik-'sā-shən\ The conversion of atmospheric nitrogen into an organic form. It can be carried out by bacteria and blue-green algae. Commercially, the most important process occurs in the roots of leguminous plants. *Rhizobium* bacteria invade the roots of the plants and form root nodules, where nitrogen fixation takes place. The plant is able to use the nitrogenous compounds produced. The relationship between the bacteria and the plant is symbiotic. The transfer of nitrogen-fixing genes into crop plants is one of the major aims of plant molecular biology, although it may turn out to be easier and as effective to induce symbiosis between *Rhizobium* and nonleguminous crop plants. *See also Azotobacter* species, nif genes, nitrogenase, *Rhizobium,* symbiosis.

nitrogenase \'nī-'träj-ə-ˌnās\ An enzyme involved in converting atmospheric nitrogen to ammonia in nitrogen-fixing organisms. The enzyme has a complex structure; it is formed from the gene products of between 17 and 20 nif genes. It comprises two main components, a dimer of an iron-containing protein (called component 1, or the Fe protein) and an $\alpha_2\beta_2$ molybdenum–iron protein (component II or the MoFe protein) that contains the active site for dinitrogen reduction.

nonideal mixing \ˌnän-ī-'dē(-ə)l 'mik-siŋ\ The situation in a reactor in which plug flow or complete mixing does not occur. Nonideal mixing must be considered when planning the mixing pattern in large-scale process reactors.

Nonidet P-40 (NP-40) \'nän-i-det 'pē 'fort-ē\ A Shell UK trade name for a nonionic detergent (an octylphenol-ethylene oxide condensate averaging 9 mol of ethylene oxide per mol of phenol). It is often used to solubilize proteins and to minimize nonspecific hydrophobic interactions in techniques such as ELISA and protein blotting.

noninvasive sensors \ˌnän-in-'vā-siv 'sen(t)-sərz\ Sensors that can monitor a change in a parameter without being in direct contact with the medium (e.g., γ-ray level detectors, ultrasonic flow meters).

nonionic surfactant \ˌnän-ī-'än-ik sər-'fak-tənt\ A detergent molecule that is uncharged in solution. These compounds usually consist of condensation products of alcohols or phenols with ethylene oxide, for example, nonidet P-40. *See also* anionic surfactant, cationic surfactant.

non-Newtonian fluid \ˌnän-n(y)ü-'tōn-ē-ən 'flü-əd\ A fluid that does not follow Newtonian fluid behavior. The viscosity of a non-Newtonian fluid varies with the shear rate (i.e., the viscosity varies with the agitation rate). Broths with polysaccharides or mycelial cells display non-Newtonian rheology. Several types of non-Newtonian fluid are recognized (e.g., Bingham plastic, Casson body, dilatant fluid, and pseudoplastic fluid).

nonrenewable resources \ˌnän-rē-'n(y)ü-ə-bəl 'rē-ṣō(ə)rs-ez\ Biomass for biotechnological processes that cannot be produced (e.g., fossil fuels, natural gas). *Compare with* renewable resources.

nopaline \'nō-pə-ˌlēn\ An opine. *See also* opines.

normal micelle \'nȯr-məl mī-'sel\ *See* micelle.

northern blotting \'nȯr-thə(r)n 'blät-iŋ\ The transfer (either by diffusion or electrophoretic transfer) of separated RNA molecules from a gel medium (e.g., agarose) to a nitrocellulose or nylon sheet where the RNA binds. The separated RNA molecules can then be further investigated (e.g., by washing the support in an appropriate probe solution). The method is analogous to Southern blotting for DNA.

nozzle-discharge centrifuge \'näz-əl-'dis(h)-ˌchärj 'sen-trə-fyüj\ A centrifuge in which the centrifuge bowl is pierced with a series of nozzles along its periphery. These nozzles are 1–3 mm in diameter and allow the continuous flow of solids as a slurry from the bowl into a surrounding container. The remainder of the construction is similar to a normal disk centrifuge. The equipment can operate on a once-through system with some recycle of solids. It will allow more extensive dewatering of the slurry up to 50% wt/wt solids, but more typical values are 20–30% wt/wt.

NTA \'en 'tē 'ā\ *See* nitrilotriacetic acid.

nuclear polyhedrosis virus \'n(y)ü-klē-ər ˌpäl-i-hē-'drō-səs 'vī-rəs\ A virus used in an insecticidal spray for the control of a sawfly, *Neoprion sertifer*, in coniferous forests. The active preparation, made from infected larvae, is produced commercially in Finland and Great Britain.

nuclease \'n(y)ü-klē-ˌās\ Any enzyme that cleaves phosphodiester bonds in either DNA or RNA. Specifically, ribonucleases (RNases) cleave RNA and deoxyribonucleases (DNases) cleave DNA. Enzymes that remove nucleotides sequentially from the ends of oligonucleotides are called *exonucleases*, whereas enzymes that cleave bonds within an oligonucleotide chain are called *endonucleases*. *See also* Bal 31, exonuclease

III, nick translation, restriction endonucleases (which play a major role in genetic engineering methodology), S1 nuclease.

nude mice \'n(y)üd 'mīs\ A genetically determined strain of mouse, lacking a thymus, characterized by a lack of hair. Because of the absence of a thymus, the mouse is depleted of T-cells and cannot react against the presence of foreign cells. Tumors from other species (*xenografts*) including humans, can be grown in these animals without rejection.

number average molecular weight \'nəm-bər 'av-(ə-)rij mə-'lek-yə-lər 'wāt\ *See* molecular weight of polymers.

nurse (feeder) callus \'nərs ('fēd-ər) 'kal-əs\ A callus mass used to stimulate plant cells to divide. If single plant cells from a suspension culture are plated at low density on semisolid agar, the cells do not divide. However, if callus masses (nurse calluses) are plated onto medium seeded with single cells, the single cells begin to divide in regions near the callus masses. This response shows a requirement for essential nutrients from the callus mass for cell division. *See also* paper raft nurse technique.

Nusselt number (Nu) \'nüs-əlt 'nəm-bər ('en 'yü)\ A dimensionless group used in heat transfer correlations:

$$Nu = \frac{hd}{k}$$

where h is the heat transfer coefficient (W/m^2K), d is the distance across the heat transfer surface (m), and k is the thermal conductivity of the fluid (W/mK)). It is a function of the Prandtl (Pr) and Reynolds (Re) numbers:

$$Nu = f(Pr, Re)$$

See also Prandtl number, Reynolds number.

nutrient agar \'n(y)ü-trē-ənt 'äg-ər\ A medium widely used for growth of nonexacting bacteria, which contains beef extract, peptones, sodium chloride, and agar. Some media manufacturers also include yeast extract in their formulations.

nutrient film technique (NFT) \'n(y)ü-trē-ənt 'film tek-'nēk ('en 'ef 'tē)\ A method for growing plants that ensures an adequate supply of both oxygen and water to the roots. Plant roots need both water and oxygen for satisfactory plant growth. Roots in soil are deprived of oxygen if there is abundant water. Conversely, good aeration is usually associated with lack of water. In NFT, plants develop their root systems in a very shallow stream of nutrient solution. The solution depth is adjusted so that the lower part of the developing root mat grows wholly in solution. The upper part projects just above the surface but remains covered by a liquid film. This exposed upper part allows for good aeration. The root mats produced by each plant are extensive and intertwined; this condition allows the plants to become self-supporting. *Compare with* hydroponics.

O

obligate \\'äb-lə-gət\\ Always required for growth, used to qualify an environmental factor (e.g., for an obligate aerobe, oxygen is essential).

obligate anaerobe \\'äb-li-gət 'an-ə-ˌrōb\\ An organism, usually a bacterium, that grows and reproduces in the absence of oxygen. In the presence of even minute traces of oxygen, it may be inhibited or killed. *See also* anaerobe, facultative anaerobe.

octopine \\'äk-tə-ˌpīn\\ An opine. *See also* opines.

OFAGE \\'ō-fāj\\ *See* pulse-field gel electrophoresis.

off-gas \\'òf 'gas\\ A term used to denote the gaseous product from a chemical or biological reaction.

off-line analysis \\'ò-f-līn ə-'nal-ə-səs\\ The analysis of a process using a technique in which a discrete sample is taken and then removed to an analytical instrument for determination. This technique is used for monitoring when continuous measurement of the parameter is not required or when suitable on-line or in-line techniques are not available. *Compare with* in-line analysis, on-line analysis.

Oldshue–Rushton contactor \\'ōld-shü–'rəsh-tən 'kän-takt-ər\\ An agitated liquid–liquid contactor consisting of a vertical shell fitted with horizontal rings, with a central opening, and vertical baffles attached to the shell walls. The rings divide the column into a series of mixing zones, and the vertical baffles improve mixing by reducing the rotational movement of the fluids imparted by the agitators. In the center of each mixing zone a flat-bladed turbine is mounted on a common shaft with the other turbines and driven by an external motor. The inlet feed distributors are placed just above the top turbine for the heavy phase and just below the bottom turbine for the light phase. Calm settling zones are situated above the top and below the bottom stator rings.

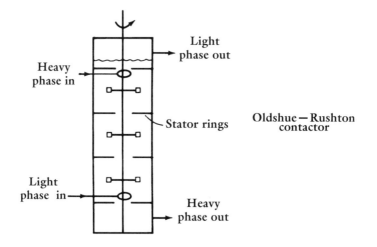

Heavy phase in → | Light phase out →

Stator rings

Oldshue — Rushton contactor

Light phase in → | Heavy phase out →

oleaginous \ō-lē-'aj-ə-nəs\ Containing or producing oils, used to describe organisms.

oligodeoxyribonucleotide synthesis \ə-'lig-ə-dē-ˌäk-sē-ˌrī-bō-'n(y)ü-klē-ō-ˌtīd 'sin(t)-thə-səs\ The chemical synthesis of chains of oligodeoxyribonucleotides for use as DNA probes, synthetic genes, and in site-directed mutagenesis. Because synthetic methods for oligodeoxyribonucleotides are more highly developed than those for oligoribonucleotides, the term *oligonucleotide* has become synonymous with *oligodeoxyribonucleotide*. The two main syntheses used are the phosphite triester method and the phosphotriester method. Also known as oligonucleotide synthesis. *See also* gene probe, phosphite triester synthesis, phosphotriester synthesis, site-directed mutagenesis.

oligomer \ə-'lig-ə-mər\ A polymeric chain of a few monomeric repeat units. The chain length need not be defined.

oligonucleotide-directed mutagenesis \ə-'lig-ə-'n(y)ü-klē-ə-ˌtīd-də-'rek-təd ˌmyüt-ə-'jen-ə-səs\ *See* site-directed mutagenesis.

oligonucleotide probe \ə-'lig-ə-'n(y)ü-klē-ə-ˌtīd 'prōb\ A short, radio- or biotin-labeled, synthetic oligonucleotide that base-pairs with a region of a gene sequence and can therefore be used to detect, by hybridization, a clone containing the gene of interest (*see also* colony hybridization) or an isolated gene or gene fragment (*see also* Southern blotting).

oligonucleotide synthesis \ə-'lig-ə-'n(y)ü-klē-ə-ˌtīd 'sin(t)-thə-səs\ *See* oligodeoxyribonucleotide synthesis.

oligosaccharide \äl-i-gō-'sak-ə-ˌrīd\ A carbohydrate comprising a small number of monosaccharide units linked together. Longer chains and branched chains of monosaccharides are called polysaccharides.

on-column injection \'ón-ˌkäl-əm in-'jek-shən\ A technique in which the sample enters the chromatography column directly from the syringe without contacting other surfaces. On-column injection is usually used with a cold sample and capillary gas chromatography columns.

on-line analysis \ˈȯn-ˌlīn ə-ˈnal-ə-səs\ The continuous measurement of an analytical parameter, whereby a sample is taken continuously via a sample line to the analytical device to assist in process control. It is not as satisfactory as in-line analysis. With the introduction of a sample line, care must be taken to ensure that a representative sample of the process is fed into the sample line and that no changes occur in the sample during its residence in the line. Finally, because of the presence of a sample line, the overall response time must be greater than with the in-line configuration. However, on-line analysis is still preferred in most cases to off-line analysis. On-line analysis is also the continuous transfer of data from a sensor to a recording device or computer. *Compare with* in-line analysis, off-line analysis.

oncogene \ˈäŋ-kō-jēn\ A cancer-causing gene. Oncogenes are mostly normal human cellular genes, but when excessively or aberrantly expressed, their gene products contribute to the process of malignancy. When an oncogene is under its normal cellular control, the term protooncogene indicates that it is a normal gene that has the potential to cause malignancy. Only when the gene is excessively or aberrantly expressed (e.g., by a change in the activity of the promoter region or by a mutation to the protooncogene that affects the activity of the gene product) does it become a true oncogene. Cellular oncogenes are referred to as c-onc to differentiate these genes from related oncogenic sequences that have been found in some viruses, which are referred to as v-onc. More than 60 human oncogenes have been identified, and each one has a three-letter acronym. Because many human oncogenes were first identified by their viral counterparts (the viruses are believed to have taken up the human oncogene sequences early in evolution), many of the acronyms are related to the virus, e.g., the oncogene *sis* is named after Simian sarcoma virus, and the oncogene *ras* is named after the rat sarcoma virus. The gene products of oncogenes are generally growth factors, growth factor receptors, signal transducers, or factors that affect gene expression by interaction with the cell nucleus.

open circular (oc) DNA \ˈō-pən ˈsər-kyə-lər (ˈō ˈsē) ˈdē ˈen ˈā\ *See* covalently closed circular DNA.

open reading frame (ORF) \ˈō-pən ˈrēd-iŋ ˈfrām (ˈō ˈär ˈef)\ A piece of nucleotide sequence with an initiation codon at on end, a series of triplet codons, and a termination codon at the other end. The sequence is potentially capable of coding for an as yet unidentified polypeptide.

operon \ˈäp-ə-rän\ A group of functionally related genes regulated and transcribed as a unit.

opines \ō-ˈpīnz\ Unusual amino acids, derived from arginine, synthesized in cells of crown gall tumors (*see Agrobacterium tumefaciens*). The ability to both induce and metabolize opines is encoded in the Ti plasmid of the bacterium *A. tumefaciens*. Infection of plant cells with *A. tumefaciens* therefore not only causes the cells to become malignant but also induces the cells to use their own metabolism to synthesize opines. These substances are of no use to the plant but can be metabolized and used as an energy source by the bacterium.

opsonization \äp-sə-nī-'zā-shən\ The coating of a target cell, e.g., bacteria or tumor, with molecules, e.g., antibodies that render the target more susceptible to phagocytosis.

optical activity \'äp-ti-kəl ak-'tiv-ət-ē\ The ability of compounds possessing one or more asymmetric atoms (chiral centers) to interact with polarized radiation and cause rotation of the plane of polarization. The degree to which this plane is rotated is a function of the structure of the compound and its concentration in solution. *See also* chirality, dextrorotatory, levorotatory.

optical density \'äp-ti-kəl 'den(t)-sət-ē\ The reduction in intensity of incident radiation on passing through an absorbent material, an obsolete term, replaced by absorbance.

optical isomerism \'äp-ti-kəl ī-'säm-ə-ˌriz-əm\ A form of isomerism in which the isomers differ in their optical activity. *See also* chirality, enantiomers.

optical rotation \'äp-ti-kəl rō-'tā-shən\ The angle through which the plane of polarized radiation of a defined wavelength is rotated on passing through a solution that contains an optically active substance. *See also* specific optical rotation.

optoelectronic sensor \äp-tō-i-lek-'trän-ik 'sen(t)-sər\ A sensor that uses optical principles coupled to an electronic transducer. For example, a sensor could consist of a detecting enzyme linked to a dye, which in turn is linked to a membrane. In the presence of a substrate, the enzyme generates a pH change that alters the color of the dye–membrane complex. This color change is recorded by using a transducer system that consists of a light-emitting diode (with wavelength equal to λ_{max} of the dye) and a photodiode.

order of reaction \'ȯrd-ər 'əv rē-'ak-shən\ The rate of a reaction can be expressed as an empirical differential rate equation that contains a factor of the form $k([A]^a)[B]^b$, where k is a proportionality constant, and a and b are constant exponents independent of concentration and time. These exponents are termed the order of reaction. Thus, in this expression, the order with respect to component A is a, of component B is b, and the total order of the reaction is $(a + b)$. The order may be integral, nonintegral, or zero. The order of reaction is determined experimentally and may differ from the actual number of molecules involved in the reaction.

ORF \'ō 'är 'ef\ *See* open reading frame.

organ culture \'ȯr-gən 'kəl-cher\ In plant tissue culture, the unlimited growth of isolated organs (such as root tips, stem tips, and leaves) in a sterile nutrient medium. Organ cultures retain their characteristic structures and grow in a manner comparable to that of their intact counterparts (*contrast with* callus culture, suspension culture). This growth allows the properties and functions of the individual organs to be studied in isolation. *See also* tissue culture.

organic pollution monitor \ȯr-'gan-ik pə-'lü-shən 'män-ət-ər\ An instrument developed for monitoring water quality and for controlling water- and wastewater-treatment processes. The monitor measures ultraviolet absorption at 254 nm, which correlates well with the total organic carbon (TOC) for a range of samples from settled sewage and effluents to raw and treated river waters.

organogenesis \\ȯr-gə-nō-'jen-ə-səs\\ The induction of root and shoot production in callus culture or living tissue (*explants*) by the addition of appropriate ratios of auxins and kinins. Such induced cultures can, under appropriate conditions, develop to form plantlets and eventually fully grown plants.

orifice meter \\'ȯr-ə-fəs 'mēt-ər\\ A device for measuring the flow rate of a fluid based on the differential measurement of pressure upstream and downstream of an orifice plate. *See also* pitot tube, venturi meter.

orifice plate \\'ȯr-ə-fəs 'plāt\\ A device for measuring fluid flow rate, consisting of a plate that contains a precisely machined hole and is fixed perpendicularly across a pipe. As fluid passes through the pipe, a pressure differential related to the square of the fluid flow rate is set up across the orifice plate.

origin of replication \\'ȯr-ə-jən 'ev ‚rep-lə-'kā-shən\\ A base sequence in DNA at which replication of DNA is initiated by DNA polymerase. In bacteria and viruses, usually only one replication occurs per genome. *See also* replicon.

osmosis \\äz-'mō-səs\\ The process by which a fluid or solvent passes through a semipermeable membrane from a solution of low solute concentration to one of higher solute concentration. The movement tends to equalize the concentrations on either side of the membrane.

osmotic pressure \\äz-'mät-ik 'presh-ər\\ The pressure required to prevent the osmotic movement of a fluid across a semipermeable membrane from a solution of low solute concentration to one of higher concentration. The osmotic pressure, a function of solute concentration, is dependent on the number of ions or molecules in solution

$$\pi = \frac{cRT}{M}$$

where π is the osmotic pressure, c is the concentration of the solute, R is the gas constant, T is the temperature, and M is the molecular weight of the solute particles.

osmotic shock disruption \\äz-'mät-ik 'shäk dis-'rəp-shən\\ A technique that uses osmosis across the outer membrane of the cell to disrupt cells and release intracellular material. Transfer of water across the outer membrane builds up pressure within the cell and causes its eventual disruption. This technique requires less energy than some of the others available (e.g., bead mills), but it tends to be less efficient and so is used mainly in the laboratory.

osmotolerant \\äz-mō-'täl-ə-rənt\\ *See* water activity.

output rate \\'aút-pút 'rāt\\ In a continuous-culture system, the quantity of cells produced in unit time.

oxidase \\'äk-sə-‚dās\\ *See* oxidoreductases.

oxidation pond \\äk-sə-'dā-shən 'pänd\\ A shallow static tank that relies on a large surface area to provide adequate aeration. Oxidation ponds are used as an aerobic treatment method for farm-animal waste slurries. The growth of algae on the surface is encouraged because released oxygen improves the system. In emergencies, sodium nitrate may be added to provide additional oxygen by nitrate respiration of some bacteria.

oxidation–reduction potential (redox potential) \äk-sə-'dā-shən-ri-ˌdək-shən pə-'ten-chəl ('rē-däks pə-'ten-chəl)\ The electrode potential of a half-cell containing a mixture of the oxidized and reduced forms of the species concerned, measured relative to a standard hydrogen or calomel electrode. The potential, a function of the concentration of the species in solution, may be calculated for any particular conditions by the Nernst equation. *See also* calomel electrode, Nernst equation, standard oxidation-reduction potential.

oxidoreductases \'äk-səd-ō-ri-'dək-ˌtās-əz\ An enzyme class covering all enzymes catalyzing oxidoreduction reactions. The substrate that is oxidized is regarded as the hydrogen donor. The systematic name is based on donor : acceptor oxidoreductase, and where possible the recommended name is dehydrogenase. The term *reductase* is used as an alternative, and *oxidase* is only used in cases where O_2 is the acceptor. For example, the systematic name for glucose oxidase is β-D-glucose : oxygen 1-oxidoreductase. Oxidoreductases are one of the six main classes (E.C. 1) used in enzyme classification. *See also* Enzyme Commission number.

oxirane coupling \'äk-sə-rān 'kəp-liŋ\ A coupling method used in the preparation of affinity columns. Bisoxiranes (bisepoxides) react readily with hydroxy- or amino-containing support matrices (e.g., Eupergit C, Sepharose) at high pH to yield derivatives with a long-chain, hydrophilic, reactive oxirane, which in turn can be reacted with nucleophile ligands (such as amines or phenols).

oxygen analyzer \'äk-si-jən 'an-ᵊl-ī-zər\ An instrument incorporating devices capable of determining oxygen gas in a gas stream or in solution. Several techniques are available, for example:

- *paramagnetic oxygen analyzer:* Oxygen is the only common gas that is paramagnetic; uncommon paramagnetic gases include nitric oxide (NO) and nitrogen dioxide (NO₂). The oxygen analyzer is specific to oxygen determination in the absence of these other gases. The technique relies on the tendency for oxygen molecules to be attracted into a strong magnetic field. This attraction, although small in itself, can be used to affect other measurable parameters, such as cooling a heated wire (magnetic wind device) or rotating a balanced dumbbell. Calibration of these devices allows conversion to oxygen concentration in the gas stream.
- *electrolyte fuel cell–oxygen electrode: See* dissolved oxygen electrode.
- *process gas chromatography:* A gas mixture can be separated into its components by gas–liquid chromatography. By suitable choice of a detector the components, including oxygen, can be determined.
- *process mass spectrometer:* Oxygen can be determined in a gas mixture by using a process mass spectrometer, which generally consists of a quadrupole instrument that gives a rapid response time with a relatively low resolution.

oxygen transfer \\'äk-si-jən 'tran(t)s-fər\\ The transfer of oxygen into solution in a fermentor, expressed in terms of the volumetric oxygen transfer coefficient $K_L a$ This coefficient is a function of the intensity of agitation, rate of aeration, gas–liquid interfacial area, and rheological properties of the fermentation broth. A number of correlations have been proposed, depending on the size of the gas bubbles (small bubbles have <2.5-mm diameter, and large bubbles have >2.5-mm diameter). In addition, in industrial fermentors, air bubbles usually form swarms or clusters. Here the correlation varies from those of single bubbles, but they are still size-dependent with the same critical diameter of 2.5 mm. *See also* surfactant.

oxygen transfer rate \\'äk-si-jən 'tran(t)s-fər 'rāt\\ The rate of oxygen transfer from an air bubble to liquid phase. It can be described by the equation:

$$\frac{dC_L}{dt} = K_L a \, (c^* - c_L)$$

where dC_L/dt is the oxygen transfer rate (mmol/dm^{3}h), K_L is the mass transfer coefficient (cm/h), a is the gas–liquid interfacial area per unit volume (cm^{-1}), c^* is the saturated dissolved oxygen concentration (mmol/dm^3), and c_L is the actual dissolved oxygen concentration (mmol/dm^3).

oxygen uptake rate (OUR) \\'äk-si-jən 'əp-ˌtāk 'rāt ('ō 'yü 'är)\\ The rate of change in dissolved oxygen content of the medium within a fermentor as the process proceeds.

oxygenic reactor \\ˌäk-si-'jen-ik rē-'ak-tər\\ A reactor in which oxygen is used in place of air to produce the dissolved oxygen required for biological processes.

oyster mushroom \\'ȯi-stər 'məsh-rüm\\ *See Pleurotus ostreatus.*

P

packed cell volume (PCV) \\'pakt 'sel 'väl-yəm ('pē 'sē 'vē)\\ The percentage volume of cells in a given volume of culture after sedimentation (packing) by low-speed centrifugation.

packed column \\'pakt 'käl-əm\\ A column filled with a loose solid matrix of rings, saddles, or other shapes made of inert material. This packing, when used as a contacting device, provides a large surface area and voidage for adsorption, coalescence, or partitioning.

PAGE \\'pāj\\ *See* polyacrylamide gel electrophoresis.

panning technique \\'pan-iŋ tek-'nēk\\ A method for purifying specific cell types by affinity chromatography. For example, anti-mouse-immunoglobulin antibodies adsorbed onto polystyrene dishes selectively bind cells that were previously treated with specific mouse monocloncal antibodies against a cell-type-specific surface antigen. Nonbound cells are washed away.

papain (E.C. 3.4.22.2) \\pə-'pā-ən\\ A thiol protease (MW 23,000) isolated from papaya latex. Its major uses are in the brewing industry (to stabilize and chillproof beer to prevent haze production) and in meat tenderization.

paper raft nurse technique \\'pā-pər 'raft 'nərs tek-'nēk\\ A method for culturing single plant cells. Single cells are isolated from cell suspensions with a needle or fine capillary. Then each cell is placed on the upper surface of a square of filter paper that is resting on an actively growing callus (nurse callus). In this way the single cell receives growth factors produced by the callus as well as growth factors from the nutrient medium. The cell divides to form small colonies, which can be subcultured onto fresh media to give callus isolates derived from single cells.

paramagnetism \\ˌpar-ə-'mag-nə-ˌtiz-əm\\ The interaction of an applied magnetic field with unpaired electrons in a substance. This interaction induces a magnetic field that reinforces the applied field so that the substance is attracted into the applied magnetic field. The strength of this interaction is proportional to the number of unpaired electrons and the temperature. This phenomenon is shown by organic free radicals, molecular oxygen, and transition metal ions and complexes (including metalloproteins). *See also* diamagnetism, ferromagnetism.

parasexual hybridization \ˌpar-ə-'seksh-(ə-)wəl ˌhī-brəd-ə-'zā-shən\ All nonmeiotic genetic recombination processes in vegetative cells. Few organisms currently used in biotechnological processes exhibit overt sexual recombination abilities, but most achieve limited recombination by parasexual mechanisms such as conjugation, mitotic recombination, protoplast fusion, transduction, and transformation.

parasite \'par-ə-ˌsīt\ An organism that lives on (ectoparasite) or in (endoparasite) another organism from which it obtains its nourishment.

partial pressure (gas) \'pär-shəl 'presh-ər ('gas)\ A measure of the amount of a certain gas in a mixture of gases. It is defined as the pressure that the gas would exert if it alone occupied the volume of the mixture of gases at the same temperature.

partially miscible substances \'pärsh-(ə-)lē 'mis-ə-bəl 'səb-stən(t)s-ez\ Fluids that, when mixed, form a single phase over a defined range of relative concentrations. Two phases occur outside the range of miscibility, and such systems may be used in the partitioning of a third component. *See also* two-phase aqueous partitioning.

partition chromatography \pär-'tish-ən ˌkrō-mə-'täg-rə-fē\ Any form of chromatography that relies on the partition of a single compound between two immiscible phases. For example differences in solubility of solutes between the stationary phase and gas phase (gas chromatography), or differences between the solubilities of solutes in the mobile and stationary phases (liquid chromatography). This term should not be applied when different forms of the compound are distributed between the two phases.

partition coefficient \pär-'tish-ən kō-ə-'fish-ənt\ The ratio of concentration of a substance in a single defined form between two immiscible or partially miscible phases. This term is not synonymous with distribution coefficient. *See also* distribution coefficient.

partition constant \pär-'tish-ən 'kän(t)-stənt\ The ratio of activity of a substance in a single defined form between two immiscible or partially miscible phases. This term is not synonymous with distribution constant. *See also* distribution constant.

partitioning effect \pär-'tish-ən-iŋ i-'fekt\ The effect on the environment surrounding an enzyme when the enzyme is immobilized on a solid support. Depending on the nature of the support material, the partitioning effect may attract or repel substrate, product, ions, inhibitors, or other molecules to the support surface and thus concentrate or deplete them in the vicinity of the enzyme. For example, attraction of positive ions (and thus hydrogen ions) will reduce the pH in the immediate vicinity of the enzyme. The kinetics of the enzyme obtained under these conditions are referred to as *inherent* kinetics. *See also* intrinsic kinetics.

pasteurization \pas-chə-rə-'zā-shən\ A heat treatment used to kill pathogenic microorganisms in food and beverages. A temperature of around 60 °C is usually used, and holding time depends on the material being treated.

PCR \'pē 'sē 'är\ *See* polymerase chain reaction.

PCV \\'pē 'sē 'vē\\ *See* packed cell volume.

PDGF \\'pē 'dē 'jē 'ef\\ *See* platelet-derived growth factor.

Peclet number (Pc) \\pə-'klā 'nəm-bər ('pē 'sē)\\ A dimensionless group used in mass transfer correlations. It is of particular use in biotechnological applications to relate the transfer of oxygen from a rising gas bubble into a fermentation broth:

$$Pc = \frac{\nu D}{D_o}$$

where ν is the velocity of a gas bubble relative to the liquid velocity (ms^{-1}); D is the bubble diameter (m), and D_o is the diffusivity of oxygen (m^2/s). *See also* Grashof number, Schmidt number, Sherwood number.

pectin \\'pek-tən\\ A complex acidic heteropolysaccharide that consists of α-1,4-D-polygalacturonide polymers. It is often complexed with two other polymers, a highly branched L-arabinan and a β-1,4-D-galactan. The carboxyl groups of the galacturonic acid units are partially esterified with methanol. Pectin, a structural component in various fruits and vegetables, becomes partially solubilized during processing of fruit juices. This change spoils the appearance and filterability of the fruit juices and reduces the yield of juice. To overcome the problem, pectinase is added to the juice during mashing. The enzyme degrades 1,4-α-D-galacturonic bonds and solubilizes the pectin, reduces the viscosity of the juice, and leads to a more stable and concentrated product. Pectinase preparations are usually a mixture of endo- and exopolymethylgalacturonidases obtained from *Aspergillus* species.

pectinases \\pek-tə-'nās-ez\\ *See* pectin.

PEG \\('peg)\\ *See* polyethylene glycol.

Pekilo process \\'präs-ҽs\\ A process for the production of single-cell protein that uses carbohydrates such as sulfite liquor or wood hydrolysates. The process is based on the fermentation of the mold *Paecilomyces varioti. See also* single-cell protein.

pellicular resins \\pə-'lik-yə-lər 'rez-ᵊnz\\ *See* ion chromatography.

penicillin acylase (penicillin amidase, E.C. 3.5.1.11) \\pen-ə-'sil-ən 'as-əl-ās (ˌpen-ə-'sil-ən 'am-ə-ˌdās)\\ An enzyme that catalyzes the deacylation of the side chain of penicillin G and leaves the penicillin nucleus, 6-aminopenicillanic acid (6-APA). 6-APA is used as the starting point for the synthesis of several semisynthetic antibiotics, such as ampicillin and cloxacillin. The immobilized enzyme, obtained from *Escherichia coli*, is used commercially in plug flow column reactors.

penicillin amidase \\ˌpen-ə-'sil-ən 'am-ə-ˌdās\\ *See* penicillin acylase.

penicillin enrichment technique \\ˌpen-ə-'sil-ən in-'rich-mənt tek-'nēk\\ A method for the isolation of bacterial auxotrophs from a mutagen-treated culture. Under normal conditions an auxotroph is at a disadvantage compared with the parental (wild-type) cells. However, because penicillin kills only growing cells, when the survivors of a mutation treatment are cultured in a medium containing penicillin and lacking the growth medium of the desired mutant, only those cells unable to grow (i.e., the desired auxotrophs) survive.

penicillin G (benzylpenicillin) \ˌpen-ə-'sil-ən gē ('ben-ˌzəl-ˌpen-ə-'sil-ən)\ The original penicillin molecule, produced commercially from *Penicillium chrysogenum*. It is the starting material for a more active range of semisynthetic penicillin molecules. *See also* 6-aminopenicillanic acid, ampicillin, penicillin acylase, peptidoglycan.

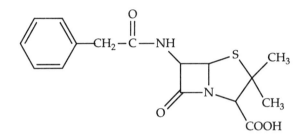

penicillinases (E.C. 3.5.2.6) \ˌpen-ə-'sil-ə-ˌnās-ez\ Enzymes that hydrolyze the β-lactam ring of the penicillin structure. Production of such enzymes by microorganisms is responsible for the resistance of the organisms to penicillins and cephalosporins. Hydrolysis of the amide bond in the ring produces penicilloic acid, which has no antimicrobial activity. β-Lactamases are used commercially to inactivate penicillin in milk and thus prevent consumer allergic reactions. Natural inhibitors of penicillinase, such as clavulanic acid, are used in drug formulations to overcome microbial resistance. Also known as β-lactamases.

Penicillium \ˌpen-ə-'sil-ē-əm\ A fungal genus belonging to the Deuteromycotina. Many widespread species may cause spoilage problems. A few species have been exploited for production of antibiotics (e.g., *P. chrysogenum*, penicillin; *P. griseofulvum*, griseofulvin; *P. utricae*, patulin), enzymes (e.g., *P. glaucum*, pectinase), and starter cultures (e.g., *P. camemberti*, *P. roquefortii*).

pentose \'pen-ˌtōs\ *See* carbohydrates.

peptidoglycan \ˌpep-təd-ō-'glī-kan\ The structural component of the bacterial cell wall, comprising linear polysaccharide chains cross-linked by short peptides. Differences in the peptidoglycan content of the cell wall of different bacteria have allowed bacteria to be categorized as either Gram-positive or Gram-negative (*see* Gram stain). The antibiotic penicillin functions by inhibiting the enzyme glycopeptide transpeptidase which is responsible for forming the peptide cross-links.

percolating (trickling) filter \'pər-kə-ˌlāt-iŋ ('trik(ə)-liŋ) 'fil-tər\ A widely used reactor design for the aerobic processing of urban and industrial wastes. The waste is passed through filter beds of clinker, stone, gravel, or plastic, where the biomass adheres to the surface of the filter material, covered only by a thin film of water. This absorbed solid is broken down by extracellular enzymes from endogenous microorganisms (mainly the *Zoogloea*) and metabolized. This process results in the release of excretory products. In this way the wastes are completely degraded to simple salts, gases, and water. Although this method is commonly used, a major problem can develop. Excessive growth of microorganisms in the filter restricts ventilation and flow. Such filters are also used to remove waste gases from industrial processes. Water containing dissolved inorganic nutrients is passed down a column. It forms a thin film on the packing material, which is covered by a biofilm of microbial flora. Waste gas is forced to rise through the column, against the water flow. Water-soluble components and oxygen are transferred to the liquid phase and then to the biolayer, from which they are eliminated by aerobic reactions. *See also* bioscrubbing.

perforated-plate column \\'pər-fə-ˌrāt-əd-'plāt 'käl-əm\\ *See* sieve plate column.

perfusion chromatography \\pər-'fyü-zhən ˌkrō-mə-'täg-rə-fē\\ A separation technique in which the problem associated with limited mass transfer through ion exchange and other derivatized membranes may be overcome by imposing a convective flow through the membrane by application of a pressure difference across the membrane.

perfusion reactor \\pər-'fyü-zhən rē-'ak-tər\\ A type of bioreactor particularly suited to mammalian cell cultivation. The cells are gently agitated by gas streams or stirrers and are retained in the reactor under optimal growth conditions while the cell products are removed continuously with some of the medium. Fresh medium is added to maintain growth. Specially designed separation or filtration devices are required to retain the cells in the reactor without damage.

periplasm \\'per-ə-ˌplaz-əm\\ *See* periplasmic space.

periplasmic space \\per-ə-'plaz-mik 'spās\\ The space between the cytoplasmic and outer membrane in Gram-negative bacteria. The fluid in the periplasmic space, called *periplasm*, contains a number of secreted enzymes.

peristaltic pump \\per-ə-'stȯl-tik 'pəmp\\ A device for transferring fluid by using a rotating drum with rollers attached to the circumference, rotating within a cylinder. A flexible tube is positioned between the drum and the containing cylinder so that rotation of the drum causes the rollers to squeeze the tube. The roller pressure carries forward discrete volumes of the fluid contained in the tube. These pumps have the advantage that the fluid has no contact with moving parts of the pump and is contained within a tube. The flow rate can generally be varied easily between limits set by the rotational speed of the drum. In addition, the system is easy to sterilize. However, the fluid flow is not generally pulse-free, the amplitude of the pulses depends on the number of rollers, and the pump is not able to transfer fluid against a significant back pressure.

permeate \\'pər-mē-ˌāt\\ The portion of material that traverses a membrane. The opposite of retentate.

permeation chromatography \\pər-mē-'ā-shən ˌkrō-mə-'täg-rə-fē\\ *See* gel filtration.

permselective membrane \\pərm-sə-'lek-tiv 'mem-ˌbrān\\ *See* semipermeable membrane.

peroxidase \\pə-'räk-sə-ˌdās\\ An enzyme that breaks down hydrogen peroxide to oxygen and water. The presence of the enzyme is detected by coupling the release of oxygen to the oxidation of a hydrogen donor such as tetramethylbenzidine (TMB) or *o*-phenylenediamine, (OPD), which results in a colored product. The horseradish enzyme is often used in enzyme-linked assays such as ELISA or the immunodetection of protein blots. It can also be used to hydroxylate aromatic compounds. At 0 °C, *p*-substituted aromatics are hydroxylated at the meta position and vice versa.

persistence \pər-'sis-tən(t)s\ The ability of a substance to remain in a particular environment in an unchanged form. Persistent compounds are not degraded by microorganisms. *Compare with* recalcitrance.

perstraction \pər-'strak-shən\ A membrane separation process similar to pervaporation in which the low pressure side uses a purge liquid rather than a vacuum to remove the permeate. Separation of the permeate from the purge liquid is carried out by conventional distillation, allowing the purge liquid to be recycled. *Compare with* membrane distillation, pervaporation.

pervaporation \pər-ˌvap-ə-'rā-shən\ A process in which components of a mixture are separated by the use of a membrane between the feed and product streams. The product side operates at a reduced pressure to produce a vapor on this side. The membranes have an asymmetric structure, with a homogeneous skin supported on a microporous substructure. The driving force for the process is a partial pressure gradient of $10^{-3}-10^5$ Pa across the membrane. The separation processes involve both solubility and diffusivity of the components within the polymeric matrix. The process is used for the separation of organic products from aqueous solution, especially with azeotropic mixtures. *Compare with* membrane distillation, perstraction.

PFGE \'pē 'ef 'jē 'ē\ *See* pulse-field gel electrophoresis.

pH auxostat \'pē 'ách 'ók-sə-stat\ An auxostat in which the controlled parameter is culture pH.

pH electrode \'pē 'āch i-'lek-ˌtrōd\ *See* combined electrode.

phage \'fāj\ Any virus that specifically infects a bacterial cell. Such viruses comprise a protein coat (*capsid*) surrounding genetic material (usually DNA, but sometimes RNA) that is injected into the cell on infection. Phages are used in genetic engineering to introduce new genetic material into a cell (i.e., they are used as vectors). *See also* lysogenic infection, M13, λ-phage.

λ-phage \'lam-də 'fāj\ A bacteriophage containing double-stranded DNA that infects *Escherichia coli*. Derivatives of λ-phage are used extensively as cloning vectors. The λ-genome is 49 kilobase pairs long. *See also* λ-insertion vector, λ-replacement vector.

phage display antibodies \'fāj dis-'plā 'ant-i-ˌbäd-ē\ A powerful, bacteriophage-based screening system for selecting and producing specific F_{ab} antibody fragments. The repertoire of antibody variable-region genes is isolated (from B-cells) by polymerase chain reaction using primers specific for conserved regions of the antibody gene families and then cloned into the gene encoding the minor coat protein (gene 3) of the filamentous bacteriophage fd. The fusion protein created, consisting of the antibody fragment at the N-terminus of the coat protein, is incorporated into the phage particle (a phage display antibody).

A large library of phages is therefore created, each phage displaying a specific antibody. Each recombinant phage genome contains the DNA encoding the specific antibody displayed on its surface. In effect, each phage mimics a B-cell. The phage particle carrying the required antibody gene can then be selected directly using the binding property of this expressed protein. To select the antibody of interest, the library of phage antibodies is passed over antigen bound to a solid surface, such as a Sepharose column. Phages that display antibody fragments specific for the antigen are retained on the column. These specific phages can be eluted and used to infect *E. coli* to give stable clones. The phage antibody can be both analyzed and used directly from the culture supernatant as a reagent in techniques such as ELISA. This method has a number of advantages. Many millions (or billions with recent refinements) of potential binding molecules can be screened using large phage libraries compared with only the several thousand molecules screened using a traditional hybridoma fusion. The overall process is considerably shorter when compared with traditional hybridoma technology. More important, it provides a convenient route for producing human monoclonal antibodies for therapeutic use, which otherwise are technically extremely difficult to produce using conventional technology.

phagocytosis \\fag-ə-sə-'tō-səs\ The process whereby either particulate matter or cells are ingested, and often metabolized, by certain cells (e.g., leukocytes) and microorganisms. This term is often incorrectly used synonymously with *pinocytosis*.

Phanerochaete chrysosporium \fan-'i(ə)r-ō-kēt 'kris-ə-ṣpō-rē-əm\ *See* lignin.

phase diagram \'fāz 'dī-ə-ˌgram\ A diagram that represents the conditions of equilibrium between various parts of a system separated by definite boundary surfaces under defined experimental conditions. Of particular use with partially miscible systems to determine the relative concentrations under which the system separates into two liquid phases. *See* binodal curve.

phase ratio \'fāz 'rā-shē-ō\ The ratio of one phase (solvent) to another (feed) in, for example, liquid–liquid extraction. It is usually expressed in terms of volume ratio.

phase-transfer catalysis \'faz-ˌtran(t)s-fər kə-'tal-ə-səs\ *See* ion-pair partitioning.

PHB \'pē 'āch 'bē\ *See* poly-β-hydroxybutyrate.

phenol extraction \'fē-ˌnōl ik-'strak-shən\ A commonly used method for deproteinizing a cell extract as a first step in the purification of DNA or RNA. Phenol, or a 1 : 1 mixture of phenol and chloroform, is mixed gently with the cell extract; then the layers are separated by centrifugation. Precipitated protein forms as a white mass at the interface between the aqueous and organic phases, with the DNA and RNA in the aqueous phase.

phenotype \'fē-nə-ˌtīp\ The physical characteristics or behavior of a cell or organism.

phenyl boronates \'fēn-ᵊl 'bōr-ō-ˌnā(t)s\ *See* boronates.

Phillips and Johnson tube \'fil-əps 'and 'jŏn-sən 't(y)üb\ A device used for the on-line monitoring of oxygen in a fermentor. It consists of a coil of polytetrafluoroethylene immersed in the culture medium, through which a stream of pure nitrogen is passed. Oxygen diffuses through the polymer wall into the nitrogen stream at a rate proportional to its partial pressure in solution. Analysis of the mixed gas stream can then be made outside the fermentor by appropriate methods.

phosphatase \\'fäs-fə-ˌtās\\ Any enzyme (esterase) that catalyzes the hydrolysis of monophosphate esters. *See also* alkaline phosphatase.

phosphite triester synthesis (phosphoramidite method) \\'fäs-ˌfīt trī-'es-tər 'sin(t)-thə-səs (ˌfäs-fər-am-i-ˌdīt 'meth-əd)\\ One of the two main methods (the other is phosphotriester synthesis) for synthesizing oligonucleotides. The method involves linking nucleoside-3′-*O*-(*N,N*-dialkylamino)phosphoramidite monomers. These monomers are used because of their resistance to hydrolysis and air oxidation during the synthesis of the oligonucleotide, which pose a problem with other nucleoside phosphites. After the synthesis, the blocking groups are removed.

phosphoramidite method \\ˌfäs-fər-am-i-ˌdīt 'meth-əd\\ *See* phosphite triester synthesis.

phosphorescence \\ˌfäs-fə-'res-ᵊn(t)s\\ A process similar to fluorescence, in which radiation is emitted from a molecule after adsorption of radiation of a defined wavelength. Phosphorescence differs from fluorescence in that radiation continues to be emitted after the exciting radiation is removed. *Compare with* fluorescence. *See also* bioluminescence.

phosphotriester synthesis \\ˌfäs-fō-trī-'es-tər 'sin(t)-thə-səs\\ One of the two main methods (the other is phosphite triester synthesis) for synthesizing oligonucleotides. The method involves chemically linking suitably protected nucleotides in which the internucleotide phosphodiester bond is protected by esterification with a third group. After the synthesis, the protecting groups are removed.

photoautotroph \\ˌfōt-ō-'ȯt-ə-ˌtrōf\\ An autotroph that uses light as its energy source for synthesizing biological molecules from carbon dioxide (e.g., plants and algae). *See also* cyanobacteria, purple bacteria.

photocorrelation spectroscopy (dynamic light scattering) \\ˌfōt-ō-kȯr-ə-'lā-shən spek-'träs-kə-pē (dī-'nam-ik 'līt 'skat-ər-iŋ)\\ A technique for determining molecular size that uses the fluctuations in scattered radiation caused by molecules moving relative to one another in a moving fluid. These fluctuations are the result of constructive and destructive interference of the scattered radiation from each molecule and vary as the relative positions of the molecules alter. The rate of fluctuation is related to the hydrodynamic size of the molecules, which provides a means of identification of molecular species. This technique has been used as a detector for liquid chromatography.

photolysis \\fō-'täl-ə-səs\\ The cleavage of water by the light reaction of photosynthesis. The production of an energy source (hydrogen) from abundantly available water by photolysis is currently a major area of research in biotechnology.

photosynthesis \\ˌfō-tō-'sin(t)-thə-səs\\ The process by which green plants, algae, and some bacteria use light to split water and transfer a free electron, which is then used in biosynthetic processes.

photosynthetic conversion efficiency \fōt-ō-sin-'thet-ik kən-'vər-zhən i-'fish-ən-sē\ The percentage of the total solar radiation falling on a given area in a given time that is converted into harvestable organic material. It follows that the harvestable biomass is a measure of photosynthetic efficiency. Mean annual figures are 0.5–1.5% in temperate areas and 0.5–2.5% for subtropical crops.

phytohemagglutinins \'fī-tō-'hē-ma-glüt-ə-nins\ *See* lectins.

pI \'pē 'ī\ *See* isoelectric point.

picometer \'pē-kō-ˌmēt-ər\ *See* angstrom.

pinocytosis \pin-ə-sə-'tō-səs\ The process of engulfment of external solid and liquid matter by a cell. The ingestion of minute droplets of fluid by various types of eukaryotic cells (e.g., macrophages, amoeba). *See also* phagocytosis.

pitot tube \'pē-tō 't(y)üb\ A device for measuring the flow rate of a fluid based on differential pressure measurement. Other devices for measuring flow rates based on this principle include orifice meters and venturi meters. *See also* orifice meter, venturi meter.

plant-growth substance \'plant-'grōth 'səb-stən(t)s\ An organic substance, either produced within a plant or synthetically produced and introduced into a plant. At low concentrations it promotes, inhibits, or qualitatively modifies growth, or it affects patterns of differentiation. Such compounds include abscisic acid, auxins, cytokinins, ethylene, and gibberellins. Also known as plant hormone.

plant hormone \'plant 'hȯr-ˌmōn\ *See* plant-growth substance.

plaque \'plak\ An area of clearing or reduced growth in a bacterial lawn caused by the lysis of cells by infecting phage particles.

plasma cell \'plaz-mə 'sel\ A terminally differentiated B-lymphocyte that produces large amounts of secreted antibody. One of the many cell types that make up the white cell component of blood. *See also* B-lymphocytes.

plasmalemma \ˌplaz-mə-'lem-ə\ The membrane of a protoplast. It is the only barrier between the external environment and the interior of the cell.

plasmids \'plaz-mədz\ Small (MW 1×10^6 to about 200×10^6) molecules of circular, double-stranded DNA that replicate independently of chromosomal DNA, found in certain bacteria. Plasmids can be isolated easily by lysing bacterial cells and separating plasmid DNA from chromosomal DNA by cesium chloride gradient centrifugation in the presence of ethidium bromide. In particular, plasmids have found considerable use as cloning vectors. However, those used currently are not naturally occuring forms but forms that have been extensively modified in vitro to give properties useful for cloning. Plasmids are generally identified by a code of the form pAB123, where p stands for plasmid, AB are initials identifying the worker or laboratory responsible for isolating the plasmid, and the number is the laboratory's code for that particular plasmid (e.g., pBR322, pSC101, pMB1, pUK230). *See also* col plasmids, compatibility, episomes, relaxed replication, resistance plasmids, virulence plasmids.

plasmin \\'plaz-mən\\ A human serum enzyme responsible for the dissolution of blood clots. The enzyme degrades insoluble fibrin in the blood clot to soluble peptides, a process known as *fibrinolysis*. The inactive precursor (*zymogen*) of this enzyme is plasminogen, which is present at a concentration of about 50 mg/100 mL of serum. Plasminogen is activated when it is converted to plasmin by the cleavage of a single peptide bond. This activation can be achieved by a number of enzymes such as the kidney enzyme urokinase, streptokinase (produced by hemolytic streptococci), and tissue plasminogen activator. Such activating enzymes have therapeutic potential for dissolving blood clots in heart attack victims.

plasminogen \\plaz-'min-ə-jən\\ *See* plasmin.

plasminogen activator \\plaz-'min-ə-jən 'ak-tə-ˌvāt-ər\\ Any molecule that activates plasminogen to give the active enzyme plasmin. *See also* plasmin, streptokinase, tissue plasminogen activator, urokinase.

plasmodesmata \\ˌplaz-mə-'dez-mət-ə\\ Fine channels that pass through the walls of adjacent plant cells joining one to another. They are thought to be one route by which viruses spread within plants.

plasticity \\pla-'stis-ət-ē\\ One of the fundamental parameters of rheology. When a force is applied to a fluid, flow cannot occur until the applied shear stress exceeds the yield stress of the material. The higher the yield stress, the greater the plasticity.

plasticizer \\'plas-tə-ˌsī-zər\\ A substance incorporated into a polymer to alter such properties as workability or flexibility. Plasticizers generally consist of high-molecular-weight compounds such as phthalates, polyglycols, and phosphate esters. They may be leached out of the polymeric material during use and hence contaminate fluids in contact with items such as peristaltic pump tubing and plastic containers.

plastid \\'plas-təd\\ An organelle found in plant cells that is enclosed in a double membrane. Plastids include *amyloplasts,* which act as reserves of starch; *chloroplasts,* which are the site of photosynthesis; *chromoplasts,* which accumulate pigments such as carotenoids; and *proteoplasts,* which act as reserves of protein. Plant-cell DNA is found in three compartments, namely the nucleus, mitochondria, and plastids. Although the introduction of DNA into the nucleus is the most common route for genetically engineering crops for improved agronomic traits, the potential exists for the engineering of plastid-encoded genes to improve plants. Transformation of plastids could lead to stable integration of transforming DNA that is transmitted to the seed progeny.

plastome \\'plas-tōm\\ The genome of a plastid. Plastomes are the genetic material (DNA) found in chloroplasts. Also known as cpDNA or ctDNA.

plate and frame filter \\'plāt 'and 'frām 'fil-tər\\ A system used for filtration and membrane processing that consists of a sandwich structure of alternating supports (plates) for the membrane or filter cloth and frames for the slurry or product. The assembly is relatively inexpensive and easy to dismantle for cleaning, but it has high operating costs. It can be difficult to provide safe containment for the overall system. *Compare with* centrifuge.

plate column \\'plāt 'käl-əm\\ An extraction or distillation column constructed with a packing made of individual plates or trays to aid coalescence and distribution. *See also* bubble-cap plate, sieve plate column.

plate efficiency \\'plāt i-'fish-ən-sē\\ A measure of the performance of an adsorption or distillation column. If perfect contacting occurred at a tray or plate in such a column, then the plate efficiency would be 100% and the two product streams would be at equilibrium. However, in practice the plate efficiency is less than 100%, so more contacting devices are required to achieve the desired separation or adsorption than theoretically calculated. The difference (expressed as a percentage) is known as the plate efficiency.

plate (theoretical) \\'plāt (ˌthē-ə-'ret-i-kəl)\\ *See* theoretical plate.

platelet-derived growth factor (PDGF) \\'plāt-let-di-ˌrīvd 'grōth 'fak-tər ('pē 'dē 'jē 'ef)\\ A protein of MW 30,000 comprising two polypeptide chains (A chain—124 amino acids, B chain—140 amino acids) first purified from blood platelets. PDGF is a potent mitogen for cells of mesenchymal origin such as smooth muscle cells and fibroblasts, and as such has potential for use as a wound-healing agent.

pleiotrope \\'plī-ə-ˌtrōp\\ A gene or mutation that affects more than one characteristic of the phenotype.

pleiotropic \\plī-ə-'trōp-ik\\ Of or referring to a single change or activity that influences more than one characteristic of a cellular phenotype.

***Pleurotus ostreatus* (oyster mushroom)** \\plù-'rō-təs ä-'strē-a-təs (ȯi-stər 'məsh-rüm)\\ An edible mushroom that has been traditionally grown in eastern Asia and is now being cultivated on a limited scale in Europe.

ploidy \\'plȯid-ē\\ The number of chromosomes in a cell; similar to karyotype. *Euploid* means having the correct number; *aneuploid* means having an abnormally high or low number, with the individual chromosomes not in their normal proportions; and *polyploid* means having a number increased by an integral factor.

PLOT \\'plät\\ *See* porous-layer, open-tubular column.

plug flow \\'pləg 'flō\\ A fluid flow in which no axial mixing occurs and thus all the fluid elements have exactly the same residence time. *See also* laminar flow, turbulent flow.

plug flow column reactor \\'pləg 'flō 'käl-əm rē-'ak-tər\\ A biochemical reactor that comprises columns packed with immobilized biocatalyst particles. Substrate solution is passed into the column, where it is exposed to a high enzyme concentration, and product molecules are flushed out of the other end. Product concentration in the reactor is therefore kept to a minimum. This procedure reduces the chances of product inhibition, but substrate inhibition can be a problem.

plug flow reactor \\'pləg 'flō rē-'ak-tər\\ *See* plug flow.

Podbielniak extractor \\päd-'bēl-nē-ak ik-'strak-tər\\ A centrifugal countercurrent liquid extractor used for antibiotic recovery from fermentation broths. Flow rates in excess of 100,000 dm^3/h are possible with the largest extractors.

point mutation \'póint m(y)ü-'tā-shən\ An alteration in the DNA base sequence that results in replacement of a single base by another. As a result of this mutation, a single amino acid may be changed in the protein sequence that is coded for by the region of DNA containing the mutation.

POL \'pōl\ The region of RNA in a retrovirus that codes for RNA polymerase. Other retroviral genes include ENV and GAG.

polishing \'päl-ish-iŋ\ The removal of the final traces of contaminant from a liquid to produce a clean product, as in polishing filtration or polishing ion exchange.

pollen culture A technique in which immature pollen is induced to divide and generate tissue, either on solid media or in liquid culture. Microspores are released from anther tissue and cultured on an appropriate medium, where they divide and generate tissue that can develop to produce a mature plant (*see* anther culture). At present the technique has only proved successful in a few species, and the closely related technique of anther culture is more generally used. These techniques allow the generation of haploid plants.

poly(A) tail \'päl-ē-ā 'tā(ə)l\ A sequence of 200–300 polyadenylic acid residues added posttranscriptionally to the 3'-end of most eukaryotic mRNAs. *See also* poly(U)-Sepharose.

polyacrylamide gel \päl-ē-ə-'kril-ə-ˌmīd 'jel\ A gel formed by the polymerization of acrylamide monomers in the presence of a comonomer *N,N'*-methylenebisacrylamide (bis), which acts as a cross-linking agent. Polymerization is induced by the addition of free radicals. Polyacrylamide gels are commonly used as a support in polyacrylamide gel electrophoresis of DNA and proteins. Polyacrylamide gels are also used as a gel filtration medium, and ion-exchange derivatives of the gel are also available. By reaction with suitable compounds (e.g., production of the aminoethyl or hydrazide derivative), polyacrylamide gels can be converted to solid carriers suitable for binding ligands. *See also* polyacrylamide gel electrophoresis.

polyacrylamide gel electrophoresis (PAGE) \päl-ē-ə-'kril-ə-ˌmīd 'jel i-ˌlek-trə-fə-'rē-səs ('pāj)\ Electrophoresis in polyacrylamide gels used to analyze both protein and nucleic acid mixtures. As samples are electrophoresed through the gel, molecules separate according to size because of the sieving properties of the gel. The pore sizes in the gel, and hence the molecular-weight range of molecules that can be separated in a given gel, are determined by the acrylamide and bis concentrations. Polyacrylamide gel electrophoresis is most frequently used during DNA sequence determination and in protein analysis by SDS polyacrylamide gel electrophoresis. *See also* isoelectric focusing.

polyadenylic acid \'päl-ē-ad-ᵊn-ˌil-ik 'as-əd\ A chain of adenylic acid residues. In biological systems, this acid is found at the 3' end of many eukaryotic RNA molecules. *See* poly(A) tail.

polyclonal antiserum \\'päl-ē-klōn-ᵊl ant-ī-ˌsir-əm\\ *See* antiserum, B-lymphocytes.

polydentate ligand \\ˌpäl-i-'den-ˌtāt 'lig-ənd\\ A molecule (ligand) capable of binding to a metal ion by more than one donor site. More precise terms such as bi-, tri-, tetra-, penta-, and hexadentate are used to indicate binding at two to six sites.

polyelectrolyte \\'päl-ē-ə-'lek-trə-ˌlīt\\ A macromolecule containing many ionizable groups that may be anionic or cationic, strong or weak acids, or bases. The term is usually used for soluble materials such as proteins or nucleic acids.

polyethylene glycol (PEG) \\ˌpäl-e-'eth-ə-ˌlēn 'glī-ˌkȯl\\ A polymer of ethylene glycol, sold as different molecular weights of polymer (e.g., PEG 4000, PEG 6000). It is used in particular to fuse protoplasts and cells (e.g., in the formation of hybridomas), in two-phase aqueous systems, and to precipitate phage particles during their isolation.

polygalacturonase (E.C. 3.2.1.15) \\ˌpäl-i-ga-lak-'tùr-ə-ˌnās\\ A pectin-degrading enzyme associated with the cell wall of fruits. The solubilization of pectin by this enzyme has been shown to be involved in the fruit-softening process. Genetically engineered tomatoes with lower levels of polygalacturonase have been produced (the so-called "Flavr Savr" tomato) by the introduction of an antisense gene and shown to be less susceptible to softening and rotting than non-engineered tomatoes.

poly-β-hydroxybutyrate (PHB) \\ˌpäl-i-bā-tə-hī-ˌdräk-sē-'byü-ti-ˌrāt ('pē 'āch 'bē)\\ A thermoplastic polymer of β-hydroxybutyrate esters, accumulated by a range of organisms as an energy reserve. It is prepared from *Alcaligenes eutrophus,* which can accumulate up to 70% of its dry weight as PHB. Cells are ruptured, and PHB is extracted with a halogenated hydrocarbon. Because it is biodegradable, this polymer has a range of potential uses (e.g., as a medical suture).

polyketides \\ˌpäl-i-kē-ˌtīdz\\ Secondary metabolites that can be derived mainly or exclusively from poly(β-keto acids). The general structure is $(\text{—(CHRCO)}_n\text{—})$. The subunits may be —$CH_2CO$—(acetate unit), or —$CH(CH_3)CO$—(proprionate unit), or —$CH(C_2H_5)CO$—(butyrate unit). A typical example is erythromycin, which is a glycoside of a polyketide aglycone. Major subgroups include flavonoids, tetracyclines, and macrolide antibiotics. Also known as acetogenins.

polylinkers \\ˌpäl-i-ˌliŋk-ərz\\ Synthetic double-stranded oligonucleotides that contain a number of different restriction sites. They are introduced into vectors to generate new and more versatile vectors with an increased number of potential restriction sites. The modified vectors can therefore be used to clone DNA fragments generated by a range of restriction enzymes.

polymer \\'päl-ə-mər\\ An assembly of a large number of repeat units (monomers) that may be represented as chains, sheets, or three-dimensional structures.

polymerase chain reaction (PCR) \\'päl-ə-mə-ˌrās 'chān rē-'ak-shən ('pē 'sē 'är)\\ An enzymatic method for selectively amplifying a specific nucleotide sequence several millionfold in a few hours. This ability to amplify as little as a single molecule of DNA to provide sufficient material for analysis, cloning or other operations has had a revolutionary effect on the study of molecular genetics and related disciplines. The method relies on the use of *Taq* polymerase, a thermostable DNA polymerase from the thermophilic bacterium *Thermus aquaticus*. *Taq* polymerase remains active at the high temperatures used to denature double-stranded DNA during each cycle of the PCR and thus obviates the need to add fresh DNA polymerase after each cycle. The use of a programmable heating block allows full automation of the process. The method requires two oligonucleotide primers that base-pair to opposite strands of the sequence to be amplified. The PCR cycle involves three steps:

(1) denaturation of the target double-stranded DNA by heating to 95 °C in the presence of the two primers,

(2) cooling to 55 °C to allow primer annealing, and

(3) raising the temperature to 70 °C. At this temperature, the DNA polymerase is active and carries out primer extension.

This denaturation, annealing, and strand synthesis cycle takes approximately 5–6 min, and a typical PCR protocol uses about 25 cycles. The diagram shows that after each denaturation step the original, as well as the newly synthesized DNA strands, serve as templates in subsequent cycles. The amount of DNA is therefore doubled at each cycle. Extension of a primer hybridized to an original template strand yields DNA copies of indefinite length, but after 3 cycles, a discrete DNA fragment is produced flanked at either end by the two primers. In further cycles this sequence increases exponentially in number and comes to predominate. The applications of PCR are legion. Applications to biotechnology include amplifying trace samples for forensic analysis, amplification of fetal DNA for genetic analysis, and amplification of bacterial and viral DNA to allow rapid identification of disease organisms using appropriate probes.

A number of modifications of the basic PCR procedure have been developed in recent years. These modifications include.

(1) *Alu PCR*: This method uses PCR primers that pair with two Alu sequences in inverse orientation. This pairing permits direct amplification of sequences between the Alu repeats in the presence of complex backgrounds such as in cosmid, lambda, or yeast artificial chromosome clones.

(2) *Hotstart PCR*: Most DNA samples ready for amplification have some single-stranded DNA present even before the denaturation step. Primers can bind to these single-stranded segments at low stringency, and since *Taq* polymerase has some activity at room temperature, the single strand region can begin to be extended before the reaction proper has started. Hotstart PCR overcomes this artifact production by leaving out a critical reagent (such as the enzyme or magnesium) from the reaction until it has reached denaturation point.

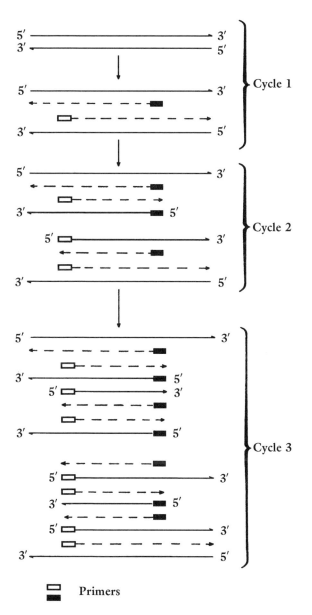

□   Primers
■

polymerase chain reaction

(3) *Inverse PCR*: This method uses end sequences in outward orientation as PCR primers in order to amplify unknown sequences adjacent to known sequences.

(4) *Touchdown PCR*: This method increases specificity by initiating synthesis at high annealing temperatures, which permits only perfectly matched primer-template hybrids to form. Once copies of the target sequence have begun to accumulate over the first few cycles the temperature can be dropped because previous products form the major template.

polymixin \\,päl-i-'mik-sən\\ One of a group of cyclic peptide antibiotics produced by various species of *Bacillus*, such as *B. polymyxa*. The basic structure is a cyclic heptapeptide with a three-residue peptide side chain that terminates in a fatty acid residue, which is either 6-methyloctanoic acid or isooctanoic acid. At least five, and often six of the ten amino acids found in polymixins are α,γ-deaminobutyric acid. Polymixins are active against many Gram-negative bacteria and act by damaging the outer membrane and by increasing the permeability of the cytoplasmic membrane to small molecules. Most Gram-positive bacteria and most fungi are resistant, but some strains are susceptible, e.g., *Candida*. Polymixins have been used for human treatment and as additives in animal feeds.

polynucleotide kinase \\,päl-i-'n(y)ü-klē-ə-ˌtīd 'kī-ˌnās\\ An enzyme, prepared from *Escherichia coli* infected with T4 phage, that adds phosphate groups to free 5'-terminal hydroxyl groups in DNA molecules.

polyploid \\'päl-i-ˌplȯid\\ *See* ploidy.

polysaccharides \\,päl-i-'sak-ə-ˌridz\\ Polymers built from monosaccharide units (e.g., amino sugars, hexoses, pentoses). Polysaccharides produced by fermentation have a range of industrial applications. *See also* alginate, dextran, polytran, succinoglucan, xanthan gum, zanflo.

polytran \\'päl-i-ˌtran\\ A linear β-1,3-glucan (polysaccharide) with a D-2-glucopyranose group linked β-1,6 to every third or fourth residue. Polytran is produced as an exopolysaccharide by the fungus *Sclerotium glucanicum* grown in submerged culture and is used in oil recovery, ceramic glazes, seed coatings, and printing inks. Also known as scleroglucan.

poly(U)-Sepharose \\,päl-i-yü-sef-ə-'rōs\\ An affinity matrix used to purify mammalian mRNA. It comprises poly(uridylic) acid covalently bound to agarose. Most mammalian mRNA molecules contain a poly(A) tail of 200–300 polyadenylic acid residues at their 3'-end. This tail region hydrogen-bonds to the poly(U) by complementary base-pairing and thus binds the mRNA to the column, whereas other molecules pass through the column unhindered.

poly(vinylidene difluoride) (PVDF) membrane \\'päl-i-vī-nil-ə-ˌdēn-ə-ˌflu̇(-ə)r-īd ('pē 'vē 'dē 'ef) 'mem-ˌbrān\\ A Teflon-like material used for protein blotting. Transferred proteins bind to the membrane by hydrophobic interactions, and contaminants can be washed from the membrane by rinsing with water. Because of the inert nature of the membrane, membranes containing transferred proteins can be placed directly into automated sequence machines for sequence analysis of the protein. *See also* Edman degradation.

polyvirus \\'päl-i-ˌvī-rəs\\ The largest, most widely distributed, and economically important group of plant viruses. Polyviruses account for more than 25% of the viruses known to infect plants.

porous glass or silica \\'pōr-əs 'glas 'ȯr 'sil-i-kə\\ Support materials for chromatography, usually treated with a silanizing reagent to minimize adsorption by residual hydroxyl groups on the glass or silica. Porous glass is also used as a medium for exclusion chromatography.

porous-layer, open-tubular (PLOT) column \\'pōr-əs-'lā-ər 'ō-pən-'t(y)ü-byə-lər ('plät) 'käl-əm\\ A capillary gas chromatography column that has a porous layer on the inner wall. This porosity may be achieved either chemically to increase the inner surface area or by a deposit of a porous solid on the inner wall. In both cases the tube wall serves as a support for the liquid stationary phase or behaves as the stationary phase itself. *See also* support-coated, open-tubular column; wall-coated, open-tubular column.

postcolumn derivatization \\'pōst-ˌkäl-əm ˌdi-'riv-ət-ə-zā-shən\\ *See* amino acid analysis, reaction chromatography.

posttranslational modification \\ˌpōs(t)-tran(t)s-'lā-shən-ᵊl ˌmäd-ə-fə-'kā-shən\\ Any change that occurs to the protein polypeptide chain during or after its synthesis at the ribosome. Commonly occuring posttranslational modifications include the covalent addition of polysaccharide chains to form a glycoprotein, the phosphorylation of specific serine or threonine residues, and the cleavage of a signal sequence once the protein has crossed a membrane. More than 200 postsynthetic modifications of proteins have been detected in living systems. Posttranslational modifications can affect both the antigenicity and the half-life upon injection of a protein and are thus important considerations when cloning proteins for therapeutic uses. Because the animal cell is the natural site for posttranslational modifications (bacteria, for example, cannot glycosylate proteins), the use of animal cells in culture is often the first choice for the expression of foreign genes. *See also* glycoform, microheterogeneity isoelectric focusing.

potable \\'pōt-ə-bəl\\ Fit to drink, used to describe liquid, e.g., water that has been suitably purified. Potability also distinguishes fermentation ethanol from the synthetic product.

potentiometric sensor \\pə-ˌten-ch(ē)ə-'me-trik 'sen-sȯ(ə)r\\ A type of sensor that gives, as an output, a potentiometric difference; for example, many types of electrodes, including ion-selective and glass.

power input (agitation) \\'pau̇(-ə)r 'in-pu̇t (ˌaj-ə-'tā-shən)\\ A measure of the energy applied to a system, as in the stirrer or agitator of a fermentor. It is measured by a wattmeter attached to the stirrer motor, torsion dynometers, or strain gauges on the inside of the hollow stirrer shaft. The power input into an agitated vessel [P(W)] is related to the power number (N_p) and the Reynolds number (Re) by the equation

$$N_p = c\mathrm{Re}^x$$

or

$$\frac{P}{\rho N^3 D_i^5} = c\left(\frac{\rho N D_i^2}{\eta}\right)^x$$

where N is the impeller speed (s^{-1}), D_i is the impeller diameter (m), ρ is the density of fluid (kg/m^3), η is the fluid viscosity (kg/ms), c is a constant that depends on the vessel geometry, and x is an exponent with the value -1 under laminar flow conditions and 0 under turbulent flow.

Thus the equation can be simplified. For laminar flow

$$P = c\eta N^2 D_i^3 \quad (\mathrm{Re} < 10)$$

For turbulent flow

$$P = c\rho N^3 D_i^5 \quad (\mathrm{Re} > 10^4)$$

These equations are for ungassed systems, whereas most fermentations are aerated. Aeration decreases power consumption because a liquid containing suspended air bubbles is less dense than an unaerated one. An empirical relationship between the gassed power consumption (P_g) and ungassed power consumption (P), which is applicable over a wide range of turbulent operating conditions, exists:

$$P_g = k\left(\frac{P^2 N D_i^3}{Q^{0.56}}\right)^{0.45}$$

where Q is the volumetric air flow rate and k is a constant relating to fermentor geometry. *See also* power number.

power number (agitation) \\'paủ(-ə)r 'nəm-bər (ˌaj-ə-'tā-shən)\\ A dimensionless group used with the Reynolds (Re) and Froude (Fr) groups to describe agitation. It is defined as.

$$N_p = \frac{P}{\rho N^3 D_i^5}$$

where P is the agitator power (W), N is impeller speed (s^{-1}), D_i is the impeller diameter (m), and ρ is fluid density (kg/m^3). The power number is related to the Reynolds and Froude numbers by the expression.

$$N_p = c(\mathrm{Re})^x(\mathrm{Fr})^y$$

where c is a constant dependent on vessel geometry but independent of vessel size and x and y are exponents.

The Froude number is only of importance if a gross vortex exists and the relationship then becomes

$$N_p = c(\mathrm{Re})^x$$

See also Froude number, power input, Reynolds number.

Prandtl number (Pr) \ 'nəm-bər ('pē 'är)\ An important dimensionless group in heat transfer between a fluid and vessel wall:

$$Pr = \frac{C_p \eta}{k}$$

where C_p is the heat capacity (J/kgK), η is the viscosity (kg/m • s), and k is the thermal conductivity of the fluid (W/mK). *See also* Nusselt number.

precoat filter \'prē-ˌkōt 'fil-tər\ A layer of filter aid applied to a filter cloth before carrying out a filtration to minimize clogging (blinding) of the filter cloth and thus maintain a high rate of filtration. It may also be applied to a drum or rotary filter fitted with a scraper discharge. *See also* rotary vacuum filter.

precolumn derivatization \'prē-ˌkäl-əm der-ə-'vā-shən\ *See* amino acid analysis, reaction chromatography.

precursor \pri-'kər-sər\ A compound that is formed before, and can be converted into, the product of interest.

preflashing (prefogging) \'prē-ˌflash-iŋ (prē-'fóg-iŋ)\ The exposure of photographic film to a brief flash of low-intensity light just before its use in fluorography. This treatment sensitizes the film to low-intensity light emissions.

pressure leaf filters \'presh-ər 'lēf 'fil-tərz\ Intermittent batch filters, often referred to by their trade names, e.g., Metafilters. They are particularly suitable for processing large volumes of liquids with a low solids concentration.

pressure release devices \'presh-ər ri-'lēs di-'vīs-ez\ Equipment designed to prevent dangerous high pressure from occuring in such vessels as reactor vessels. They may be directly activated, e.g., spring-loaded valves designed to open at a set pressure, indirectly activated valves that open on receipt of a signal, or, bursting disks, which rupture at a predetermined pressure.

primary cell culture \'prī-mer-ē 'sel 'kəl-chər\ A freshly isolated culture of cells derived directly from a particular organ, tissue, or the blood of an organism. Free cells can be obtained by treatment of tissue pieces with the proteolytic enzyme trypsin (trypsinization) or perfusion of the tissue with the proteolytic enzyme collagenase. Primary cell cultures are usually heterogeneous and have a low growth fraction, but they are representative of the cell types in the tissue from which they were derived. Subculturing of the primary culture gives rise to a secondary culture.

primary metabolites \prī-ˌmer-ē mə-'tab-ə-lītz\ The metabolites formed during the log phase (tropophase) of microbial culture. *See also* tropophase.

primer \'prī-mər\ A short (~10) oligonucleotide that base-pairs (anneals) to a region of single-stranded template oligonucleotide. Primers are necessary to form the starting point for reverse transcriptase to copy adjacent sequences of mRNA, or for DNA polymerase to produce complementary-strand synthesis with single-stranded cDNA. Primers are also a necessary requirement for the polymerase chain reaction.

probe \'prōb\ *See* gene probe.

Procion dyes \\'prō-ṣī-ən 'dīz\ An ICI Chemicals and Polymers trade name for monochloro- or dichlorotriazine dyes that are used extensively in dye–ligand chromatography, e.g., Procion blue, Procion red, Procion green.

productivity \prō-ˌdək-'tiv-ət-ē\ A measure of the production of biomass per unit time in a microbial culture. It is measured as grams of biomass per cubic decimeter per hour.

prokaryote \(')prō-'kar-ē-ˌōt\ Any organism that comprises a prokaryotic cell, e.g., a bacterium. *See* prokaryotic cell.

prokaryotic cell \prō-ˌkar-ē-'ät-ik 'sel\ A cell whose genetic material is present throughout the cell and not organized into a discrete nucleus (e.g., all bacterial cells are prokaryotic). *Compare with* eukaryotic cell.

prolamins \'prō-lə-mənz\ One of the four major categories of seed storage proteins. *See also* seed storage proteins.

promoter \prə-'mōt-ər\ A nucleotide sequence found upstream of a gene that acts as a signal for the binding of RNA polymerase. In *Escherichia coli*, promoter sequences are about 10 and 35 base pairs from the start of transcription and are about 7 base pairs long. The exact nature of these sequences is not yet fully understood, but comparison of many different genes has identified sequences common to all promoters. *See* enhancer, TATA box.

prophage \'prō-ˌfāj\ Bacteriophage DNA integrated into the bacterial chromosome. *See also* lysogenic infection.

PROSITE \'prō-ṣīt\ A database of protein sequence motifs.

prosthetic group \präs-'thet-ik 'grüp\ Any tightly bound, specific nonpolypeptide unit required by a protein for its biological activity. For example, heme is the prosthetic group of hemoglobin. A protein without its characteristic prosthetic group is termed an apoprotein. When a prosthetic group in associated with an enzyme it is usually referred to as a cofactor or coenzyme. *See also* cofactor, cofactor recycling.

protease nexin I (PNI) \'prōt-ē-ˌās 'neks-in 'ī ('pē 'en 'ī)\ A potent antithrombin and antiurokinase protein of MW 44 kDa, secreted by fibroblasts and certain other extravascular cells.

α-1-protease inhibitor (α, PI) *See* α-1-antitrypsin.

proteases \'prōt-ē-ˌās-ez\ Enzymes that break down proteins by hydrolyzing specific peptide bonds. Most proteases can be classified under one of four headings, based on specific groups present at the active site that are essential to the catalytic mechanism of the enzyme. They are

- *acid (or carboxyl) proteases:* These proteases all contain a carboxyl group at the active site that is essential for activity (e.g., rennin (E.C. 3.4.23.4)).
- *metalloproteases:* These enzymes contain a metal ion, usually zinc, that is essential for enzymatic activity (e.g., carboxypeptidase A (E.C. 3.4.17.1), thermolysin (E.C. 3.4.24.4)).
- *serine proteases:* These enzymes all cleave peptide bonds by a common mechanism that involves a highly nucleophilic serine residue at the active

site that is essential for enzyme activity (e.g., trypsin (E.C. 3.4.21.4), chymotrypsin (E.C. 3.4.21.1)).

- *thiol proteases:* Proteases that have a thiol (—SH) group at the active site that is essential for activity (e.g., papain (E.C. 3.4.22.2), bromelain (E.C. 3.4.22.4), and ficin (E.C. 3.4.22.3)).

Proteases have a wide range of industrial uses. One of the most widely usedenzymes is subtilisin Carlsberg (E.C. 3.4.21.14), a serine protease produced from *Bacillus licheniformis*. This enzyme is extensively used in enzyme washing powders. Other uses of proteases include wheat gluten degradation in the baking industry, chillproofing in the brewing industry, curd formation in the cheese industry, meat tenderization, removal of hair in the tanning industry, and the recovery of silver from spent photographic films.

protein A \\'prō-ˌtēn 'ā\\ A protein produced by and purified from certain strains of *Staphylococcus aureus*. The protein binds to the F_c portion of certain classes of IgG molecule (*see* immunoglobulin G) without impairing the ability of the antibody to bind to an antigen. Protein A has found particular use in immunoassays and in purifying IgG subclasses by affinity chromatography. Protein A is bivalent (i.e., each molecule binds two IgG molecules).

protein blotting (western blotting) \\'prō-ˌtēn 'blät-iŋ ('wes-tərn 'blät-iŋ)\\ The transfer (by either diffusion or electrophoretic transfer) of separated proteins from a gel medium (e.g., polyacrylamide) to a nitrocellulose (or nylon) sheet, where the proteins bind. The proteins, once concentrated on the surface of the solid support, may be further investigated by the use of probes, such as antibodies. Such analysis of the proteins is not possible in the gel medium. Because the method is analogous to Southern blotting for DNA, the term western blotting is sometimes used.

protein engineering \\'prō-ˌtēn en-jə-'ni(ə)r-iŋ\\ Originally a term used to describe any chemical modification of a protein that resulted in an altered structure or function for the protein. Now it is used more specifically to describe the predetermined replacement by site-directed mutagenesis of specific amino acids in a protein. This replacement can result in the synthesis of a protein with enhanced characteristics, such as increased stability, increased catalytic activity, or altered substrate specificity. *See also* bispecific antibodies, chimeric antibody, site-directed mutagenesis.

Protein Identification Resource \\'prō-ˌtēn ī-ˌdent-ə-fə-'kā-shən 'rē-sō(ə)rs\\ A protein amino acid sequence database operated by the National Biomedical Research Foundation in Washington, D.C. *See also* Brookhaven Protein Data Bank, SWISS-PROT.

protein inclusion bodies (refractile bodies) \\'prō-ˌtēn in-'klü-zhən 'bäd-ēz (rē-'frak-t⁀l 'bäd-ēz)\\ Dense, insoluble protein bodies on the order of 1 μm in diameter, often formed when eukaryotic proteins are produced by the genetic modification of microorganisms. Following cell concentration (centrifugation) and disruption, inclusion bodies are separated from soluble cell protein and particulate cell debris by centrifugation. The solubilization of protein from the inclusion bodies, followed by refolding of the product to the native conformation, are critical to the recovery and purification of the required protein.

protein sequencing \\'prō-ˌtēn 'sē-kwen(t)s-iŋ\\ *See* Edman degradation.

protein synthesis *See* translation.

protein targeting \\'prō-ˌtēn 'tär-gət-iŋ\\ The attachment, by genetic manipulation, of a leader sequence to the N-terminus of a protein so that, after translation by cytoplasmic ribosomes, the protein passes through a target membrane. In this way, nuclear-coded proteins can be targeted to enter specific organelles.

proteolysis \\prōt-ē-'äl-ə-səs\\ The hydrolysis of proteins into peptides or amino acids by proteases. *See* proteases.

proteoplasts \\prōt-ē-ə-ˌplas(t)s\\ *See* plastid.

prothrombin \\(')prō-'thräm-bən\\ *See* thrombin.

protoclonal variation \\'prōt-ə-ˌklōn-ᵊl ˌver-ē-'ā-shən\\ *See* protoclone.

protoclone \\'prōt-ə-ˌklōn\\ A clone of cells generated from a single plant protoplast. Protoclones that produce regenerated plants that show differences in phenotype when compared with the parent plant are said to have undergone *protoclonal variation,* which is effectively the same as somaclonal variation.

protomer \\'prōt-ə-mər\\ An individual polypeptide chain in a multichain protein. For example, the α- or β-chain in hemoglobin. Hemoglobin consists of two identical α-chains and two identical β-chains; it is therefore a multimeric protein made up of four protomers.

protooncogene \\prōt-ə-'äŋ-kō-jēn\\ *See* oncogene.

protoplast fusion \\'prōt-ə-ˌplast 'f(y)ü-zhən\\ Fusion of two protoplasts to give a single hybrid cell containing a single fused nucleus. This fused protoplast can then regenerate a cell wall and grow as a normal cell. The method has been used to fuse different strains of filamentous fungi, yeast, streptomycetes, and bacteria (and thus produce improved strains by recombination). The same method is used to fuse plant protoplasts (*somatic fusion*) that can regenerate and grow to give full plants. Fusion methods include the use of polyethylene glycol and electrofusion.

protoplasts \\'prōt-ə-ˌplas(t)s\\ Microbial or plant cells devoid of their cell walls. They are prepared by treating cells with cell-wall-degrading enzymes (e.g., snail gut enzyme, Novo Industri A/S Denmark's Novozyme) in isotonic solution.

Pruteen process \\'prü-ˌtēn 'präs-ˌes\\ A process devised by Imperial Chemical Industries that uses extremely large-scale culture (1500 m³) to produce bacterial single-cell protein (ICI Chemicals and Polymers's Pruteen) for use as animal feed. *Methylophilus methylotrophus* is the microorganism used, with methanol as a substrate.

pseudomycelium \\ˌsüd-ə-mi-'sē-lē-əm\\ A series of microbial cells adhering together end-to-end to form a chain resembling mycelium but not a tubular filament. It is produced by some yeasts.

pseudoplastic fluid \\ˌsüd-ə-'plas-tik 'flü-əd\\ A fluid in which viscosity decreases with increasing shear rate, such that shearing thins the fluid. The relationship between the applied shear (τ) and the shear velocity (γ) follows a power law of the form

$$\tau = k\gamma^n$$

where k is the consistency index and n is the flow behavior index. For example, when n is less than 1, pseudoplastic or shear thinning is observed.

$$\gamma = a\tau^3 + c\tau$$

where a and c are coefficients. Pseudoplastic fluids generally show Newtonian behavior at high and low shear rates on either side of the power law region.

psychrophiles \\'sī-krō-ˌfīlz\\ Microorganisms that require low temperatures (i.e., 5–25 °C) for optimum growth. *Contrast with* mesophiles, thermophiles.

pullulan \\'pəl-yə-lan\\ A neutral linear homopolysaccharide produced commercially by the growth of *Aureobasidium pullulans*. It is an exopolysaccharide. Pullulan consists of glucose units polymerized into repeating maltotriose units. Within each maltotriose unit, the glucopyranose units are linked by α-1,4-glucosidic bonds. The repeating maltotriose units are linked by α-1,6-glucosidic bonds. Pullulan forms strong resilient films and fibers that can be molded and shows low oxygen permeability when cross-linked with other materials to form thin polymeric films. Cross-linking is an important feature for food-packaging materials. It is also used in the preparation of adhesives, fibers, molded articles, coatings, and films.

pullulanase (E.C. 3.2.1.41) \\'pəl-yə-la-ˌnās\\ An enzyme that acts specifically on the α-1,6-bonds of the amylopectin of starch and therefore is called a debranching enzyme. It is prepared commercially from *Klebsiella pneumoniae* and is used in conjunction with glucoamylase (which only slowly hydrolyzes α-1,6-bonds) in the saccharification of starch.

Pulmozyme \\'pəl-mō-ˌzīm\\ A Genentech trade name for a recombinant DNAse. This pulmonary drug is delivered to the lungs by an aerosol, where it degrades polymerized DNA, which otherwise increases the viscosity of purulent secretions. It is used to dissolve the phlegm of patients with acute chronic bronchitis and to treat cystic fibrosis patients. The enzyme degrades excess DNA, which is the main ingredient of sputums lining the lungs of these patients.

pulsed column \\'pəls(t) 'käl-əm\\ A contactor for liquid–liquid extraction. The column usually contains some form of coalescence and distribution devices such as plates and is fitted with an oscillating pump that provides a pulsing action on the continuous phase. This action causes the continuous phase to oscillate during its progress through the column. The oscillation applies a mild form of agitation to the system and thus improves the efficiency of the process. *See also* pulsed-plate column.

pulse-field gel electrophoresis (PFGE) \\'pəls-ˌfē(ə)ld 'jel i-ˌlek-trə-fə-'rē-səs ('pē 'ef 'jē 'ē)\\ An electrophoretic technique for separating high-molecular-weight DNA molecules. Traditional agarose gel electrophoresis can fractionate DNA fragments in the range 20 kilobase pairs or less. The introduction of PFGE means that DNA fragments in excess of 2×10^6 (2.0 M base pairs) can be separated. This technique therefore allows the separation of whole chromosomes by electrophoresis. The method basically involves electrophoresis in an agarose gel in which two electrical fields are alternately applied at different angles for a defined time period (e.g., 60 s). Activation of the first electric field causes the coiled molecules to be stretched in the horizontal plane and start to move through the gel. Interruption of this field and application of the second field force the molecule to move in the new direction. Because there is a length-dependent relaxation behavior when a long-chain molecule undergoes conformation change in an electrical field, the smaller a molecule, the quicker it realigns itself with the new field and is able to continue moving through the gel. Larger molecules take longer to realign. In this way, by continually reversing the field, smaller molecules draw ahead of larger molecules, and separation according to size is achieved. More recently a number of modified techniques have been described, but all techniques are still based on the principle of DNA reorientation. Parameters that have been modified include the electrode configuration, the polarity, and the position of the gel in the box. These modified methods include orthogonal field alternation gel electrophoresis (OFAGE), field inversion gel electrophoresis (FIGE), transverse alternating field gel electrophoresis (TAFE), contour clamped homogeneous electric field electrophoresis (CHEF), and rotating field electrophoresis (RFE). PFGE is used as an acronym for pulse field gel electrophoresis to indicate any technique that resolves DNA by continuous reorientation. FIGE and CHEF are currently the most commonly used PFGE systems. *See also* reptation.

pulsed liquid protein sequencer \'pəlst 'lik-wid 'prō-ˌtēn 'sē-kwən(t)s-ər\ *See* Edman degradation.

pulsed-plate column \'pəlst-ˌplāt 'käl-əm\ An alternative design to the pulsed column. The stack of plates is capable of vertical oscillation and thereby promotes agitation in the column in the same way as the pulsed column. Both the pulsed column and the pulsed-plate column may be designed with solid disks or modified as in the perforated-plate or sieve plate designs.

pure culture \'pyu̇(ə)r 'kəl-chər\ A culture containing a single species of organism.

purification tag \ˌpyu̇r-ə-fə-'kā-shən 'tag\ *See* affinity tailing.

purple bacteria \'pər-pəl bak-'tir-ē-ə\ Photosynthetic bacteria that operate an anoxygenic photosynthesis. In addition to light they also rely on reduced hydrogen donors, such as hydrogen sulphide, and evolve no oxygen during photosynthesis.

$$H_2S + CO_2 \xrightarrow{\text{light}} \text{cell substance} + S$$

PVDF membrane \'pē 'vē 'dē 'ef 'mem-ˌbrān\ *See* poly(vinylidene difluoride) membrane.

pyrethrin \pī-'rē-thrən\ A commercially useful insecticide isolated from the plant *Chrysanthemum cinerariaefolium*.

pyrogen \'pī-rə-jən\ Any fever-inducing substance. Historically, pyrogens are lipopolysaccharides from the cell walls of Gram-negative bacteria. Because many biotechnology products are produced in such organisms, the elimination of pyrogens from the product is extremely important. In solution the lipopolysaccharides tend to form complexes that can be removed by filtration through a membrane with a cut-off of MW 100,000. If, however, the pyrogens are in solution with detergents they will disaggregate into units of about MW 20,000; therefore, the use of a membrane with a 10,000 MW cut-off is required.

pyrolysis-gas chromatography \pī-'räl-ə-səs-'gas ‚krō-mə-'täg-rə-fē\ An analytical technique whereby a sample is thermally decomposed to produce simpler components before entering a chromatographic column.

![Q](decorative letter Q)

Q-beta replicase \'kyü 'bāt-ə 'rep-li-ˌkās\ *See* RNA replicase.

quadrupole mass spectrometer \'kwäd-rə-ˌpōl 'mas spek-'träm-ət-ər\ A configuration of mass spectrometer commonly used with process instruments for the detection and analysis of process streams, for example, gases in the exhaust from fermentors. The operating basis of the instrument is as follows. After ionization and fragmentation of the molecule, the ions are fed into the mass analyzer, which consists of four electrically conducting rods that form the corners of a rectangular box. Opposite pairs of rods are connected to an electrical supply that carries both direct current (dc) and radio frequency (rf) voltages, arranged so that at any time one pair of electrodes is positive and the other negative. Ions are injected along the axis of the rod electrodes, where they meet the alternating electrical field. This field is arranged to filter out all ions except the one that resonates at the particular applied radio frequency. These ions pass down the axis of the rods and emerge at the end of the mass analyzer, where they are detected and counted in the ion collector. Variation of the radio frequency allows the entire mass spectrum to be scanned at rates up to 1000 mass units per second, but with a lower resolution than that attainable with double-focusing instruments.

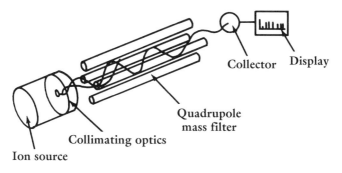

quantum yield \\'kwänt-əm 'yē(ə)ld\ The number of species produced by the absorption of a quantum of energy in radiation-induced processes such as fluorescence. *See also* fluorescence.

quaternary ammonium salts \\'kwät-ə(r)-ˌner-ē ə-'mō-nē-əm 'sȯltz\ Ionic compounds with cations derived from the reactions of tertiary amines with, for example, alkyl halides to give compounds like $R_4N^+X^-$. These compounds have many uses, e.g., cationic surfactants and ion-partition reagents.

quenching, in fluorescence \\'kwen-chiŋ 'in ˌflu̇(-ə)r-'es-ənts\ A process by which excited molecules lose their excess energy by collision with other species; the amount of fluorescent radiation emitted is thereby reduced. Oxygen is a particularly good quenching agent and therefore should be removed from solution before analysis.

quenching, in scintillation counting \\'kwen-chiŋ 'in ˌscint-ᵊl-'ā-shən 'kau̇nt-iŋ\ A process that results in a lower observed counting rate than the true disintegration rate of the analyte. In this case, the emitted photons are intercepted in the solution by, for example, quenching agents or dilution effects before reaching the photomultiplier.

quinine \\'kwī-ˌnīn\ A plant alkaloid isolated from the bark of the *Cinchona ledgeriana* tree. It is used as an antimalarial compound and as a bittering agent in the food and drink industry.

$$R$$

R_f \\'är 'ef\ The ratio of the distance traveled by a component to that traveled by the solvent front. Used in chromatography, the term characterizes components in a mixture when separated by paper or thin-layer techniques.

R-form \\'är 'fȯ(ə)rm\ The stereochemical arrangement displayed by a chiral molecule in which the substituents listed according to defined sequence rules follow a right-handed route around the chiral center. Derived from *rectus*, Latin for right. The converse is S-form. *See also* S-form, sequence rules.

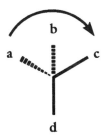

sequence rule a>b>c>d

racemate \rā-'sē-ˌmāt\ An equimolar mixture of the dextrorotatory and levorotatory optical forms of a compound that does not exhibit any overall optical activity.

radial flow chromatography \\'rā-dē-əl 'flō ˌkrō-mə-'täg-rə-fē\ A form of column chromatography that provides high-speed, preparative-scale separations of proteins and other biomolecules and can be used with either soft gel or rigid packings. The equipment consists of a short wide column with a radial liquid flow pattern from core to periphery, which allows high flow rates with a very low pressure drop.

radioautography \ˌrād-'ē-ō-ȯ-'täg-rə-fē\ *See* autoradiography.

radiolabeling \ˌrād-ē-ō 'lā-b(ə-)liŋ\ The incorporation of a radioactive isotope of an element into a chemical compound to follow the progress of the compound through a series of operations or reactions.

radioimmunoassay (RIA) \'rād-ē-ō-im-yə-nō-'as-ā ('är 'ī 'ā)\ An assay method, much used in clinical and research laboratories, that uses antibodies to detect small concentrations of substances (antigens). A known amount of radiolabeled antigen is added to the test solution, together with an antibody to the antigen. The radiolabeled and unlabeled (test) antigen compete for binding to the antibody molecules. The amount of radiolabel that binds to the antibody is determined; it is inversely proportional to the amount of test antigen. The greater the amount of test antigen present, the less radiolabeled compound will bind to the antibody, and vice versa.

raffinate \'raf-ə-ˌnāt\ The product stream that contains the lower concentration of the product when it leaves a process; normally applied to waste streams from extraction processes.

rare cutters \'ra(ə)r 'kət-ərz\ *See* low-frequency restriction enzymes.

reaction chromatography \rē-'ak-shən ˌkrō-mə-'täg-rə-fē\ A technique in which the components of a test solution are intentionally changed between injection and detection to improve the chromatography. This reaction may take place upstream of the column, for example, to produce a derivative better suited for the chromatographic procedure (precolumn derivatization) or downstream of the column (postcolumn derivatization) to produce a derivative suitable for identification by the available detector. Reaction chromatography may also be used to modify particular components to improve resolution.

Reactive Blue 2 \rē-'ak-tiv 'blü 'tü\   \rē-'ak-shən ˌkrō-mə-'täg-rə-fē\ *See* Cibacron Blue 3G-A.

read-through \'rēd-'thrü\ A situation that sometimes occurs in transcription, in which the RNA polymerase molecule continues to transcribe beyond the expected termination point.

real time \'rē(-ə)l 'tīm\ In computing, the situation in which computer calculations keep pace with actual events. For example, in computer graphics, the term is used to describe the ability to instantaneously and continuously rotate and translate a three-dimensional structure in response to an operator's commands. This effect is achieved by the rapid computation of new atomic coordinates. Such computation results in a rapid change in the graphics display that simulates motion of the molecule.

reassociation \ˌrē-ə-sō-sē-'ā-shən\ The pairing of complementary single strands of DNA to form a double helix.

recalcitrance \ri-'kal-sə-trənts\ The ability of a substance to remain in a particular environment in an unchanged form. Recalcitrant compounds are degraded poorly by microorganisms. *See also* persistence, xenobiotic.

recombinant \rē-'käm-bə-nənt\ A transformed cell containing a recombinant DNA molecule.

recombinant DNA \rē-'käm-bə-nənt 'dē 'en 'ā\ A DNA molecule formed in vitro by ligating DNA molecules that are not normally joined.

recombinant protein \rē-'käm-bə-nənt 'prō-ˌtēn\ Any protein synthesized in a recombinant by expression of a cloned gene.

recombination \ˌrē-ˌkäm-bə-'nā-shən\ Any process that helps to generate new combinations of genes that were originally present in different individuals.

recycle bioreactor \rē-'sī-kəl ˌbī-ō-rē-'ak-tər\ *See* loop reactor.

redox-mediated sensors \'rē-däks 'mēd-ē-ˌāt-əd 'sen(t)-sərz\ Sensors that use a transducer device by which electrons generated by an oxidoreductase enzyme (or enzyme system) are transferred to an electrode surface by mediators such as cytochromes or ferrocene and its derivatives.

redox potential \'rē-däks pə-'ten-chəl\ *See* oxidation–reduction potential.

reductases \ri-'dək-ˌtās-əz\ *See* oxidoreductases.

reflux \'rē-fləks\ A process by which part of a product stream from a reactor is fed back into the system to improve the performance of the process. For example, in fractional distillation, condensed liquid is returned to the top of a column, where it comes into contact with rising vapor of the components. Thus, the process aids separation of the components in the mixture. Also used in liquid–liquid extraction.

regioselective \'rē-jē-ō-si-'lek-tiv\ Descriptive of any reaction that produces or degrades a predominance of one positional or structural isomer of a compound. For example, sugars contain multiple hydroxyl groups. A regioselective enzyme recognizes specific hydroxyls only. *Compare with* stereoselective.

rejection coefficient (membranes) \ri-'jek-shən kō-ə-'fish-ənt ('mem-ˌbrānz)\ A measure of the amount of a species in a feed solution to a semipermeable membrane that does not permeate the membrane.

$$R = 1 - \frac{C_p}{C_f}$$

where R is the rejection coefficient, C_p is the concentration of permeate, and C_f is the concentration of feed. In multicomponent systems, all species have their own individual rejection coefficients.

relative viscosity \'rel-ət-iv vis-'käs-ət-ē\ The ratio of dynamic viscosity of a solution to that of the solvent, each measured at the same temperature.

relaxed plasmid \ri-'lakst 'plaz-məd\ *See* covalently closed circular DNA.

relaxed replication \ri-'lakst ˌrep-lə-'kā-shən\ Plasmid replication is said to be relaxed when it is not stringently coupled to chromosomal DNA replication. Plasmids, therefore, exhibit relaxed replication by continuing to replicate after the bacterial cell stops dividing. This activity results in the presence of multiple copies (up to ~3000 molecules) of a plasmid in each cell.

renewable resources \ri-'n(y)ü-ə-bəl 'rē-sō(ə)rs-əz\ Biomass for biotechnological processes that can be produced on a regular basis as required (e.g., straw, sugar cane, and microbial biomass). Nonrenewable resources, on the other hand, include fossil fuels, natural gas, and petroleum.

rennet \'ren-ət\ A crude commercial enzyme preparation from the fourth stomach (abomasum) of preruminant (suckling) calves, used extensively in the cheese industry for its milk-clotting activity. Its major proteolytic enzyme component is rennin (chymosin) (88–94%), but it also contains small amounts of pepsin. Because the availability of young calves for slaughter cannot match the commercial demand for rennet, rennet substitutes from microbial sources are increasingly used for the production of fermented milk products (e.g., a microbial product from *Mucor miehei*). However, such substitutes differ slightly from rennet in their clotting activity. This variation modifies cheese yield and fat retention, and proteolysis products produced during ripening can have undesirable effects on flavor in the final product. *See also* rennin.

rennin (E.C. 3.4.23.4) \'ren-ən\ A proteolytic enzyme also known as chymosin; the main component of calf rennet. Rennin or rennet is used predominantly in the production of fermented milk products because of its ability to coagulate (curdle) milk. Coagulation occurs principally as a result of the limited proteolysis by rennin of the κ-casein component of milk; it produces insoluble para-κ-casein and soluble macropeptides. Other casein fractions then precipitate as a result of exposure to calcium ions released by κ-casein hydrolysis. However, the overall process is probably more complex than these two simple steps. The use of genetic engineering to produce rennin from microbial sources has been reported. *See also* *Mucor,* rennet.

repertoire cloning \'rep-ə(r)-ˌtwär 'klōn-iŋ\ The cloning and expression of human antibody genes such that each clone produces a different human monoclonal antibody. Currently, methodologies are being developed such that the expression products from cells containing cloned human antibody genes can be examined for antigen-binding ability, and thus cells expressing specific human monoclonal antibodies can be selected. When fully developed, these methods will obviate "humanizing" mouse monoclonal antibodies (*see* chimeric antibody), and indeed specific antibodies will be produced without the need to immunize animals. Such methods will almost certainly replace established hybridoma technology. Specific functional antibody fragments have been produced and screened in bacteria, but phage display antibody technology currently seems to be the most successful form of repertoire cloning. *See also* phage display antibody.

λ-replacement vector \'lam-də ri-'plās-mənt 'vek-tər\ Modified λ-DNA that has two recognition sites for the same restriction enzyme on either side of the nonessential (or "stuffer") part of the λ-DNA molecule. The DNA fragment to be cloned (maximum length is 21 kilobase pairs) is produced by cleavage with the same enzyme and replaces the excised stuffer region. *See also* λ-insertion vector.

replica plating \'rep-li-kə 'plāt-iŋ\ A technique for producing identical patterns of bacterial colonies on a series of Petri dishes. The surface of a Petri dish that contains colonies is pressed with a cylindrical block covered with sterile velvet, which causes about 10% of the bacteria in each colony to be transferred to the fabric. The velvet disk is then pressed against the surface of a bacteria-free plate. Some of each bacterial colony are thus transferred to the new plate; a replica pattern of bacterial growth is produced on this new plate. About six replicas can be printed from a single pad in this way.

replicon \'rep-li-ˌkän\ Any DNA molecule that has an origin of replication and is therefore capable of being replicated by DNA polymerase. If exogenous DNA is added to a bacterial cell, this DNA will not be replicated when the cell divides. Such exogenous DNA has no origin of replication, and therefore only the parent cell and none of the daughter cells will contain this DNA. For DNA to be replicated and passed on to all daughter cells, it must first be attached to a suitable replicon. Such replicons are known as vectors or cloning vehicles (e.g., plasmids or bacteriophage DNA).

reporter gene \ri-'pōrt-ər ˈjēn\ A coding sequence with an easily identifiable gene product that is fused to a gene being used in gene-transfer experiments to help monitor the expression of that gene. This ligation is necessary because often it is difficult to assay the expressed gene product or indeed to distinguish between the expression of the transferred gene and that of a similar or identical endogenous gene. Reporter genes have their own start codon (AUG) and are ligated to the gene of interest such that both genes are under the control of the same promoter or gene regulatory element. Expression of the gene of interest therefore results in an equal expression of the reporter gene. Reporter genes normally code for an enzyme activity that can be easily assayed and is not encoded in the host cell genome. The assay of such an enzyme in a cell extract provides a quick biochemical test for gene activity; the amount of enzyme present is proportional to the amount of expression of the gene of interest.

The most commonly used reporter genes include β-galactosidase, chloramphenicol acetyl transferase (CAT), and luciferase. The following is a brief description of their measurement:

1. The *Escherichia coli lac Z* gene encodes the enzyme β-galactosidase (β-gal) and is frequently used as a reporter gene. Commercially available antibodies allow immunochemical localization and detection of β-gal. Colorimetric and fluorimetric assays include the cleavage of o-nitrophenol-β-D-galactoside (ONPG) to yield the chromophore o-nitrophenol; in histochemical analysis, the cleavage of 5-bromo-4-chloro-3-indolyl-β-D-galactoside (X-gal) to generate soluble indoxyl molecules, which in turn are oxidized to insoluble indigo; and the cleavage of 4-methylumbelliferyl-β-D-galactoside (MUG) to produce the highly fluorescent 4-methylumbelliferone.
2. CAT activity is measured by incubating cell lysates with chloramphenicol and radiolabeled acetyl coenzyme A. Acetylated (radiolabeled) chloramphenicol is extracted from the mixture, and the amount of acetylchloramphenicol is measured.
3. Two commercially available sources of the enzyme luciferase have been used. Bacterial luciferase oxidizes a long-chain aldehyde; the firefly enzyme

oxidizes a heterocyclic carboxylic acid (luciferin), both with the concomitant release of light. Advantages of the system include both accurate measurement of extremely low levels of light and its quantitation over many orders of magnitude. Because light does not diffuse or accumulate in situ, the source of gene expression can be localized with high resolution. *See also* entrapment vector.

repressor \ri-'pres-ər\ A protein that binds to DNA at a specific site (operator) to "switch off" a promoter and thus prevent transcription of the gene under the control of that promoter.

reptation \rep-'tā-shən\ The unwanted effect in the electrophoresis of DNA in which the movement of large DNA molecules through the gel is restrained because the diameters of the DNA molecules are almost as big as those of the pores in an agarose gel. Reptation occurs with DNA fragments larger than about 50 kilobases, and when it does, mobility becomes independent of size, and DNA cannot be separated into discrete bands. The term is derived from the phrase "reptilian stutter," which was used to describe the movement of DNA. The problem of reptation was overcome by the development of pulse-field gel electrophoresis. *See also* pulse-field gel electrophoresis.

reserpine \ri-'sər-pēn\ An antihypertensive agent prepared from the plant *Rauwolfia serpentina*.

resistance plasmids (R plasmids) \ri-'zis-tən(t)s 'plaz-mədz ('är 'plaz-mədz)\ Plasmids that carry genes for antibiotic resistance. Such plasmids confer antibiotic resistance to the host bacterium (e.g., plasmid pBR322 carries genes for tetracycline and ampicillin resistance). Such markers can be used to detect transformants.

resolution \rez-ə-'lü-shən\ The ability of a process to separate (resolve) two components with similar properties. Resolution is used in analytical processes as a measure of selectivity. For example, the resolving power of a spectrophotometer is the ability to separate two spectral lines, and the resolution of a chromatograph is the ability to separate two compounds.

$$\text{resolution} = \frac{2 \times \text{difference between peak maxima}}{\text{sum of peak widths}}$$

respiration \res-pə-'rā-shən\ The oxidation of an organic or inorganic reductant by organisms to produce energy for metabolic reactions by a process involving an electron-transport chain.

respiratory quotient \'res-p(ə-)rə-ṭōr-ē 'kwō-shənt\ The ratio between the carbon dioxide production rate and the oxygen uptake rate in a fermentor or in a Warburg or Gilson flask.

respirometer \res-pə-'räm-ət-ər\ An instrument used to measure the rate of respiration or gas exchange of a culture or particular tissue.

restriction \ri-'strik-shən\ A mechanism whereby a microorganism prevents the entry of foreign DNA (e.g., by a bacteriophage). This mechanism involves the production of specific restriction endonucleases that cleave the heterologous DNA. The microbial host protects its own DNA by a modification system, normally methylation of specific DNA bases, which prevents the restriction enzyme from cleaving the microbial DNA. *See also* restriction endonucleases.

restriction analysis \ri-'strik-shən ə-'nal-ə-səs\ Determination of the sizes of DNA fragments produced by the cleavage of a given DNA molecule (e.g., a plasmid) by a given restriction endonuclease.

restriction endonucleases \ri-'strik-shən en-dō-'n(y)ü-klē-ās-ez\ A number of enzymes derived from a wide range of prokaryotes that all cleave double-stranded DNA molecules. Different types of restriction endonucleases occur (types I–IV), but type II is of particular importance to genetic engineering. To date almost 200 type II enzymes have been isolated. The enzymes differ in the nucleotide sequence that they recognize and cleave. The sites recognized by most of the enzymes used in genetic engineering are palindromes (i.e., the 5' to 3' sequence in one strand of the DNA duplex is the same as in the complementary strand) of four, five, or six bases. Some enzymes cleave to give "blunt" or "flush" ends (e.g., *Hpa*I) or "sticky" (cohesive) ends, where a few unpaired nucleotides project from either end (e.g., *Eco*RI).

$$
\begin{array}{ccc}
\textit{Hpa}\text{I} & & \textit{Eco}\text{RI} \\
\downarrow & & \downarrow \\
5'\ \text{GTTAAC}\ 3' & & 5'\ \text{GAATTC}\ 3' \\
3'\ \text{CAATTG}\ 5' & & 3'\ \text{CTTAAG}\ 5' \\
\uparrow & & \uparrow
\end{array}
$$

Arrows indicate the sites of cleavage. The generation of sticky ends is of particular use in cloning DNA. If a plasmid and the gene to be inserted are both cleaved with the same restriction enzyme, both DNA molecules will generate the same sticky ends, which will anneal by complementary base-pairing. They can then be joined by DNA ligase to give a circular recombinant molecule that can be used to transform a bacterium.

Because so many restriction enzymes are available from a range of organisms, a standard terminology has been adopted. The genus and species names of the organism are identified by the first letter of the genus and the first two letters of the species. This system generates a three-letter abbreviation in italics (e.g., *Escherichia coli* is *Eco*). Strain or type identification is given in nonitalicized symbols (e.g., *Eco*R), and the number (sequence specificity) of the endonuclease is given in Roman numerals (e.g., *Eco*RI). The ability of restriction enzymes to cleave DNA specifically and reproducibly into discrete DNA fragments (a particular hexanucleotide sequence should occur on average only every 4^6 [or 4096] base pairs) has been an essential part of the development of genetic engineering. Also referred to as restriction nucleases or restriction enzymes. *See also* low-frequency restriction enzymes.

restriction enzymes \ri-'strik-shən 'en-zīmz\ *See* restriction endonucleases.

restriction fragment-length polymorphisms (RFLPs) \ri-'strik-shən 'frag-mənt 'leŋ(k)th ˌpäl-i-'mȯr-ˌfiz-əmz ('är 'ef 'el 'pē)\ Any differences observed between genotypes in the fragment lengths produced when DNA is digested by a restriction enzyme. RFLPs occur as a result of the creation or destruction of a restriction site due to base-pair changes, a rearrangement involving a restriction site, or internal deletion or insertion. In the following example, a certain restriction enzyme cleaves a region of DNA in genotype A as shown by the arrows; but a mutation in this region in genotype B has introduced an additional cleavage site for this enzyme. When the genomic digests are separated by electrophoresis, then probed with a probe that binds as indicated, the probe identifies fragments of different lengths in the two samples. This difference in restriction fragment lengths (polymorphism) has therefore identified a genetic difference between genotypes A and B. *See also* DNA fingerprinting.

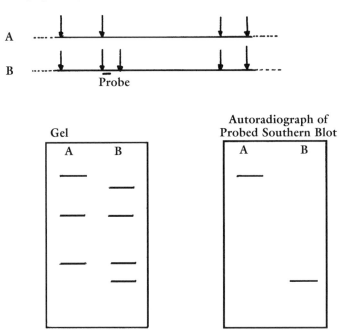

restriction map \ri-'strik-shən 'map\ A map that shows the positions of different restriction sites (i.e., nucleotide sequences that are cleaved by various restriction enzymes) in a given DNA molecule (e.g., a plasmid). *See also* restriction analysis.

restriction nucleases \ri-'strik-shən 'n(y)ü-klē-ˌās-əz\ *See* restriction endonucleases.

retentate \rē-'ten-ˌtāt\ The portion of material retained by a membrane. The opposite of permeate.

retention time \ri-'ten-chən 'tīm\ In column chromatography, the time taken for a particular component to emerge from the column; the time of injection is considered zero. This time depends on the experimental conditions used for the chromatography but can, with standard conditions or the use of internal standard compounds, be used to identify components in a mixture.

retention volume \ri-'ten-chən 'väl-yəm\ An alternative to retention time for the characterization of species emerging from a chromatographic column. To measure retention volume, the volume of eluant required to elute a particular component is recorded. *See also* breakthrough profile.

retroelement \re-trō-'el-ə-mənt\ Any genetic entity that uses reverse transcriptase in its life cycle (e.g., a retrovirus).

retrogradation \re-trō-grā-'dā-shən\ The formation of insoluble precipitates that are difficult to hydrolyze or redissolve because of the association of poorly soluble linear dextrins present in starch solution of low dextrose equivalent (DE) hydrolysates. *See also* starch.

retroviruses \'re-trō-vī-rəs-əz\ A family of small viruses with RNA genomes of 5,000–10,000 nucleotides. Retroviruses have the unique property of replicating via the formation of a nuclear DNA intermediate, unlike most other viruses with RNA genomes, which replicate in the cytoplasm. The genome comprises only three genes, one of which codes for the enzyme reverse transcriptase, which transcribes the single-stranded viral RNA to a double-stranded DNA intermediate. This DNA is then integrated into the chromosomal DNA of the host cell as a "provirus" and is stably inherited during subsequent cell divisions. One type B and all of the type C virus genera are oncogenic, particularly in animals (e.g., feline leukemia virus and bovine leukemia virus). Diseases in humans are caused by human T-cell leukemia virus (HTLV). Two strains, HTLV-1 and HTLV-2, are associated with T-cell malignancies; because only a small proportion of infected people develop the disease, other factors may be involved. Certain retroviruses have considerable potential for use in human gene therapy. Because retroviruses integrate within target cell genomes, integration of a modified retroviral vector carrying a therapeutic gene ensures that the gene is stable and propagated when the target cells divide. Retroviruses are relatively easy to manipulate and can carry quite large DNA inserts. *See also* reverse transcriptase.

reverse micellar extraction \ri-'vərs mī-'sel-ər ik-'strak-shən\ *See* micellar extraction.

reverse micelle \ri-'vərs mī-'sel\ *See* micelle.

reverse osmosis (hyperfiltration) \ri-'vərs äs-'mō-səs (͵hī-pər-fil-'trā-shən)\ A membrane process for the separation of a solute from solution. The process operates by the application of an external pressure greater than the osmotic pressure, which forces the solvent to pass through the membrane against its normal osmotic flow. Thus, solvent molecules are removed from the concentrated solution. The process operates at high pressure, 1500–8000 kPa, and uses an asymmetric membrane with pore sizes less than 25 μm. Uses include the concentration of dilute solutions of salts, sugars, and amino acids; purification and recovery of water; and desalination.

reverse transcriptase (E.C. 2.7.7.49) \ri-'vərs tran-'skrip-͵tās\ An enzyme prepared from avian myeloblastosis virus that synthesizes single-strand DNA from an RNA template. The normal process of transcription in a cell is the synthesis of RNA from a DNA template, hence the term *reverse* transcriptase. Such enzymes are unique to RNA viruses (retroviruses), which need to produce a DNA copy of their genetic materials (RNA) upon infection of a cell. The DNA copy thus directs the in-cell synthesis of protein and RNA necessary for the synthesis of new viral particles. The isolated enzyme is used extensively in genetic engineering methodology to synthesize in vitro complementary DNA (cDNA) from messenger RNA. *See also* retroviruses.

reversed-phase chromatography (RPC) \ri-'vərst-'fāz ˌkrō-mə-'täg-rə-fē ('är 'pē 'sē)\ A chromatographic method in which the support matrix (e.g., silica) is modified to replace hydrophilic groupings with large hydrophobic alkyl carbon chains. This modification allows the separation of compounds according to their hydrophobicity. The more hydrophobic compounds bind most strongly to the support and are eluted later than the less hydrophobic compounds. Strongly bound components can be eluted by applying a gradient of a competing hydrophobic solvent, such as methanol or acetonitrile. The supports most commonly used for RPC of proteins or peptides have alkyl groups of either 8 (C_8) or 18 (C_{18}) carbon molecules. This method is commonly carried out by high-performance liquid chromatography (HPLC), although thin-layer chromatography may also use reversed-phase media.

Reynolds number (Re) \'ren-əl(d)z 'nəm-bər ('är 'ē)\ A dimensionless number used in fluid dynamics to determine the flow regime. It is a relation between the fluid properties of the system and the flow velocity:

$$\mathrm{Re} = \frac{LV\rho}{\mu}$$

where L is the linear dimension of the fluid channel (m), V is the linear velocity (m/s), ρ is the density of the fluid (kg/m^3), and μ is the fluid viscosity (kg/ms). The Reynolds number defines the transition between laminar and turbulent flow as velocity increases. Its value depends on the channel geometry and ranges from 2000 to 3000 for a circular pipe. Generally, for laminar flow $\mathrm{Re} < 2300$; and for turbulent flow, $\mathrm{Re} > 10^4$. In agitated fermentors Re can be based on the root mean square fluid velocity, V_{rms}, and the average diameter of gas bubbles, $\overline{D}$:

$$\mathrm{Re} = \frac{\overline{D}V_{rms}}{\mu}$$

The Reynolds number in the vicinity of an impeller can also be considered by an impeller Reynolds number, Re_i:

$$\mathrm{Re}_i = \frac{\rho\, ND_i^2}{\mu}$$

where N is the rotational speed of the impeller (rpm) and D_i is the impeller diameter (m). In a stirred tank, if $\mathrm{Re}_i > 10^4$, then turbulent flow exists; if $\mathrm{Re}_i < 10$, then laminar flow is present.

RFE \\('är 'ef 'ē)\\   *See* pulse-field gel electrophoresis.

RFLP \\('är 'ef 'el 'pē)\\ *See* restriction fragment-length polymorphisms.

rheodestructive fluid \\rē-ə-di-'strək-tiv 'flü-əd\\ A fluid that undergoes irreversible breakdown when subjected to an applied shear stress. The fluid exhibits a decrease of viscosity with time that, when the stress is removed, does not return to its original value and may indeed continue to decrease slightly.

rheogram \\'rē-ə-ˌgram\\ The pictorial representation or graph illustrating a rheological relationship (e.g., shear stress as a function of shear rate).

rheology \\rē-'äl-ə-jē\\ The science of deformation and flow of matter that gives rise to the fundamental parameters of elasticity, plasticity, and viscosity.

rheopectic liquid \\rē-ə-'pek-tik 'lik-wəd\\ A fluid that rapidly exhibits an increase in apparent viscosity when subjected to a constant shear rate.

Rhizobium \\rī-'zō-bē-əm\\ A genus of bacteria involved in the process of nitrogen fixation in leguminous plants. Rhizobia are introduced deliberately into soil sown with leguminous plants to encourage root-nodule formation. *See also* nitrogen fixation.

rhodamine isothiocyanate \\'rōd-ə-ˌmēn īs-ō-thī-ō-'sī-ə-ˌnāt\\ A reagent commonly used to label antibodies fluorescently, which are then used in immunofluorescent labeling of cells. The antibody, which works against a specific cell component, can be used as a label to map the distribution of that component in the specimen. When the specimen is illuminated by light of the appropriate wavelength under a light microscope, the component is seen because it is fluorescent. Fluorescein isothiocyanate is used similarly.

RIA \\'är 'ī 'ā\\ *See* radioimmunoassay.

riboflavin \\rī-bə-'flā-vən\\ A vitamin (vitamin B_2) produced industrially in submerged culture from *Eremothecium ashbyii* (grown on carbohydrate-free media with lipid as an energy source), *Ashbya gossypii* (grown on glucose, sucrose, or maltose), *Candida* sp. (grown on sugar solutions), or *Clostridium acetobutylicum* (grown on grain mash or whey with a low iron content). In the purified form, it is used for human nutrition and therapy; in the crude concentrated form, it is used as an animal feed supplement.

ribonuclease (RNase) \\rī-bō-'n(y)ü-klē-ˌās ('är-'en-ˌās)\\ *See* nuclease.

ribosomal RNA (rRNA) \\rī-bə-'sō-məl 'är 'en 'ā ('är 'är 'en 'ā)\\ *See* ribosome.

ribosome \\'rī-bə-ˌsōm\\ A particle comprising RNA and protein found in all cells. Each cell contains many thousand ribosomes. The ribosome is the site of protein synthesis, in which the code in a molecule of messenger RNA (mRNA) is translated into a protein sequence. Ribosomes are classified on the basis of their rate of sedimentation, which is measured in Svedberg units (S). Prokaryotic ribosomes have rates of 70 S and are made up of subunits of 50 and 30 S (the units are not additive). The 50 S subunit comprises two RNA molecules (5 and 23 S RNA) and approximately 34 proteins. The 30 S unit comprises one RNA molecule (16 S RNA) and about 21 proteins. Eukaryotic ribosomes are larger (80 S) and are made up of 60 and 40 S subunits. The 60 S subunit comprises three RNA molecules (5, 5.8, and 28 S RNA) and approximately 45 proteins. The 40-S subunit comprises one RNA molecule (18 S RNA) and approximately 33 proteins. Many antibiotics (e.g., streptomycin, chloramphenicol, and erythromycin) exploit the structural differences between prokaryotic and eukaryotic ribosomes being able to selectively inhibit bacterial protein synthesis by binding to bacterial ribosomes.

ribozymes \rī-bə-zīmz\ Short RNA molecules capable of inducing a site-specific hydrolytic reaction in target RNA. The reaction depends on the tertiary structure of the RNA–RNA complex. The development of ribozymes is based on the initial observation that several RNA molecules undergo self-splicing. Ribozymes have potential uses as tools in molecular biology and as inhibitors of specific gene expression in the treatment of diseases. For example, they could be used to break down oncogene mRNA or to defend plants and animals against viral infection.

ricin \'rīs-ᵊn\ A protein consisting of two nonidentical subunits, derived from the plant *Ricinis communis*. It is toxic to mammalian cells because of its ability to inactivate ribosomes. It is one of a number of proteins being studied as a component of an antibody-toxic conjugate for use in chemotherapy. *See also* bispecific antibodies, chimeric antibody.

rising-film evaporator \'rī-ziŋ-'film i-'vap-ə-ˌrāt-ər\ A device for concentration of solutions, similar in operation to a falling-film evaporator in that the fluid forms a thin film on the surface of the heat exchanger tubes. However in this case, the overall flow of the liquid is upward, with the energy supplied by the system operating under vacuum. Circulation of the concentrated fluid is via a gravity return to the feed vessel. Also known as climbing-film evaporator.

RNA polymerase \'är 'en 'ā 'päl-ə-mə-ˌrās\ An enzyme that synthesizes RNA by using DNA as a template. Three different enzymes in the nuclei of human cells carry out transcription: RNA polymerase I, II, and III. RNA pol. I transcribes three of the ribosomal RNA (rRNA) genes, RNA pol. II transcribes protein genes into mRNA, and RNA pol. III transcribes the genes for the other ribosomal RNA and all the genes for tRNA. In bacteria, a single polymerase enzyme synthesizes all three types of cellular RNA. *See also* transcription.

RNA replicase \'är 'en 'ā 'rep-li-ˌkās\ An enzyme that uses single-strand RNA as a substrate to produce a complementary single-stranded RNA molecule. The parent and daughter strands are forced apart as they are synthesized, and then both serve as templates for further rounds of replication. RNA replicases have been isolated from *Escherichia coli*, RNA phages, and plant RNA viruses. The best example of these enzymes is Q-beta replicase.

RNA splicing \'är 'en 'ā 'splī-siŋ\ *See* introns.

roller bottles \'rō-lər 'bät-ᵊlz\ Bottles used for growing cells. The design provides increased surface area for the growth of anchorage-dependent mammalian cells. The simplest design involves a continuously rolled cylindrical vessel that contains the cells and growth medium. This design makes almost all the internal surface available for cell growth, even though only ~20% is covered by medium at any one time. Other methods exist to increase the surface area further within a roller bottle. These methods include incorporating a spirally wound plastic film and packing the bottle with a cluster of small parallel glass tubes separated by silicone spacer rings.

root nodules \'rüt 'näj-ü(ə)lz\ *See Leguminosae*, nitrogen fixation, *Rhizobium*.

rosette technique \rō-'zet tek-'nēk\ A method for separating cells by affinity chromatography. For example, anti-mouse IgG is fixed on ox red blood cells using chromic chloride. The cell population to be fractionated is treated with specific mouse monoclonal antibodies against surface antigens of the cells to be purified and is added to the red blood cells. This procedure causes the formation of rosettes. Bound cells are recovered by selective lysis of the blood cells with ammonium chloride.

rotameter \'rōt-ə-ˌmēt-ər\ A flow-measuring device. It operates on the principle of variable area for flow, thus maintaining a constant differential pressure across the meter.

rotary-disk biological contactor \'rōt-ə-rē-'disk ˌbī-ə-'läj-i-kəl 'kän-tak-tər\ A contactor used in the treatment of wastewaters and effluents that consists of a series of plastic disks, 3–4 m in diameter. Up to 40% of the disk surface is immersed in a tank. On rotation at 2–5 rpm, the disks pick up liquid from the tank and bring it into contact with the air as a thin film. Microbes on the disk surface oxidize any organic compounds present in the liquid film. As microbial growth increases, the shearing effect of the liquid in the tank tends to remove the excess growth from the disks, which then can be removed from the waste stream by conventional settling tanks. The contactor is very popular for treating industrial wastes because of its small size and low energy requirements. However, problems exist with highly contaminated wastes because of excessive growth and insufficient oxygen to meet demand. Typical loading rates for this design are 13 g of biological oxygen demand $(BOD)/m^2/day$ for domestic waste and partial treatment of 400 g of $BOD/m^2/day$ for industrial waste.

rotary vacuum filter \'rōt-ə-rē 'vak-yüm 'fil-tər\ A device for filtration that consists of a horizontal drum covered by a cloth (filter cloth). The drum rotates in a bath continuously fed by a process stream. Vacuum applied to the inside of the drum draws the filtrate through the cloth into the drum, from which it is continuously removed. The filter cake (precipitate) is retained on the cloth, from which it can be removed periodically. Drum filters can be designed to allow in situ washing of the precipitate on the filter, followed by dewatering by increasing the vacuum. Various operating procedures may be used to maintain the rate of filtration, including the use of a precoat on the filter cloth, to minimize blinding, or to improve the filterability of slimy precipitates. The filter cake may be removed continuously from the filter by the application of a scraper or knife, which can be set to remove the cake once the desired cake depth has been reached. An alternative is a string discharge, in which a series of belts runs around the drum and a separate cylinder. These strings continuously lift off the cake as they detach from the drum.

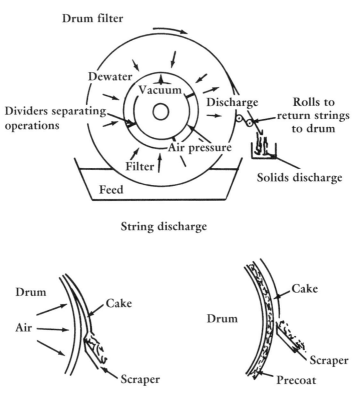

Drum filter

Dewater

Vacuum

Discharge

Rolls to return strings to drum

Dividers separating operations

Air pressure

Filter

Feed

Solids discharge

String discharge

Drum

Cake

Air

Scraper

Cake

Drum

Scraper

Precoat

Scraper discharge alternative procedures

rotary vacuum filter

rotating biological contactor \\'rō-tāt-iŋ ˌbī-ə-'läj-i-kəl 'kän-tak-tər ('är 'dē 'sē)\\ *See* rotary-disk biological contactor.

rotating-disk contactor (RDC) \\'rō-tāt-iŋ-'disk 'kän-tak-tər\\ This agitated contactor is used for liquid–liquid extraction. It consists of a vertical shell fitted with a series of horizontal flat plates with a central opening. In the middle of the compartments formed by these stator rings, disks are mounted on a common vertical shaft. These disks, which have a diameter less than that of the ring opening, are rotated with the shaft by a motor. Grids are installed above the top stator ring and below the lowest ring to provide calm settling zones, and the feed inlets are mounted tangentially so that the flow patterns in the column are not disturbed. In other designs of RDC (asymmetric RDC), the ring opening and disks are offset from the center.

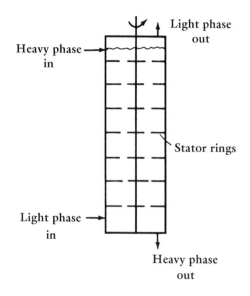

rubber \'rəb-ər\ *See* latex.

runaway plasmid \'rən-ə-ˌwā 'plaz-mid\ A plasmid that has lost its ability to control its copy number at increased growth temperatures. The plasmids are present in the cell at low copy numbers (10–25) at the permitted temperature (30 °C), but increasing the temperature to 40 °C causes an increase of the copy number to as much as several thousand by uncontrolled plasmid replication. Such plasmids are necessary for the high expressions of cloned genes that code for products lethal to the cell at high concentrations. Cells are grown normally at the permitted temperature, then shifted to the nonpermitted higher temperature. Protein synthesis continues at the higher temperature for several hours after the increase in copy number, allowing overproduction of the cloned gene. This runaway replication is ultimately lethal to the cell.

S

S-form \'es 'fȯ(ə)rm\ The stereochemical arrangement displayed by a chiral molecule in which the substituents listed according to defined sequence rules follow a left-handed route around the chiral center. Derived from *sinister*, Latin for left. The converse is R-form. *See also* R-form, sequence rules.

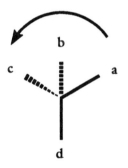

sequence rule a>b>d>d

S1 nuclease \'es 'wən 'n(y)ü-klē-ˌās\ An endonuclease, usually isolated from the fungus *Aspergillus oryzae*, that cleaves only single-stranded DNA. It is used to cut the hairpin formed during the synthesis of double-stranded DNA from cDNA and in conjunction with exonuclease III to create deletions in cloned DNA molecules.

saccharides \'sak-ə-ˌrīdz\ *See* carbohydrates.

saccharification \sə-ˌkar-ə-fə-'kā-shən\ The hydrolysis of a polysaccharide to give a sugar solution. For example, the process whereby dextrin derived from starch is broken down to dextrose (glucose). *See also* starch.

Saccharomyces cerevisiae \ˌsak-ə-rō-ˈmī-sēz ser-ə-ˈvē-sē-ē\ *See* baker's yeast.

salting out \ˈsȯl-tiŋ ˈau̇t\ A process by which the aqueous solubility of a compound, usually organic (e.g., proteins), can be reduced by increasing the concentration of neutral salts.

sand filter \ˈsand ˈfil-tər\ A type of percolating filter or depth filter in which sand is used as the filtering medium.

saprophytic \sap-rə-ˈfit-ik\ Able to live on dead or decaying organic matter.

scale-up (factor) \ˈskā-ˌləp (ˈfak-tər)\ Translation of a process to a larger scale. Scale-up involves the variation of parameters required to ensure optimum performance at the increased scale. The design of a commercial-scale reactor cannot be accomplished solely by a theoretical approach; at least some laboratory data are required on the reaction involved in the process. Thus, although the behavior of a population of microorganisms should be invariant to scale, the behavior of factors affecting the surrounding environment is less easy to predict (e.g., oxygen transfer and shear behavior). Thus, one approach to scale-up uses an extensive series of systematic scale-up and scale-down experiments to ascertain the effect of process variables on the product yield. This approach has the inevitable consequence of considerable costs in both time and equipment. An alternative approach uses a combination of modeling techniques based on an understanding of the kinetic behavior of the processes involved with practical experience gained from knowledge of the observed behavior of the process under actual process conditions.

Scheibel contactor \ˈshī-bəl ˈkän-tak-tər\ A type of column contactor in which the countercurrent liquid phases make contact in mixing zones agitated by rotating turbines on a common central shaft. The mixed phases then enter calming zones, which may be filled with wire-mesh packing to separate the flow patterns of the mixing zones. These calm zones minimize loss of efficiency by back-mixing and eliminate the rotatory motion imparted to the liquids by the impeller. Once in the calming zones, the two phases separate. The light phase and heavy phase move in opposite directions into other mixing zones.

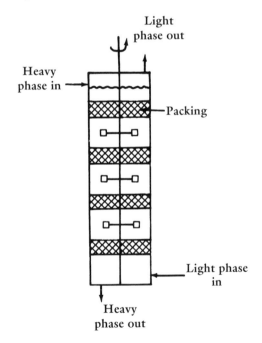

Schmidt number (Sc) \'shmit 'nəm-bər ('es 'sē)\ A dimensionless group associated with convective mass transfer. For example, in the mass transfer of oxygen:

$$Sc = \eta/\rho D_o$$

where η is the viscosity (kg/ms), ρ is the density (kg/m³) of the continuous phase, and D_o is the diffusivity of oxygen (m²/s). *See also* Grashof number, Peclet number, Sherwood number.

scleroglucan \ˌskler-ə-'glü-ˌkan\ *See* polytran.

scopolamine \skə-'päl-ə-ˌmēn\ A plant alkaloid isolated from *Datura stramonium*, used as an antihypertensive agent.

SCOT \'skät\ *See* support-coated, open-tubular column.

SCP \'es 'sē 'pē\ *See* single-cell protein.

screening \'skrē-niŋ\ Any selective procedure used to identify specific organisms or metabolites in large populations.

scrubbing \'skrəb-iŋ\ The process of removing impurities from an extract phase by making contact with another solution phase such that the impurities are preferentially re-extracted from the solvent phase. *See also* bioscrubbing.

SDS \'es dē 'es\ *See* sodium dodecyl sulfate.

SDS polyacrylamide gel electrophoresis (SDS PAGE) \'es 'dē 'es ˌpäl-ē-ə-'kril-ə-ˌmīd 'jel i-ˌlek-trə-fə-'rē-səs ('es dē 'es 'pāj)\ An analytical electrophoretic method for studying a protein or protein mixtures. The proteins to be separated are first denatured by boiling in a solution of the detergent sodium dodecyl sulfate (SDS). Mercaptoethanol is also present to reduce any disulfide bridges. Negatively charged SDS molecules bind at an average of one SDS molecule for every two amino acids; thus, each protein is opened out into a rod-shaped molecule covered in negative charges. The proteins are then electrophoresed through a polyacrylamide gel. Because the charge-to-mass ratio for each protein is the same, the proteins enter the gel at a uniform rate. As they pass through the gel, however, the proteins begin to separate as a result of the sieving effect of the gel pores. Smaller proteins are able to travel more freely than larger proteins. After electrophoresis, the separated protein bands are visualized by using an appropriate protein stain, normally Coomassie brilliant blue, although the more sensitive silver stain can also be used. The presence of a single band on an SDS gel is good evidence of protein purity. Because the method separates proteins by size, the method can also be used to obtain molecular-weight data. A mixture of standard proteins of known molecular weights is run on the same gel, and a calibration curve of log MW versus distance moved is produced. The distance moved by the unknown protein is also measured, and the molecular weight is determined from the calibration curve. *See also* isoelectric focusing, polyacrylamide gels.

secondary cell culture \'sek-ən-ˌder-ē 'sel 'kəl-chər\ A cell culture obtained by the repeated in vitro passage of primary cells. Most secondary cell cultures survive only for a limited number of cell divisions before dying. However, some continue to grow (i.e., they transform) and establish a continuous *cell line* that is capable of being passaged indefinitely in vitro. *See also* primary cell culture.

secondary fermentations \'sek-ən-,der-ē ,fər-mən-'tā-shənz\ Fermentations, other than the major fermentation, that occur during the production of fermented milk products (e.g., cheese, buttermilk, sour cream, and yogurt). The major fermentation is the bacterial breakdown of lactose to lactic acid, usually by streptococci and lactobacilli. Other reactions that may occur, either during the main fermentation or postfermentation, produce the distinctive flavors of individual products. These reactions are referred to as secondary fermentations; for example, propionic acid fermentation in milk leads to the typical flavor and hole formation (as the result of carbon dioxide production) in Swiss cheese.

secondary metabolites \'sek-ən-,der-ē mə-'tab-ə-,lītz\ Metabolic intermediates or products not essential to growth and life of the producing organism. They are normally produced during the idiophase (stationary phase) of microbial culture and are synthesized from several general metabolites by a wider variety of pathways than those available for general metabolism. The occurrence of a vast range of secondary metabolites with sometimes extremely complicated structures and often without obvious functions has led to many hypotheses about the function of secondary metabolism. Although there is no consensus, microbial secondary metabolites definitely benefit mankind. For example, the majority of microbial antibiotics are secondary metabolites.

sedimentation \,sed-ə-mən-'tā-shən\ The process of settling of particulate matter. Four types of settling are recognized:

- *compression settling:* \kəm-'presh-ən 'set-liŋ\ The particles are present at such a concentration that a structure is formed, and further settling occurs only by compression caused by the weight of particles on top.
- *discrete particle settling:* \dis-'krēt 'pärt-i-kəl\ In systems with a low solids content, individual particles settle without any significant interaction with each other.
- *flocculant settling:* \'fläk-yə-lənt 'set-liŋ\ In a dilute suspension, particles flocculate and coalesce so that their mass is increased, and thus the settling rate is increased.
- *hindered settling:* \'hin-dərd 'set-liŋ\ In some suspensions, particle concentration is such that interparticle forces are sufficient to hinder the settling of neighboring particles, and the total mass settles as a single unit with a solid–liquid interface at the top. This behavior is typical of a fermentation broth.

sedimentation field-flow fractionation \,sed-ə-men-'tā-shən 'fē(ə)ld-'flō ,frak-shə-'nā-shən\ *See* field-flow fractionation.

sedimentation potential \,sed-ə-men-'tā-shən pə-'ten-chəl\ The potential difference caused by sedimentation of particles in a gravitational field or in a centrifuge as measured between identical electrodes at different levels in a gravitational field or at different distances from the center of rotation in centrifuges. The potential is positive if the lower electrode or the electrode at the periphery of the centrifuge is negative. Also known as the Dorn effect.

seed \\'sēd\\ A microbial culture used as a starter for further cultures.

seed storage proteins \\'sēd 'stōr-ij 'prō-ţēnz\\ The proteins of seeds that are used during germination to provide nitrogen for the developing seedling. They are classified into four major categories, based on their solubility properties: albumin, globulins, glutelins, and prolamins. Seed crops such as cereals and grain legumes play an important role in human and animal nutrition. However, most seed crops contain limited amounts of amino acids that are nutritionally essential to humans. Most cereals are deficient in lysine, legumes are deficient in sulfur-containing amino acids, and rice has an overall low protein content. The improvement of the nutritional quality of seed storage proteins is therefore one of the aims of plant molecular biology.

selectable marker \\sə-'lek-tə-bəl 'mär-kər\\ Any gene carried by a vector that confers a recognizable characteristic on a cell containing the vector, e.g., antibiotic resistance genes on plasmids.

selection \\sə-'lek-shən\\ Any means of identifying a clone that contains a desired recombinant DNA molecule. The term is also used to describe the detection and isolation of microbial strains with the required characteristics after mutation of the organism. *See also* reporter gene, selectable marker.

selection enrichment \\sə-'lek-shən in-'rich-mənt\\ *See* enrichment culture.

self-transmissible plasmids \\self-tranz-'mis-ə-bəl 'plas-ˌmədz\\ Plasmids containing transfer (*tra*) genes that allow the production of pili involved in conjugal transfer of plasmid DNA. Laboratory-used plasmids have their *tra* genes deleted to prevent the possibility of self-transmission of recombinant plasmids if they should escape from the laboratory environment. *See also* mobilization genes.

semipermeable membrane \\sem-i-'pər-mē-ə-bəl 'mem-brān\\ A membrane that is permeable to some substances but not others. The selectivity can be achieved by pore size or by the presence of charged species in the membrane. Also known as permselective membrane.

sense strand \\'sen(t)s 'strand\\ The strand of duplex DNA that is transcribed into mRNA.

separation factor \\sep-ə-'rā-shən 'fak-tər\\ The degree of separation achieved by a particular process or system, defined as the ratio of the concentration of the components (C_a/C_b) in the product phase (x) to those in the feed phase (y).

$$\text{separation factor} = \frac{(C_a/C_b)_x}{(C_a/C_b)_y}$$

Sephacryl \\sef-ə-'kril\\ Pharmacia trade name for cross-linked dextran polymer beads used as a gel-filtration medium. *See also* dextran, gel filtration.

Sephadex \\sef-ə-'deks\\ Pharmacia trade name for cross-linked dextran polymer beads used as a gel-filtration medium. *See also* dextran, gel filtration.

Sepharose \sef-ə-'rōs\ Pharmacia trade name for cross-linked agarose beads used as a gel-filtration medium. *See also* gel filtration.

sequence rules \'sē-kwən(t)s 'rülz\ A series of rules for naming chiral forms. Ligands attached to a chiral center are placed in a conventional order of preference to determine the nomenclature for the chiral forms, R- and S-.

    Subrule 1: Ligands with higher atomic numbers are preferred to those with lower atomic numbers (e.g., Br > Cl > C > H). When the donor atom in ligands is the same, then the next atom in the ligand is considered (e.g., CH_2—Br > CH_2—H).

    Subrule 2: Higher mass numbers are preferred to lower mass numbers, which enable isotopic labeling to be considered.

    Subrule 3: If cis–trans isomerization occurs, then Z > E. This complex subject continues to evolve, and for the latest information, the IUPAC recommendations and *Chemical Abstracts* nomenclature should be consulted.

sequence-tagged site (STS) \'sē-kwən(t)s-'tagd 'sīt ('es 'tē 'es)\ A short, single-copy DNA sequence that characterizes mapping landmarks on the genome and can be detected by polymerase chain reactions. The region of a genome can be mapped by determining the order of a series of STSs.

sequencing gel \'sē-kwən(t)-siŋ 'jel\ A long (up to 1 m), thin (0.2–0.5 mm) polyacrylamide slab gel, which has the resolving power to separate single-stranded DNA fragments that differ by one nucleotide. The gel, which includes urea, is kept hot to prevent base-pairing. Such gels are used to separate radiolabeled DNA fragments in DNA-sequencing methodologies. *See also* dideoxy sequencing, Maxam–Gilbert method.

sequestering agents \si-'kwes-tər-iŋ 'ā-jəntz\ Compounds that are able to form complexes with metal ions in solution and thus prevent their activity as simple cationic species. Ethylenediaminetetraacetic acid (EDTA), for example, is effective for many metal ions and can be used (1) to ensure that the metal ions may be held in a form appropriate for chemical reaction, (2) to prevent the deleterious catalytic effects of traces of the metal ion, (3) to analyze metal ions, and (4) to prepare culture media.

serine proteases \'se(ə)r-ēn 'prōt-ē-ās-əz\ One of the four classifications of proteases. All serine proteases cleave peptide bonds by a common mechanism, which involves a highly nucleophilic serine residue at the active site. The serine protease most commonly used in industry is the microbial enzyme subtilisin Carlsberg, which is used extensively in enzymatic washing powders. Other serine proteases include chymotrypsin, elastase, thrombin, and trypsin.

serum-free media \'sir-əm-'frē 'mē-dē-ə\ *See* fetal calf serum.

sex pilus \'seks 'pī-ləs\ A fimbria-like structure (tube) that is formed by one of the two mating types during bacterial conjugation, through which the genetic material is transferred.

sexual hybridization \'seksh-(ə-)wəl ˌhī-brəd-ə-'zā-shən\ The combining of haploid nuclei from opposite mating types in one cell. The nuclei ultimately fuse to form the diploid nucleus, which will then undergo meio is. During meiotic division, a rearrangement and reorganization of the chromosomes results in recombination of the genetic element.

shear \\'shi(ə)r\\ The movement of a layer of fluid relative to parallel adjacent layers in the fluid.

shear rate \\'shi(ə)r 'rāt\\ The rate of change of shear (i.e., velocity gradient) in a fluid.

shear stress \\'shi(ə)r 'stres\\ The force per unit area that causes the movement of a fluid by the process of shearing.

Sherwood number (Sh) \\'shər-wüd 'nəm-bər ('es 'āch)\\ A dimensionless group associated with convective mass transfer. The Sherwood number relates the mass-transfer coefficient (k_c) with the characteristic length of the system (L) and the diffusivity (D) of the solute, A, in the solvent, B:

$$Sh = \frac{k_c L}{D_{AB}}$$

Thus, for the transfer of oxygen from a gas bubble to the liquid phase of a fermentor:

$$Sh = k_L \frac{D}{D_o}$$

where k_L is the liquid-side mass-transfer coefficient (m/s), D is the bubble diameter (m), and D_o is the diffusivity of oxygen (m/s). *See also* Grashof number, Peclet number, Schmidt number.

shikonin \\shē-kōn-ēn\\ A bright red naphthoquinone compound used as a dye. Traditionally used in Japan as a medicine because of its antibacterial and anti-inflammatory properties. Although originally extracted from the root of the plant *Lithospermum erythrorhizon*, it is now produced commercially by plant-cell culture.

Shine–Dalgarno sequence \\'shīn ˌdal-ə-'gär-nō 'sē-kwən(t)s\\ A base sequence of about six to eight nucleotides located on mRNA, lying about four to eight nucleotides upstream from the initiation codon (AUG). The sequence acts as a ribosome-binding site at the first step of translation by base-pairing with a complementary base sequence on the 16S rRNA chain and ensures that the mRNA is translated in the correct reading frame.

shotgun cloning \\'shät-'gən 'klō-niŋ\\ A cloning strategy in which genomic DNA is cleaved into fragments with a restriction endonuclease. These fragments are then inserted into a vector, and the vector is used to transform bacteria. The product is a large number of clones (library) that between them contain all or most of the genes present in the original genome.

shuttle vector \\'shət-ᵊl 'vek-tər\\ Any vector that can replicate in more than one organism. For example, plasmid pJDB219 is made up of part of the 2-μm circle of yeast and the entire bacterial plasmid pBR322 and can replicate and be selected for in both yeast and *Escherichia coli*.

sialic acid \\sī-'al-ik 'as-əd\\ *See* neuraminic acid.

siderophores \\'sid-ə-rə-ˌforz\\ Iron-complexing compounds released by microorganisms to solubilize and transport iron into the microorganism. Iron is a trace element and plays an important role in nutrition. In an aerated nutrient solution, iron is present as weakly soluble iron(III) compounds. The siderophores solubilize the iron and transport it into the cells.

sieve plate column \\'siv 'plāt 'käl-əm\\ A column contactor used in mass-transfer processes such as adsorption, distillation, and liquid–liquid extraction, which consists of a stack of perforated plates that aid contact between the gas–liquid, vapor–liquid, or liquid–liquid phases. Also referred to as a perforated plate column. *See also* bubble-cap column, packed column, pulsed column, pulsed-plate column.

signal sequence \\'sig-nᵊl 'sē-kwən(t)s\\ A sequence of about 15–30 amino acids, mainly hydrophobic, found at the N-terminus of proteins that are exported from cells. As this N-terminal sequence is synthesized at the ribosome, it attaches to the endoplasmic reticulum (to give "rough" endoplasmic reticulum). The protein is synthesized directly into the interstices of the endoplasmic reticulum. The signal sequence is removed by specific proteolytic cleavage after it has passed through the membrane and therefore does not appear in the final secreted protein. Similar signal sequences are thought to exist to direct proteins across membranes between cell compartments. *See also* protein targeting.

silage \\'sī-lij\\ Plant material preserved by natural lactic acid fermentation and used as animal feed (fodder). To prepare silage, fodder plants are fermented in the absence of air. The lactic acid that is formed by the fermentation preserves the fermented material.

silanization \\ˌsī-lə-na-'zā-shən\\ The conversion of —OH groups, which could act as adsorption sites, commonly on silica or glass stationary chromatographic phases, to give the inactive —O—SiR_3 grouping by reaction with, for example, alkylsilicon halides R_3SiCl.

silica gel \\'sil-i-kə 'jel\\ An amorphous form of hydrated silica formed by precipitation or chemical hydrolysis. When dehydrated, it forms hard granules that are chemically and physically inert but highly hygroscopic. This property leads to their main use in drying solvents and gases. The presence of —OH groups provides the active sites for adsorption and thus allows the use of silica gel as a stationary phase for chromatography.

silicone \\'sil-ə-ˌkōn\\ Polymeric organosilicon derivatives containing the —Si—O—Si— links. Depending on the nature of the polymeric chain, the products can be oils, greases, resins, or rubbers. They are chemically inert and immiscible with water, and as oils they have high flash points. One use is as an antifoaming agent.

silver stain \\'sil-vər 'stān\\ A highly sensitive method for detecting proteins on polyacrylamide gels. The method is based on the reduction of silver ions to metallic silver at the site of the protein on the gel. Protein bands stain black. Traditionally, polyacrylamide gels have been stained for protein by using Coomassie brilliant blue, which detects protein bands containing as little as about 0.1 μg of protein. The silver stain, which extends the sensitivity about 100-fold, is able to detect less than 1 ng of protein in a gel band.

silyl compounds \\'sil-il 'käm-ˌpaůndz\\ Compounds that contain the —SiH₃ or —SiR₃ group.

simian virus 40 (SV40) \\'sim-ē-ən 'vī-rəs 'fȯrt-ē ('es 'vē 'fȯrt-ē)\\ A mammalian DNA virus used as the basis for a number of cloning vectors. It has potential use in transforming animal cells, in which it follows either a lytic or lysogenic cycle, depending on the species from which the cell is derived.

single-cell protein (SCP) \\ˌsiŋ-gəl-'sel 'prō-ˌtēn ('es 'sē 'pē)\\ The microbial biomass component of a fermentation process that is produced commercially for human food or animal feed. After harvest, the biomass is processed to a suitable food product. The name indicates its microbial origin, thus distinguishing it from proteins originating from higher multicellular plants and animals. Sources of SCP include Pekilo protein, an animal feed product produced in Finland by the growth of the fungus *Paecilomyces varioti* on stripped sulfite waste liquor, and mycoprotein, a human food product produced by the British company Rank–Hovis–MacDougall by the growth of the fungus *Fusarium graminaerum* on glucose syrup. Other processes, which provided sources such as Toprina, an animal feed product produced by British Petroleum by the growth of yeast on *n*-alkane, and Pruteen, an animal feed product produced by ICI by the growth of *Methylophilus methylotrophus* on methanol, are now only of historical interest because these processes are no longer financially viable. The nucleic acid content of SCP should not exceed 3% of the dry weight; a greater nucleic acid content would lead to gout and gouty arthritis because abnormal levels of uric acid in the blood are caused by nucleic acid catabolism.

single-chain antibody \\ˌsiŋ-gəl-'chān 'ant-i-ˌbäd-ē\\ An antibody formed by genetically linking the genes for the variable regions of an immunoglobulin heavy and light chain by a DNA segment encoding a synthetic linker. It should be possible to produce such antibodies in bacteria, thereby reducing the cost of antibody production. The perceived advantages of such antibodies are that they should be less immunogenic in vivo (and thus more

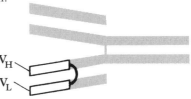

Single chain antibody. V_H and V_L region peptides joined by a synthetic linker. Shadow indicates relative size of normal antibody.

amenable for use in human therapy) and that their small size should give them the ability to penetrate tissues in the body normally restrictive to larger molecules.

single-stranded DNA (ssDNA) \\ˌsiŋ-gəl-'stran-dəd 'dē 'en 'ā ('es 'es 'dē 'en 'ā)\\ DNA that consists of only one strand of nucleotides rather than two base-paired strands found in the double helix form of DNA. Any method that denatures (melts) DNA results in the production of ssDNA.

site-directed mutagenesis \\ˌsīt-də-'rek-təd myüt-ə-'jen-ə-səs\\ A technique for producing a single mutation in a DNA sequence. With this technique, a single codon can be modified in a predetermined manner to cause the specific replacement of one amino acid by another in the protein gene product. On the basis of a knowledge of the relationship between the protein's structure and its function, amino acid substitutions can lead to a modified (engineered) protein that has increased stability, increased catalytic activity, or altered substrate specificity. Although potentially a very exciting development in biotechnology, the applications of this technique are restricted at present by our limited knowledge of the factors relating protein structure to function in many of the proteins of commercial interest. Indeed, the technique is mainly used in basic research into an understanding of the factors relating protein structure and function.

size-exclusion chromatography   \'sīz-iks-'klü-zhən ˌkrō-mə-'täg-rə-fē\   *See* gel filtration.

sizing   \'sī-ziŋ\ The application of adhesive starch to fabrics to increase strength during weaving.

slime layer   \'slīm 'lā-ər\ An ill-defined layer of polysaccharide exterior to the cell wall in some bacteria. *See also* capsule, glycocalyx.

slops   \'släps\ The residual liquids from fermentations in the manufacture of products such as citric acid, alcohol, or acetone. For example, the slops are the byproduct from which the alcohol has been removed by distillation. Slops contain mainly residual sugars and organic acids, as well as fermenting microorganisms, and are often used as cattle feed.

slow flow   \'slō 'flō\ *See* creep flow.

slug flow   \'sləg 'flō\ A type of two-phase flow, particularly in gas–fluid systems, in which the gas forms large bubbles with dimensions similar to the cross section of the fluid flow. This process occurs in flow-through tubes in which the gas segments the fluid into discrete volumes (slugs) as in some forms of automatic analyzer equipment and gas-lift devices and in fluidized beds when aggregative fluidization occurs. *See* gas lift.

snail gut enzyme   \'snā(ə)l 'gət 'en-ˌzīm\ A commercial preparation of snail digestive juice used to degrade cell walls. It is used to prepare protoplasts and sold as gluculase, helicase, or sulfatase.

sodium azide (NaN₃)   \'sōd-ē-əm 'ā-ˌzīd ('en 'ā 'en 'thrē)\ A compound that is often used as a bacteriostatic agent (at ~0.02%) in buffers and solutions that are stored for any length of time and in buffers used in chromatographic columns, where bacterial contamination can cause clogging. Not to be used in solutions containing horseradish peroxidase because this enzyme is inhibited by sodium azide. Its bacteriostatic effect is caused by interference with the electron-transport chain.

sodium dodecyl sulfate (SDS)   \'sōd-ē-əm 'dō-də-sil 'səl-ˌfāt ('es 'dē 'es)\ $CH_3(CH_2)_{10}CH_2OSO_3Na$. An anionic detergent used to denature and solubilize proteins. Also known as sodium lauryl sulfate. *See also* SDS polyacrylamide gel electrophoresis.

sodium lauryl sulfate   \'sōd-ē-əm 'lor-əl 'səl-ˌfāt\ *See* sodium dodecyl sulfate.

solid-bowl scroll centrifuge \ˌsäl-əd-ˈbōl ˈskrōl ˈsen-trə-fyüj\ A type of centrifuge used for continuous solid–liquid separations. Feed is passed through the spindle of an Archimedian screw within a rotating solids bowl. Solids settling on the walls of the bowl are scraped to the conical end of the bowl and discharged. Clarified liquor is discharged from the other end of the bowl. Also known as a decanter centrifuge.

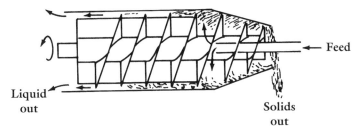

Feed

Liquid
out

Solids
out

solid-phase extraction \ˌsäl-əd-ˈfāz ik-ˈstrak-shən\ A technique for the concentration or purification of a substance whereby a solid substrate, appropriately derivatized, is contained in a short column (about 5 cm^3) such as hypodermic syringe barrels. The dilute solution containing the desired solute is passed through the column, and any nonabsorbed species are washed off the support. The absorbed solute may then be eluted with a small volume of liquid for direct analysis or recovery. Several such columns derivatized for commonly required species are commercially available.

solid-substrate fermentation \ˈsäl-əd-ˈsəb-ˌsträt fər-mən-ˈtā-shən\ A process in which an organism is allowed to grow on a solid substrate (e.g., the Koji process). The solid is extracted to recover the desired microbial product after the optimum time for growth and production has been reached. Growth may take place in shallow trays or in large heaps, which may be turned either by hand or mechanically in rotating drums, or in tanks with blown air and mechanical stirring.

solvent power \ˈsäl-vənt ˈpaủ(-ə)r\ *See* critical fluid.

somaclonal variation \ˌsō-mə-ˈklōn-ᵊl ˌver-ē-ˈā-shən\ All types of variation that occur in plants regenerated from culture cells or tissues. *See also* somatic variants.

somatic \sō-ˈmat-ik\ Of or relating to the vegetative (nonsexual) stages of a eukaryotic life cycle.

somatic cell \sō-ˈmat-ik ˈsel\ Any eukaryotic cell other than a germ cell (egg or sperm in mammals, ovule or pollen in plants).

somatic embryogenesis \sō-ˈmat-ik ˌem-brē-ō-ˈjen-ə-səs\ *See* embryogenesis.

somatic fusion \sō-ˈmat-ik ˈf(y)ü-zhən\ *See* protoplast fusion.

somatic hybridization \sō-ˈmat-ik ˌhī-brəd-ə-ˈzā-shən\ The fusing of two protoplasts or cells with different genotypes to produce hybrids that contain various mixtures of nuclear and cytoplasmic genomes.

somatic variants \sō-ˈmat-ik ˈver-ē-ən(t)s\ Plant-cell clones that differ from the parent cell. Although plants derived by vegetative means (clones) are usually like the parent plant, not all clones are genetically identical. Clones that differ significantly from

The Language of Biotechnology

the parent are called somatic variants or sports and are said to show somaclonal variation. Occasionally sports become established as important new varieties (e.g., the nectarine). *See also* clone.

somatotropin \sō-ˌmat-ə-'trō-pən\ *See* human growth hormone.

sonication \ˌsän-ə-'kā-shən\ The use of ultrasonic energy to disrupt cells. The application of ultrasonics to a liquid causes areas of compression and rarefaction to occur. Cavities form in the areas of rarefaction, which rapidly collapse as the area changes to one of compression. Bubbles produced in the cavities are therefore compressed to several thousand atmospheres, and on their collapse shock waves are formed. These shock waves are thought to be responsible for cell damage. Generally only used for laboratory-scale work.

souring \'sau(ə)r-iŋ\ A drop in the pH of a process caused by the production of organic acids. In anaerobic digestion, souring caused by the overproduction of fatty acids results in an inhibition of methanogenesis.

Southern blotting (Southern transfer) \'səth-ərn 'blät-iŋ ('səth-ərn 'tran(t)s-ˌfər)\ A method for transferring separated DNA fragments (e.g., a restriction enzyme digest) from an agarose gel to a solid support such as a nitrocellulose or nylon membrane. The method is named after E. M. Southern, who developed the technique. The DNA is first denatured in the gel by soaking the gel in sodium hydroxide. The denatured DNA is transferred from gel to membrane by placing first the membrane and then a pad of adsorbent material on the gel. Buffer is allowed to soak through the gel, carrying the DNA from the gel to the membrane, where the DNA binds. The membrane therefore contains a replica of the DNA bands separated in the gel. The membrane can then be washed in a solution of an appropriate radiolabeled DNA or RNA probe, which binds to the DNA fragment of interest by complementary base-pairing. This binding can be detected by autoradiography.

soybean \'sȯi-ˌbēn\ A bean from the plant *Glycine max. (L) Merr.* The seed contains a high concentration of protein (40%) and oil (20%) and supplies about a quarter of the world's fats and oils and about two-thirds of the protein-concentrate animal feeds.

spacer arm (spacer group) \'spā-sər 'ärm ('spā-sər 'grüp)\ Usually a short alkyl chain 4–12 carbon units long. When an affinity adsorbent is constructed for affinity chromatography, it is often necessary for ligands to be held at some distance from the matrix. This distance obviates problems of steric hindrance between the ligate and matrix that would otherwise prevent correct binding. The ligand is therefore linked to the matrix via a spacer arm.

sparger \'spärj-ər\ A device for introducing air or oxygen into a reactor by way of an orifice that generates bubbles. The size of the bubbles and the velocity at which they rise in the fermentor determine the rate of oxygen transfer. *See also* bubble column fermentor.

specific growth rate \spi-'sif-ik 'grōth 'rāt\ *See* dilution rate, Monod kinetics.

specific ion electrode \spi-'sif-ik 'ī-än i-'lek-trōd\ *See* ion-selective electrode.

specific optical rotation \spi-'sif-ik 'äp-tə-kəl rō-'tā-shən\ A measure of the rotation of the plane of polarization of plane-polarized light by an optically active substance. It is defined as the following:

$$[\alpha]_\lambda^T = \frac{\alpha V}{lm}$$

where α is observed rotation, V is volume of sample, l is path length, and m is mass of solute or sample if it is a pure liquid. As the value of the rotation is dependent on wavelength of the radiation, λ, and temperature, T, these must be specified and are normally the sodium D line and 20 or 25 °C, respectively.

specific viscosity \spi-'sif-ik vi-'skäs-əd-ē\ The difference between the viscosity of a solution or suspension and that of the solvent or continuous phase in which the species is dissolved or suspended, all measured at the same temperature.

spheroplast \'sfir-ə-plast\ A bacterial or yeast cell that has been largely, but not entirely, freed of its cell wall, usually by enzymatic digestion, and that retains an intact cytoplasmic membrane.

spinner bottle (spinner culture vessel) \'spin-ər 'bät-ᵊl ('spin-ər 'kəl-chər 'ves-əl)\ A magnetically stirred culture vessel for the suspension culture of plant or animal cells. The vessel has a magnetic bar, suspended a few millimeters from the bottom of the vessel, and is placed on a magnetic stirrer, which drives the magnetic bar. Spinner vessels are available in sizes from 0.2 to 20 dm³.

spirally wound module \'spi-rə-lē 'waund 'mäj-u(ə)l\ A configuration for a membrane in which a sandwich of membranes and spacers is wound into a spiral. The module produces a high surface area (800–1000 area-to-volume ratio) with both low investment and low operating costs. However, the module suffers from poor control of fouling and concentration or polarization effects. The module is used for both reverse osmosis and ultrafiltration when low fouling is expected.

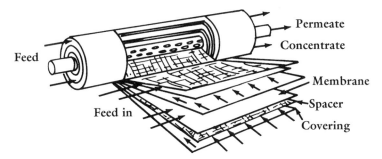

splenocytes \'splē-nō-ˌsīts\ Lymphocytes isolated from the spleen.

splicing \'splī-siŋ\ *See* exons, introns.

split injection \'split in-'jek-shən\ A capillary gas chromatographic technique in which the sample size is adjusted to suit the column requirements by splitting off the major fraction of sample vapors in the inlet so that only a minor portion (e.g., about 0.1%) of the sample enters the column. *Contrast with* splitless injection.

splitless injection \\'split-les in-'jek-shən\\ A modification of split injection in which the sample is not split for the first 0.5–4 min of sampling, and thus the whole sample enters the column. Splitting is restored after a defined time, and the remainder of the sample in the inlet is purged. In this technique as much as 99% of the sample enters the column. *Contrast with* split injection.

spontaneous mutation \\spän-'tā-nē-əs myü-'ta-shən\\ *See* mutation.

sport \\'spȯ(ə)rt\\ Any organism (or cell) with unusual characteristics produced by natural mutation. A mutant. *See also* somatic variants.

spray column \\'sprā 'käl-əm\\ The simplest type of column contactor in which one phase is introduced as a spray via a distributor to pass through the continuous phase in an unhindered pattern. An inefficient type of contactor, used only for the most favorable of processes.

spray drying \\'sprā 'drī-iŋ\\ A method for the large-scale recovery of dry solids from a concentrated aqueous solution or suspension. The solute or suspension is atomized in a drying chamber into which a stream of heated air is introduced. The moisture evaporates into the air stream, and the dried powder is collected. Reduced pressure is also sometimes applied to increase the rate of solvent evaporation.

stacked-plate reactor \\'stakt-'plāt rē-'ak-tər\\ A reactor designed for the large-scale growth of mammalian cells. Circular glass or stainless steel plates are fitted vertically ~5 mm apart on a central shaft. The shaft is immersed in growth medium and can be stationary, with an airlift pump for mixing, or can revolve on a vertical or horizontal axis. The large surface area provided by the plates is ideal for the growth of anchorage-dependent cells. A 200-dm^3 reactor can give a surface area of 2×10^5 cm^2.

stage (theoretical) \\'stāj (ˌthē-ə-'ret-i-kəl)\\ *See* theoretical plate.

standard oxidation-reduction potential (E^o) \\'stan-dərd ˌak-sə-'dā-shən-ri-'dək-shən pə-'ten-chəl ('ē 'ō)\\ The potential of a redox couple when the activity of the oxidized form is equal to the activity of the reduced form of the species. Thus, from the Nernst equation:

$$E = E^o + \frac{RT}{nF} \ln \frac{a_{ox}}{a_{red}}$$

when $a_{ox} = a_{red}$, then $E = E^o$. E is the observed electrical potential, E^o is the standard potential, R is the gas constant, T is the absolute temperature, n is the number of electrons involved in the redox couple, F is the Faraday constant, and a is the activity of the species involved.

starch \\'stärch\\ One of the major nutritional storage molecules in plants. It has two forms: amylose, which is unbranched and consists of glucose molecules in α-1,4 linkages, and amylopectin, which consists of chains of α-D-glucose with one α-1,6 linkage about every 30 α-1,4 linkages. The main sources of industrial starch are maize, wheat, and potato. These starches are hydrolyzed enzymatically on an industrial scale. Following liquefaction (i.e., dispersing and rupturing of the starch granules into aqueous solution by heating in the presence of a thermostable

α-amylase), the starch is progressively hydrolyzed to dextrose (glucose) by the use of enzymes such as α-amylase (from *Bacillus subtilis*), amyloglucosidase (from *Aspergillus niger* or *A. oryzae*), and pullulanase to give a solution with a high (~96) dextrose equivalent. This stage after liquefaction is known as saccharification. The glucose syrup product can be further processed to produce high-fructose syrups.

start codon \\'stärt 'kō-ˌdän\\ *See* initiation codon.

start-up \\'stärt-ˌəp\\ The commencement of a process. The process conditions may vary from those found for continuous running and tend toward steady-state operating conditions with time.

starter culture \\'stärt-ər 'kəl-chər\\ Microbial inocula used to inoculate milk for the preparation of a range of milk products, such as butter, cheeses, and yogurt. The cultures, which may be bacterial or fungal strains and are used as pure or mixed cultures, have been selected for their ability to produce (1) lactic acid for curd production and a low pH value to prevent spoilage, (2) metabolites that give desirable flavors, or (3) enzymes that mature the dairy product.

starter rotation \\'stärt-ər rō-'tā-shən\\ In cheese making, several cultures with susceptibility to different bacteriophage strains are used in rotation as inocula. This rotation reduces the likelihood that a particular bacteriophage will become established in a cheese-manufacturing plant and halt production.

static head \\'stat-ik 'hed\\ The pressure in a fluid due to the head or level of a fluid above the reference point.

static maintenance reactor \\'stat-ik 'mānt-nən(t)s rē-'ak-tər\\ A reactor design for the growth of mammalian cells. Some mammalian cells secrete products maximally when in an active growth phase, and others produce better when quiescent. The reactor is designed for the quiescent type of cells.

stationary phase \\'stā-shə-ˌner-ē 'fāz\\ The point at which growth ceases in a microbial culture. As log phase proceeds in a microbial culture, toxic byproducts tend to accumulate, and the substrate (nutrient) becomes increasingly depleted until exhausted. Growth of the microorganism therefore slows down (*deceleration phase*) and finally ceases, at which point the culture is said to have reached the *stationary phase*. Eventually, the culture enters the *death phase,* and the number of viable cells declines.

stationary phase (chromatography) \\'stā-shə-ˌner-ē 'fāz (ˌkrō-mə-'täg-rə-fē)\\ The fixed component in partitioning process across which the mobile phase passes. For example, paper, gel, silica (thin-layer chromatography), column packing (gas chromatography, high-performance liquid chromatography).

Staverman reflection coefficient \\'stā-ver-ˌmən ri-'flek-shən ˌkō-ə-'fish-ənt\\ *See* Darcy law.

steam seal \\'stēm 'sel\\ The use of a continuous flow of steam condensate to maintain sterility (e.g., around a sample line).

stem cells \\'stem 'selz\\ Proliferating eukaryotic cells that can both maintain their own numbers by self-renewal divisions and give rise to differentiated progeny.

stepped-feed aeration \\'stept-'fēd a(-ə)r-'ā-shən\\ *See* activated sludge process.

stepwise elution \\'step-ˌwīz ē-'lü-shən\\ A chromatographic procedure in which the composition of the mobile (eluant) phase is changed in steps during elution to change a parameter such as polarity, thereby varying parameters such as retention times or resolution of solutes on the column. *See also* gradient elution, isocratic elution.

stereoisomers \\ˌster-ē-ō-'ī-sə-mərz\\ Forms of a chemical compound (isomers) that vary from one another in the spatial orientation of the atoms in the molecule. Various types of stereoisomers can be distinguished, such as optical isomers (which give rise to optical activity) and geometrical isomers (which depend on the nature of the groupings and spatial configurations and have names like *cis* and *trans*, *syn* and *anti*, and *mer* and *fac*).

stereoselective \\'ster-ē-ə-si-'lek-tiv\\ Descriptive of any reaction that produces or degrades a predominance of one stereoisomer. When two stereoisomers are mirror images (i.e., enantiomers), then the term *enantioselective* is used. *Compare with* regioselective.

sterilization criterion \\ˌster-ə-lə-'zā-shən krī-'tir-ē-ən\\ *See* del factor.

sticky ends \\'stik-ē 'endz\\ The ends of a double-stranded DNA molecule from which single-stranded nucleotides extend. *See also* restriction endonucleases.

stillage \\'stil-ij\\ *See* slops.

stirred-tank reactor \\'stərd-'tank rē-'ak-tər\\ A type of fermentor in which agitation is provided by a mechanically rotated stirrer or turbine.

stoichiometry \\ˌstói-kē-'äm-ə-trē\\ In a chemical system, the stoichiometry of the reactants is defined as the relative number of atoms or molecules taking part in a chemical reaction. In a biological system, because of its more complex nature and the uncertainty of the chemical composition of the components, a mass-based rather than a mole-based approach is used. This mass balance can be related to a number of parameters, such as total mass, carbon content, and chemical oxygen demand.

Stokes law \\'stōks 'ló\\ An equation that relates the velocity of a particle in a fluid to the size of the particle and the fluid viscosity. It is commonly used in determining the rate of sedimentation of particles suspended in a fluid by gravity:

$$V_g = d^2 g(\rho_P - \rho_L)/18\eta$$

where V_g is the rate of sedimentation under gravity (m/s); d is the particle diameter (m); g is the gravitational constant (m/s^2); ρ_P and ρ_L are the density of the particle and fluid, respectively (kg/m^3); and η is the viscosity of the fluid (kg/ms).
    The equation can be modified for sedimentation in a centrifugal field:

$$V = d^2 \omega^2 r(\rho_P - \rho_L)/18\eta$$

where ω is the angular velocity of the centrifuge (s^{-1}), and r is the radial position of particle (m).

Stokes law tube \\'stōks 'lȯ 't(y)üb\\ A device used to remove particulate matter from fluid streams before sampling for on-line analysis. Particles that enter the sampling line decelerate under gravity and fall back into the main fluid flow, leaving a particle-free sample for analysis.

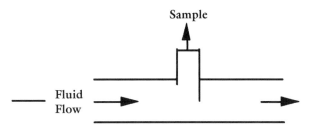

Stokes law tube

strain gauge \\'strān 'gāj\\ A device for monitoring the extension or compression of an elastic material. An electrical resistance strain gauge is normally a grid of metal foil, usually constantan, deposited on a thin sheet of plastic material. Gauges are generally used with resistances of the order of 50–500 Ω. Strain gauges are used in load cells to monitor weight and also in fermentor agitator shafts to measure power uptake.

streamline flow \\'strēm-līn 'flō\\ *See* laminar flow.

streptavidin \\ˌstrep-tə-'vī-dən\\ A protein from *Streptomyces avidinii* that has the same biotin-binding properties as avidin. Unlike avidin, it is not a glycoprotein and thus is said to cause less nonspecific binding.

streptokinase \\ˌstrep-tō-'kī-ˌnās\\ A serine protease produced by hemolytic streptococci and capable of activating plasminogen. It is used in the treatment of diseases requiring the dissolution of blood clots. *See also* plasmin, tissue plasminogen activator.

Streptomyces \\ˌstrep-tə-'mī-ˌsēz\\ A genus of the actinomycetes that produces branching mycelia and aerial hyphae-bearing chains of spores. They are common in the soil and are involved in the breakdown of a variety of organic compounds. Streptomycetes produce more than 60% of the known antibiotics, including streptomycin and the tetracyclines. A number of *Streptomyces* species are being used in genetic studies, and more than 100 genetic markers have been identified in *S. coelicolor*.

stringency \\'strin-jən-sē\\ Conditions under which nucleic acid hybridizations are carried out (e.g., probing Southern blots). Conditions of high stringency (i.e., high temperature and low salt strength) allow base-pairing only between nucleotide sequences with a very high frequency of complementary base sequences. However, to use a probe to detect a nucleotide sequence that has a lower frequency of complementary base sequences, the condition of low stringency (low temperature and high salt strength) must be used to allow these sequences to base-pair.

STS \\'es 'tē 'es\\ *See* sequence-tagged site.

stuffer fragment \'stəf-fər 'frag-mənt\ *See* λ-replacement vector.

subculture \'səb-ˌkəl-chər\ A culture obtained by the subdivision of a culture and the transfer of one or more of these subdivisions into fresh culture media.

submerged culture \səb-'mərjd 'kəl-chər\ Culture conditions in which the fermentation broth is agitated at sufficient rate to ensure equal distribution of the organisms throughout the medium and to ensure that concentration gradients of nutrients or oxygen do not occur and lead to local variations in growth conditions.

subtilisin Carlsberg \səb-'til-ə-sən 'kärlz-bərg\ A serine protease produced from *Bacillus licheniformis*. The enzyme is used extensively in enzymatic washing powders; more than 500 tons of pure enzyme is produced annually.

succinoglucan \suk-'sin-ō-glü-kan\ β-1,3-glucan (a polysaccharide) isolated from cultures of *Alcaligenes faecalis* var. *myxogenes*. It consists of glucose, galactose, and succinic acid in a ratio of 7:1:1.5.

sucrase \'sü-ˌkrās\ Another name for invertase.

sugar \'shüg-ər\ The common name for sucrose. Sucrose is obtained commercially from sugar cane and sugar beet. It is a disaccharide of glucose and fructose, which can be hydrolyzed to its constituent sugars by the enzyme invertase. *See also* disaccharide.

sulfatase \'səl-fə-ˌtās\ *See* snail gut enzyme.

sulfhydryl (thiol) proteases \səlf-'(h)i-drəl ('thī-ȯl) 'prōt-ē-ˌās-əz\ One of the four classifications used for proteases, based on the mechanistic action at the active site. Sulfhydryl proteases have a functional thiol (—SH) group at their active site. Many of them, such as bromelain, ficin, and papain are isolated from plant sources. *See also* proteases.

supercoiled plasmid \'sü-pər-ˌkȯi(ə)ld 'plaz-məd\ *See* covalently closed circular DNA.

supercritical extraction \sü-pər-'krit-i-kəl ik-'strak-shən\ The extraction of a substance using a supercritical fluid (e.g., carbon dioxide) as the solvent. The process, which takes place under a pressure greater than atmospheric, has the advantage that release of this pressure causes rapid volatilization of the solvent. The process usually leaves the extracted material uncontaminated by solvent residues. This advantage, together with the relatively low temperatures involved during extraction, makes the process particularly suitable for foodstuffs and flavors.

supercritical fluid \sü-pər-'krit-i-kəl 'flü-əd\ Fluid with properties of both gases and liquids, behaving like gases in filling containers and like liquids with densities of $0.1-1$ g/dm^3 and the ability to dissolve substances. They exist when a substance is above its critical temperature and pressure, but these conditions alone do not always produce supercriticality. *See also* critical fluid.

supercritical fluid chromatography \sü-pər-'krit-i-kəl 'flü-əd ˌkrō-mə-'täg-rə-fē\ A chromatographic system that uses a supercritical fluid as a mobile phase with either a capillary gas chromatograph or packed-column high-performance liquid chromatography (HPLC) equipment. A number of supercritical fluids can be used (e.g., ammonia, carbon dioxide, and nitrous oxide), and their solvent properties can be modified by varying the pressure or by the addition of small amounts of polar liquids. The whole system operates at higher-than-atmospheric pressure to retain the fluid in its liquid state. The advantages include rapid separations with clean products uncontaminated by solvents. On release of pressure, these mobile phases rapidly evaporate and thus facilitate the linking of the chromatograph to other analytical equipment.

superinfection \\ˌsü-p(ə-)rin-ˈfek-shən\\ The infection by a virus of a host previously infected by another virus. Not to be confused with *double infection,* in which two viruses invade the host simultaneously.

Superose \\ˈsü-pər-ōs\\ Pharmacia trade name for cross-linked agarose beads used as a gel-filtration medium. The beads are much smaller ($10-30$-μm diameter) and physically stronger than the softer Sephadex beads. Whereas large-scale Sephadex columns have to be packed as segmented columns to avoid compression of the gel medium and consequent blocking of the column, the much more rigid Superose beads can be packed in much larger columns.

support-coated, open-tubular (SCOT) column \\sə-ˈpȯ(ə)rt-ˌkōt-əd ˈō-pən-ˈt(y)ü-byə-lər (ˈskät) ˈkäl-əm\\ A capillary chromatography column in which the stationary phase is coated onto a support material distributed over the column inner wall. This column generally has a higher capacity than a wall-coated open-tubular column with the same average film thickness. *See also* porous-layer, open-tubular column; wall-coated, open-tubular column.

supported liquid membrane extraction \\sə-ˈpȯ(ə)rt-əd ˈlik-wəd ˈmem-ˌbrän ik-ˈstrak-shən\\ A modification of conventional liquid–liquid extraction that provides a simple method of selective recovery and concentration of solute from dilute solutions. The supported liquid membrane is formed by adsorbing an appropriate organic extractant phase on a porous hydrophobic polymeric support, typically a microfiltration membrane. This membrane can be used as a physical barrier between appropriate feed and strip solutions because the thin layer of extractant in the pores of the support acts as a liquid membrane that allows transport of a selected solute between feed and strip phases (*see* diagram). The characteristics of the system can be arranged such that the concentration of the solute in the strip phase can be many times that of the feed.

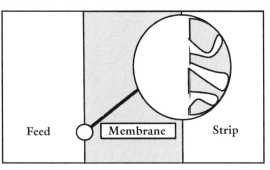

The disadvantage is that the membrane flux is quite small because of the finite size of the liquid membrane, but the flow sheet is much simpler than that for emulsion liquid membrane extraction, and conventional microfiltration module technology can be used. *Compare with* emulsion liquid membrane extraction.

surface-active agent (surfactant) \\'sər-fəs 'ak-tiv 'ā-jənt (ˌsər-'fak-tənt)\\ A compound that consists of a polar (hydrophilic) head group and a nonpolar (lipophilic) tail. Common head groups include sulfonic, sulfate, carboxylic acid, and hydroxyl groups, whereas the lipophilic groups generally consist of carbon chains and rings, e.g., sodium lauryl sulfate:

$$CH_3CH_2CH_2CH_2CH_2CH_2CH_2CH_2CH_2CH_2CH_2CH_2SO_4 - Na^+$$

lipophilic tail   polar head

At a water–oil interface, the hydrophilic groups tend to move into the water phase, but complete solution is inhibited by the nonpolar organic tail, which remains in the oil. Thus, the observed surface-active properties are a balance between these hydrophilic and lipophilic properties, i.e., the hydrophilic–lipophilic balance (HLB). *See also* hydrophilic–lipophilic balance.

surface activity \\'sər-fəs ak-'tiv-ət-ē\\ The variation of the surface tension of a liquid by the presence of a solute. As a result of the tendency of the free energy of a surface to decrease, the concentration of a solute on the surface may differ from that in the bulk phase. The component with the lower surface tension will tend to congregate at the surface because in this way the surface energy is reduced. Thus, if a given solute lowers the tension at a particular interface, then the concentration of the solute in the interface will be greater than that in the bulk phase (i.e., the solute will concentrate at the interface). Such solutes are said to display surface activity. *See also* antifoam agents, surfactant.

surface fermentation \\'sər-fəs ˌfər-mən-'tā-shən\\ The growth of an organism on the surface of a static fermentation liquor, where the microorganism grows as a raft floating on the surface of the culture and the required product is collected from the aqueous medium.

surfactant \\sər-'fak-tənt\\ Generally, a surface-active compound (i.e., a compound with low surface tension that tends to congregate at a liquid–liquid interface). These compounds generally contain both hydrophilic (polar) and hydrophobic (nonpolar) groupings; they may carry an overall positive (cationic), negative (anionic), or neutral (nonionic) charge. They may be used to disperse oils in aqueous systems, to transfer charged compounds from an aqueous to organic phase (as in ion-partition extraction), or to increase the gas–liquid interfacial area in a fermentor. In a fermentor, however, the presence of surfactant molecules at the interface may also increase the mass-transfer resistance and reduce the mobility of the interface. But the increase of interfacial area generally outweighs the other effects. Therefore, the oxygen-transfer coefficient normally increases with increasing concentrations of surfactants at low overall concentrations. *See also* surface-active agent.

surfactant liquid membrane extraction \\sər-'fak-tənt 'lik-wəd 'mem-ˌbrān ik-'strak-shən\\ *See* emulsion liquid membrane extraction.

suspension culture \sə-'spen-chən 'kəl-chər\ The growth of cells in submerged liquid culture.

- *animal cells:* \'an-ə-məl 'selz\ Suspension culture is the preferred method for scaling up mammalian cell cultures. However, only hematopoietic and transformed cells grow naturally in suspension culture because most other mammalian cells show anchorage dependence. Fortunately, a number of anchorage-dependent cells can be selected or adapted to grow in suspension culture, although a few will not survive in suspension at all. Because of the fragile nature of mammalian cells, they are best grown under conditions of minimum shear stress (e.g., in airlift fermentors). The encapsulation of cells in gelatin, alginate, or agarose, has also been used to protect cells from mechanical stress in other large-scale culture equipment (e.g., stirred fermentors). *See also* microcarrier.
- *plant cells:* \'plant 'selz\ Suspension cultures are usually initiated by transferring established callus tissue from a semisolid medium to a liquid medium. The liquid medium is then agitated and gives rise to a heterogeneous population of cells made up of a mixture of free-floating single cells and cell aggregates. Unlike mammalian cells, plant cells are not attachment-dependent. Single-cell suspensions from suspension culture can be plated out in semisolid media in a manner much like that used routinely in microbiology. Cell colonies that grow up (calliclones) can be transferred to new Petri dishes and grown to form a callus. These callus cultures can then be induced (embryogenesis or organogenesis) to generate complete plants. *See also* paper raft nurse technique.

Svedberg unit (sedimentation) \'sfed-berg 'yü-nət (ˌsed-ə-men-'tā-shən)\ A unit of sedimentation equal to a sedimentation coefficient of 1×10^{-13} s.

sweet sorghum \'swēt 'sor-gəm\ A tall grass with high sucrose content in the stems. It is widely grown for cattle forage but has been used for ethanol production by fermentation.

SWISS-PROT \'swis 'prōt\ A protein and peptide database sequence with detailed annotation.

symbiosis \ˌsim-bē-'ō-səs\ An association of dissimilar organisms to their mutual advantage, such as the association of nitrogen-fixing bacteria with leguminous plants (e.g., peas or beans). The bacteria inhabit nodules on the plant roots and synthesize nitrogenous compounds from nitrogen in the air. These nitrogenous compounds become available to the plant for metabolism. The bacteria obtain carbohydrates and other food substances from the plant.

synergism \'sin-ər-jiz-əm\ The interaction of substances such that the additive properties of two components are greater than the sum of the effects of the two components taken separately. *Compare with* antagonism.

synergist \'sin-ər-jəst\ A compound that by itself has little activity, but when combined with another compound enhances the activity of the latter.

syngeneic \ˌsin-jə-'nē-ik\ Of or relating to genetically identified individuals of the same species. The term is used in reference to tissue transplants.

synthases \'sin(t)-thə-ˌsēz\ *See* lyases.

synthetic medium \sin-'thet-ik 'mēd-ē-əm\ *See* defined medium.

syntrophy \'sin-trō-fē\ Nutritional interdependence of two organisms.

T

T-DNA \'tē 'dē 'en 'ā\ *See Agrobacterium tumefaciens.*

T4-DNA ligase \'tē-fȯr 'dē 'en 'ā 'lī-gāse\ *See* DNA ligase.

T4-DNA polymerase \'tē-fȯr 'dē 'en 'ā 'päl-ə-mə-ˌrās\ A DNA polymerase, isolated from bacteriophage T4, that also has $5' \rightarrow 3'$ exonuclease activity.

T-lymphocytes \'tē 'lim(p)-fə-ˌsītz\ Thymus-derived lymphocytes; blood cells that originate in the bone marrow but migrate to the thymus (hence, "T"), where they mature before migrating to the spleen and lymph nodes. Unlike B-lymphocytes, they do not secrete a single major protein product but carry out their functions either by direct contact with other cells or by producing lymphokines (cytokines) that have powerful biological effects on other cells and are active at very low concentrations. The roles of T-lymphocytes in the body are numerous and include involvement in delayed type hypersensitivity, graft rejection, and protection against viral and fungal infections.

TAFE \'tē 'ā 'ef 'ē\ *See* pulse-field gel electrophoresis.

tailing in chromatography \'tā(e)l-iŋ 'in ˌkrō-mə-'täg-rə-fē\ Asymmetry of a chromatographic peak, as shown by an elongation of the tailing edge.

tailings in hydrometallurgy \'tā(e)l-iŋz 'in ˌhī-drō-'met-ᵊl-ˌər-jē\ The solid residues remaining after processing that are normally deposited in heaps. Tailings heaps may be reworked for the metal resources they contain, often with the aid of microbial leaching.

taka-diastase \'täk-ə-'dī-ə-stās\ *See* Koji.

tangential-flow filtration \tan-ˈjen-chəl-'flō fil-'trā-shən\ *See* crossflow filtration.

***Taq* polymerase** \'tak 'päl-ə-mə-ˌrās\ *See* polymerase chain reaction.

TATA box \'tä 'tä 'bäks\ A nucleotide sequence, approximating to TATA T^A_AT, found about 30 bases upstream of the start of transcription of all genes in eukaryotes. This promoter sequence seems to have the function of accurately aligning the initiating RNA polymerase at the start site. *See also* enhancer.

taxonomy \tak-'sän-ə-mē\ The classification of organisms into groups based on similarities of structure or origin.

telomeres \'tel-ə-ˌmi(ə)rz\ The two ends of a linear chromosome. The sequences at the ends of the chromosomes are necessary to allow chromosomes to replicate. DNA fragments not containing these telomeres (e.g., a circular plasmid cleaved to give a linear molecule) are unable to be replicated.

temperature-sensitive (ts) mutation \'tem-pə(r)-ˌchù(ə)r-'sen(t)-sət-iv myü-'tā-shən\ A mutation in a gene that results in a gene product that is only functional within a certain temperature range. For example, many laboratories use λ-phage with a temperature-sensitive mutation in the gene cI, which is one of the genes responsible for maintaining the phage in the integrated state. At 30 °C, the gene is functional and lysogeny occurs, but at 42 °C the gene product is inactive and lysogeny cannot occur. Cultures are therefore grown at 37 °C and then induced to produce extracellular phages by transferring to 42 °C.

template \'tem-plət\ Any polynucleotide that acts as a substrate for the synthesis of a complementary nucleic acid strand (e.g., mRNA acts as a template for reverse transcriptase in the synthesis of cDNA).

terminal transferase (terminal deoxynucleotidyl transferase) \'tərm-nəl 'tran(t)s-fər-ˌās ('tərm-nəl 'dē-äk-sī-ˌn(y)ük-lē-ō-ˌtid-əl 'tran(t)s-fər-ˌās)\ An enzyme isolated from calf thymus tissue that adds one or more deoxynucleotides to the 3′-terminus of a DNA molecule. The enzyme is used for homopolymer tailing of DNA molecules for subsequent cloning.

Tet^R \'tet 'är\ Abbreviation for tetracycline resistance gene.

tetracyclines \ˌte-trə-'sī-ˌklēnz\ A family of broad spectrum antibiotics, based on the naphthacene skeleton produced by species of *Streptomyces*. They inhibit protein synthesis in prokaryotes by binding to ribosomes. Clinically useful tetracyclines include chlortetracycline (aureomycin), oxytetracycline (terramycin), and tetracycline (tetramycin).

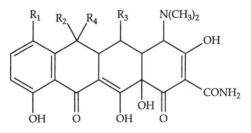

in oxytetracycline, R_1 = H, R_2 = CH_3, R_3 and R_4 = OH

TGF \'tē 'jē 'ef\ *See* transforming growth factor.

thaumatin \\'thau-mə-ˌtin\ An intensely sweet-tasting, nonnutritive protein from the fruit of the African plant *Thaumatococcus danielli*. Aqueous solutions of thaumatin have an intense sweetness about 1600 times that of a 10% sucrose solution. It is sold under the trade name Talin and is used in the food and drink industry.

theoretical plate \\ˌthē-ə-'ret-i-kəl 'plāt\ The number of individual equilibria equivalent to the performance of a column used for a particular process. Thus, a 10-plate extraction column achieves the same separation performance as 10 successive contacts between the solute and solvent. *See also* height equivalent to a theoretical plate.

theoretical stage \\ˌthē-ə-'ret-i-kəl 'stāj\ *See* theoretical plate.

thermal gradient field-flow fractionation \\'thər-məl 'grād-ē-ənt 'fē(ə)ld-'flō ˌfrak-shə-'nā-shən\ *See* field-flow fractionation.

thermistor \\'thər-ˌmis-tər\ A device used to monitor temperature. Changes in temperature are detected by a change of resistance within a semiconductor.

thermistor-based sensors \\'thər-ˌmis-tər-bāst 'sen(t)-sərz\ Sensors that use a thermistor to monitor small temperature changes generated in biochemical reactions. Temperature changes in the range 0.01–0.001 °C are usually recorded.

thermodynamic activity \\ˌthər-mō-dī-'nam-ik ak-'tiv-ə-tē\ A thermodynamic parameter that measures the active concentration (a) of a substance in a given chemical system, in contrast to the molecular concentration (c). These two terms are related by the activity coefficient (f) such that:

$$a = fc$$

where f, a dimensionless parameter with a wide range of positive values, approaches unity in dilute solutions. However, in concentrated solutions f may become very large, thus increasing the active concentration over the molecular concentration. Activity is the parameter measured by electrochemical sensors such as selective ion electrodes and pH electrodes. *See also* distribution constant, partition constant.

thermogravimetric analysis (TGA) \\ˌthər-mō-grav-ə-'me-trik ə'nal-ə-səs ('tē 'jē 'ā)\ An analytical technique based on the continuous recording of weight loss of a sample as a function of temperature. It can also be used in the derivative mode, differential thermal analysis (DTA). Both techniques are used as rapid and simple methods of characterization and analysis of both simple compounds and complex mixtures as found in pharmaceutical formulations.

thermophiles \\'thər-mə-ˌfīlz\ Microorganisms capable of growth at temperatures above about 55 °C. *See also* mesophiles, psychrophiles.

thickening \\'thik-(ə-)niŋ\ The process of concentrating suspended solids and separating them from the contained liquor, generally carried out under gravity.

thimerosal \\thī-'mər-ə-ˌsal\ Sodium ethylmercurithiosalicylate. It is often used as an antimicrobial agent in buffers and solutions that are stored for any length of time and in buffers used in chromatographic columns in which bacterial contamination can cause clogging or degradation of column materials.

thin-layer chromatography (TLC) \'thin-ˌlā-ər ˌkrō-mə-'täg-rə-fē ('tē 'el 'sē)\ A chromatographic separation developed from paper chromatography in which the stationary phase is a thin layer of a powdered adsorbent, usually silica or alumina, supported by a glass or aluminum sheet. Such plates are commercially available in a number of forms depending on the nature of the solid phase and type of binder. The chromatogram is developed by placing the plate spotted with the analyte mixture in a small amount of the mobile phase, which is usually an organic solvent or mixture of solvents. The mobile phase rises up the vertical plate by capillary action, and separation of the components occurs by partition between the solid support material and the mobile phase. After developing and drying the plate, the spots may be visualized in a number of ways selective for the type of compound or submitted to UV radiation, under which the organic compounds often fluoresce.

High-performance thin-layer chromatography (HPTLC) is a development of TLC that uses specialized adsorbent phases bonded onto the backing plate. Various adsorbents are commercially available for affinity chromatography or reversed-phase chromatography. High performance is also achieved by having an adsorbent of a regular particle size.

Thiobacillus ferrooxidans \thī-ō-bə-'sil-əs ˌfer-ō-'äks-i-danz\ *See* microbial leaching.

thiol proteases \thī-ól 'prōt-ē-ˌās-ez\ One of the four possible classifications of proteases. Thiol proteases have a thiol (—SH) group at the active site that is essential for activity. *See also* bromelain, ficin, papain.

thixotrophy \thik-'sä-trə-fē\ A time-dependent reversible behavior of a fluid under an applied shear stress, such that on application of a shearing force the fluid becomes less viscous. When the shear force is removed, the viscosity returns to its original value.

thixotropic fluid \thik-sə-'trō-pik 'flü-əd\ A fluid that, when subjected to a constant shear stress, such as agitation, exhibits a reduction in apparent viscosity with time.

thrombin (E.C. 3.4.21.5) \'thräm-bən\ A blood coagulation enzyme responsible for the conversion of soluble fibrinogen to insoluble fibrin, which forms part of blood clots. Thrombin is a serine protease and is present in the serum as the inactive form prothrombin. Prothrombin is converted into thrombin as a result of tissue damage. *See also* fibrin, hirudin.

thrombolytic \thräm-bə-'lit-ik\ Any compound that initiates dissolution of a blood clot (*thrombus*). *See*, for example, eminase, plasmin, streptokinase, tissue plasminogen activator, urokinase.

thrombus \'thräm-bəs\ *See* thrombolytic.

thylakoids \'thī-lə-ˌkoidz\ Membranous structures, shaped like flattened sacs, found in chloroplasts. A pile of these sacs is a *granum*. The thylakoid membranes contain the chlorophyll molecules and other components of the energy-transducing machinery necessary for photosynthesis.

thymidine kinase (tk) gene \'thī-mə-dēn 'kī-ˌnās ('tē 'kā) 'jēn\ A widely used selectable marker for transfection studies in eukaryotic cells. *See* selectable marker.

ti (tumor-inducing) plasmid \'tē 'ī ('t(y)ü-mər-in-'d(y)üs-iŋ) 'plaz-məd\ *See Agrobacterium tumefaciens.*

time-of-flight mass spectrometer \'tīm-əv-'flīt 'mas spek-'träm-ət-ər\ A simple mass spectrometer that operates on the varying time taken for ions to reach the detector. A pulse of ions is accelerated by a voltage gradient to a given kinetic energy. Because the ions have the same kinetic energy, ions with different masses travel at diferent velocities, the lower the mass of the ion the greater its velocity. If time zero is taken when the pulse of ions is generated, the time taken for ions to reach the detector (time-of-flight) indicates the ion's mass. Thus a measurement of ion signal against time produces a mass spectrum.

The mass spectrometer is compact and has a linear configuration, with ions traveling in a straight line from source to detector. It has a high sensitivity, and ions with molecular weights above 2×10^5 can be detected. However, the resolution is relatively low because of slight variations in initial kinetic energy imparted to the ions in the source. This resolution can be improved (up to 2000) by the introduction of an ion mirror. This ion mirror or reflector consists of a series of perforated plates held at different potentials to stop and reverse the ions. Because the ions with slightly different kinetic energies penetrate the mirror to different depths, on reversal they are gathered together and thus focused at the detector, hence improving the resolution of the instrument.

tissue culture \'tish-ü 'kəl-chər\ The process whereby small pieces of living tissue (*explants*) are isolated from an organism and grown aseptically in a defined or semidefined nutrient medium. Originally the term was used to describe the culture of whole fragments of explanted tissue. Currently the term covers both *organ culture* (in which a piece of tissue or embryonic organ explant is grown and retains tissue architecture, cell interaction, and histological and biochemical differentiation) and *cell culture* (in which explant tissue is dispersed enzymatically or mechanically and propagated as a cell suspension or attached monolayer).

tissue plasminogen activator (t-PA or TPA) \'tish-ü plaz-'min-ə-jən 'ak-ti-vāt-ər ('tē 'pē 'ā)\ A mammalian enzyme that converts plasminogen into its active form, plasmin. Plasmin then dissolves fibrin, which is the major component of blood clots. The enzyme has proven to be particularly successful clinically for dissolving blood clots in heart attack victims. It is currently produced through recombinant DNA technology by a number of biotechnology companies. t-PA's promise lies in the specificity with which it dissolves blood clots but does not cause widespread bleeding in the patient. Currently approved plasminogen activators such as streptokinase and urokinase do not bind to fibrin, so they activate plasminogen throughout the bloodstream. This activation leads to systemic degradation of blood proteins, such as fibrinogen, which causes internal bleeding as well as blood clot dissolution. *See also* plasmin.

tk gene \'tē 'kā 'jēn\ *See* thymidine kinase gene.

TNFα \'tē 'en 'ef 'al-fə\ *See* tumor necrosis factor.

top yeasts \'töp 'yēst\ *See* brewing yeasts.

Toprina \tō-prī-na\ *See* single-cell protein.

toroidal chromatography \tȯ-'rȯid-ᵊl ˌkrō-mə-'täg-rə-fē\ *See* countercurrent chromatography.

torque \\'tȯ(ə)rk\\ A force that results from the rate of increase of the moment of momentum for rotating bodies, the magnitude of which is the product of the momentum and the perpendicular distance of the point of application from the axis of the rotating body.

tortuosity \\ˌtȯr-chə-'wäs-ət-ē\\ The nature of the pore structure within a solid, taking account of the deviation of these pores from a straight-line configuration. A system that follows a straight line is given a tortuosity of unity; the higher the value, the greater is the path length followed by the pore channel.

torula yeast \\'tȯr-(y)ə-lə 'yēst\\ A contaminant in some milk products that is responsible for unwanted alcohol fermentation.

total dissolved solids (TDS) \\'tōt-ᵊl diz-'älvd 'säl-ədz ('tē 'dē 'es)\\ The total mass of material in a filtered solution after removal of the solvent by drying at 105 °C. The units are milligrams per cubic decimeter or parts per million. Also known as total nonfilterable residue.

total filterable residue \\'tōt-ᵊl 'fil-t(ə-)rə-bəl 'rez-ə-ˌd(y)ü\\ *See* total suspended solids.

total nonfilterable residue \\'tōt-ᵊl 'nön-fil-t(ə-)rə-bəl 'rez-ə-ˌd(y)ü\\ *See* total dissolved solids.

total organic carbon (TOC) \\'tōt-ᵊl ȯr-'gan-ik 'kär-bən ('tē 'ō 'sē)\\ A measure of the carbon (organic) content of a sample, especially in the case of water, where it may be used as a measure of pollution. The technique used for its determination involves pyrolysis of the sample in oxygen, followed by analysis of the evolved carbon dioxide.

total oxygen demand \\'tōt-ᵊl 'äk-si-jən di-'mand\\ The amount of oxygen consumed in the combustion of organic compounds in a sample at 900 °C.

total suspended solids \\'tōt-ᵊl sə-'spend-əd 'säl-ədz\\ The total dried solid material retained by a standard fiberglass filter after filtration of a well-mixed sample. Alternatively, it is the dried solid material separated by centrifugation. These residual solids are dried at 103–105 °C, weighed, and recorded as milligrams per cubic decimeter or parts per million. Also known as total filterable residue.

totipotent \\tō-'tip-ət-ənt\\ Able to develop into a complete and differentiated organism from a part. For example, callus cells can be induced to form either embryos (*embryogenesis*) or roots and shoots (*organogenesis*), which then develop into full plants.

touchdown PCR \\'təch-ˌdoún 'pē 'sē 'är\\ *See* polymerase chain reaction.

tower bioreactor \\'taü(-ə)r ˌbī-ō-rē-'ak-tər\\ An elongated tubular vessel for carrying out fermentations. There is no mechanical agitation of the culture. Air is introduced at the bottom of the tower, and mixing is produced by the rising bubbles. This method produces very little shear effect on the organisms.

trans-acting region \\tran(t)s-'akt-iŋ 'rē-jən\\ Any region of DNA that encodes a diffusible product that influences the activity of genes on a different DNA molecule. *Contrast with* cis-acting region.

trans configuration \\'tran(t)s kən-ˌfig-(y)ə-'rā-shən\ A form of stereoisomer in which similar groups are on opposite sides of a double bond. *Contrast with* cis isomers.

trans isomers \\'tran(t)s 'ī-sə-ˌmərz\ A configuration of geometric isomers in which similar groups are on opposite sides of a double bond. *Contrast with* cis isomers. *See also* E-isomers (the IUPAC preferred term).

transcription \tran(t)s-'krip-shən\ The process whereby a molecule of RNA is synthesized by the enzyme RNA polymerase using DNA as a template. The process involves complementary base-pairing. In this way the genetic information (*protein sequence*) encoded in the DNA is transferred to an RNA molecule, which can diffuse into the cytoplasm. The encoded message in the RNA can then be used to direct the synthesis of the protein by the process of translation, which occurs at the ribosome. *See also* RNA polymerase.

transducer \tran(t)s-'d(y)ü-sər\ A device that converts a measurable parameter into a signal, commonly an electrical signal whose magnitude is related to the magnitude of the original parameter. *See also* biosensor.

transduction \tran(t)s-'dək-shən\ The transfer of genetic material from one cell to another by means of a viral vector and the subsequent incorporation of this genetic material (by recombination) into the genome of the recipient cell.

transesterification \ˌtran(t)s-e-ˌster-ə-fə-'kā-shən\ The reaction of an alcohol with an ester. The reaction is favored in almost anhydrous organic solvents, in which competition from hydrolysis is minimized.

transfection \tran(t)s-'fek-shən\

(1) The introduction of DNA into cultured cells.
(2) The introduction of purified phage DNA molecules into a bacterial cell.

transfer (tra) genes \\'trans(t)s-fər ('tē 'är 'ā) 'jēnz\ *See* self-transmissible plasmids.

transfer RNA (tRNA) \\'tran(t)s-fər 'är 'en 'ā ('tē 'är 'en 'ā)\ A major class of RNA molecule that is involved in protein synthesis. The function of tRNA molecules is to carry amino acids to a messenger RNA (mRNA) template bound to a ribosome, where the amino acids are linked, in a specific order dictated by the code on the mRNA, to produce a protein. Each tRNA contains both a site for the attachment of a specific amino acid and a site (the *anticodon*) that recognizes the corresponding three-base codon on the mRNA. *See also* anticodon, initiation codon, translation.

transferases \\'tran(t)s-fər-ˌās-ez\ An enzyme classification covering all enzymes that transfer a group (e.g., a methyl group) from one compound (the donor) to another compound (the receptor). The systematic name is based on donor : acceptor group transferase, and the recommended name is acceptor group transferase and donor group transferase. In many cases, the donor is a cofactor (*coenzyme*) charged with the group to be transferred. Transferases are one of the six main classes (E.C. 2) used in enzyme classification. *See also* Enzyme Commission number.

transformant \tran(t)s-'fȯr-mənt\ *See* transformation.

transformation \tran(t)s-fər-'mā-shən\

(1) Any alteration in the apparent growth properties, morphology, or behavior of a cell in culture.

(2) The acquisition of new genes in a cell after the incorporation of nucleic acid (usually double-stranded DNA) into the cell (e.g., the introduction of a plasmid into a bacterial cell). The bacterial cell is then said to have been transformed and is referred to as a *transformant*.

transformation efficiency \tran(t)s-fər-'mā-shən i-'fish-ən-sē\ In a transformation experiment, the number of transformants produced per microgram of cloning-mixture DNA.

transformation frequency \tran(t)s-fər-'mā-shən 'frē-kwen-sē\ A measure of the proportion of cells in a population that are transformed in a given experiment.

transformation processes \tran(t)s-fər-'mā-shən 'präs-es-əz\ The use of micro-organisms to catalyze the conversion of a compound into a structurally similar (and usually more commercially useful) compound. For example, the production of vinegar (the conversion of ethanol into acetic acid) is a well-established transformation process.

transforming growth factors \tran(t)s-'fó(ə)r-miŋ 'grōth 'fak-tərz\ Growth factors that have potential uses as wound-healing agents. TGFα is a peptide of 50 amino acid residues, which bind the same receptor as EGF. TGFβ is a protein comprising two identical subunits of 112 amino acids. Two forms, TGFβ1 and TGFβ2 have been identified, which have >70% sequence homology.

transgenic animal \tran(t)s-'jen-ik 'an-ə-məl\ An animal whose genetic composition has been altered to include selected genes from other animals or species by methods other than those used in traditional animal breeding. Methods include microinjection of DNA into the zygote, the introduction of DNA by retroviral infection, and DNA transfection into cultured embryonic stem cells that are injected into the blastocyst.

transgenic plant \tran(t)s-'jen-ik 'plant\ A plant whose genetic composition has been altered to include selected genes from other plants or species by methods other than those used in traditional plant breeding. In particular, many cultures have been modified using recombinant DNA technology to introduce resistance to herbicides, viruses, insects, and bacterial and fungal pathogens. Insect resistance has been achieved by both the introduction of the *Bacillus thuringiensis* endotoxin gene and the expression of protease inhibitor genes. Strategies to introduce virus resistance include protection by overexpression of viral coat protein genes, expression of antisense gene constructs, and the use of ribozymes to cut and inactivate viral RNA. Resistance to fungal pathogens is being attempted by the expression of both chitinase genes and β-1,3-glucanases, which should degrade the hyphal walls of invading fungi. Resistance to bacterial pathogens has been achieved by expressing the antibacterial peptide cecropin and the expression of lysozyme, which degrades the peptidoglycan bacterial cell wall. Herbicide resistance has been introduced by amplification of the target gene, introduction of a mutant, herbicide-insensitive gene, and introduction of a gene for an enzyme that can inactivate or degrade the herbicide. *See also* antisense RNA, *Bacillus thuringiensis*, cecropin.

translation \tran(t)s-'lā-shən\ The process whereby the information encoded in an RNA molecule is used to synthesize a protein. Translation occurs by a ribosome binding to the mRNA molecule and reading the triplet code. For each codon, a specific transfer RNA molecule (tRNA) is bound to the appropriate amino acid. tRNA molecules therefore provide the relevant amino acids at the site of protein synthesis. *See also* anticodon.

transmission (%*T*) \tran(t)s-'mish-ən (pər-'sent 'tē)\ A measure of the amount of radiation transmitted by a sample, used in particular for IR spectrophotometry, and defined by the equation

$$\%T = \frac{100I}{I_0}$$

where I is the transmitted radiation intensity and I_0 is the incident radiation intensity. Related to absorbance by

$$A = 2 - \log \frac{1}{T}$$

transmittance \tran(t)s-'mit-ᵊn(t)s\ In spectrophotometry, the relative amount of radiation transmitted by a sample:

$$T = \frac{I}{I_0}$$

where I is the transmitted radiation intensity, I_0 is the incident radiation intensity, and T is the transmittance.

transplastome \tran(t)s-'plas-ˌtōm\ A transformed plastid genome.

tray column \'trā 'käl-əm\ *See* plate column.

triazine dyes \'trī-ə-ˌzēn 'dīz\ *See* dye–ligand chromatography.

trickling filter \'trik-(ə-)liŋ 'fil-tər\ *See* percolating filter.

triose \'trī-ōs\ *See* carbohydrates.

triple helixes \'trip-əl 'hē-lik-səz\ A potential form of therapy in which an oligonucleotide targets the major groove of double-stranded DNA, shutting down transcription and in some cases translation. *Compare with* antisense RNA and aptamer as related forms of treatment.

triplet code \'trip-lət 'kōd\ The method whereby details of protein structure are encoded in DNA. Each possible group of three nucleotides codes for a specific amino acid. This DNA code is converted into protein at the ribosome via the processes of transcription and translation. *See also* codon, transcription, translation.

triplet codon \'trip-lət 'kō-ˌdän\ *See* codon.

trisacryl \'tris-ə-ˌkril\ Synthetic gels formed by the polymerization of *N*-acryloyl-2-amino-2-hydroxymethyl-1,3-propanediol. The resulting polymer has three hydroxymethyl groups and one alkylamide group per repeating unit. It can be used as a gel-filtration medium, or as an ion-exchange and affinity chromatography support.

Triton X-100 (isooctylphenoxypolyethoxyethanol) \'trīt-ᵊn 'eks 'wən 'hən-dred ('ī-sō-ˌäk-tīl-fe-'näk-sī-'pä-lī-'ēth-äks-ī-'eth-an-ol)\ A Rohm and Haas trade name for a nonionic surfactant used to dissolve cytoplasmic membranes without causing the denaturing of proteins.

tRNA: \'tē 'är 'en 'ā\ *See* transfer RNA.

trophophase \'trō-fə-ˌfāz\ The log or exponential phase of a culture, during which the sole products of metabolism are either essential to growth (such as amino acids, nucleotides, proteins, lipids, and carbohydrates) or are the byproducts of energy-yielding metabolism (such as ethanol, acetone, and butanol). The metabolites produced during the trophophase are referred to as primary metabolites. The term trophophase therefore describes the behavior of microbial cells based on their metabolism, whereas the terms *log phase* or *exponential phase* describe the kinetic behavior of the cells.

trypsinization \ˌtrip-sin-ə-'za-shən\ The generation of primary-culture cells by disaggregating tissue through the use of a trypsin solution.

tubular bowl centrifuge \'t(y)ü-byə-lər 'bōl 'sen-trə-ˌfyüj\ A type of centrifuge in which the bowl is suspended from the top and hangs free, with only a loose guide at the bottom. Feed enters at the bottom through a stationary nozzle, under sufficient pressure to jet it upward into the bowl. Particles collect on the inside walls of the bowl and must be harvested intermittently. The liquid leaves through the top of the bowl into a stationary casing and exits via a discharge nozzle at high velocity. A 10-cm-diameter centrifuge can handle 5–45 dm³/min at a speed of 15,000 rpm to generate a centrifugal force of 14,000 *g*.

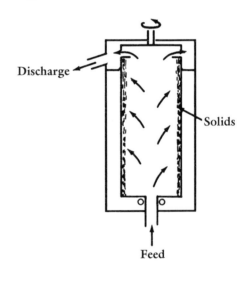

tubular membrane module \'t(y)ü-byə-lər 'mem-brān 'mäj-ü(ə)l\ A configuration of membranes used in cross-flow, micro-, and ultrafiltration processes with high solids content. The membranes are formed into tubes 1–2 cm in diameter supported by rigid porous shells. This design results in a low membrane area to tube length ratio (i.e., 20–30 m^{-1}) and high investment and operating costs. However, the system does allow control of membrane fouling and concentration polarization.

tumor necrosis factor (TNFα) \'t(y)ü-mər nə-'krō-səs 'fak-tər ('tē 'en 'ef 'al-fə)\ A human protein currently being used in clinical trials for cancer chemotherapy. The protein has been shown in animal models to affect tumor vascularization leading to central necrosis and tumor regression. *See also* cytokines.

turbidity \tər-'bid-ət-ē\ A measure of the amount of light scattered by the presence of microscopic suspended matter in a fluid. It may be determined by a turbidimeter, in which a collimated light beam is passed through the sample and the turbidity is measured by the apparent increase in the absorbance resulting from scattering of the light. *See also* nephelometry.

turbidostat \tər-'bid-ə-ˌstat\ A continuous-culture vessel in which fresh medium flows in response to an increase in turbidity of the culture. *See also* continuous culture.

turbulent flow \'tər-byə-lənt 'flō\ Fluid flow in which fluid particle motion at any point varies rapidly in direction and magnitude. It is characteristic of fluid motion at high levels of mixing and high Reynolds number. The converse of laminar flow. *See also* laminar flow, Reynolds number.

turbulent shear \'tər-byə-lənt 'shi(ə)r\ Shear resulting from turbulent flow. In turbulent flow, short-term hydrodynamic forces give rise to turbulent eddies, which can generate higher stresses than in laminar, or time-average shear. Plant and animal cells are susceptible to turbulent eddies, and detrimental effects have been shown to arise with eddy size below 100 mm. Eddy length (L) is given by the following equation:

$$L = \left(\frac{\mu/\rho^3}{P}\right)^{0.75}$$

where μ is viscosity (kg/ms), ρ is density (kg/m^3), and P is the power dissipation per unit mass.

turnover number \'tər-ˌnō-vər 'nəm-bər\ In enzymology, the number of substrate molecules processed by one enzyme molecule per second. Typically, enzymes have turnover numbers of the order of 10^4, but some enzymes have values as high as 10^8.

two-phase \'tü-ˌfāz\ Of or referring to a system of two immiscible fluids, e.g., air–water, oil–water, commonly used in extraction and purification of organic compounds. *See also* liquid–liquid extraction, two-phase aqueous partitioning, two-phase reactions.

two-phase aqueous partitioning \'tü-ˌfāz 'ak-wē-əs pər-'tish-ən-iŋ\ The distribution or partitioning of a substance between two partially miscible phases largely composed of water. A typical phase system is polyethylene glycol–dextran–water, which separates into two phases, one mostly polyethylene glycol–water and the other mostly dextran–water. These systems are used widely for the partitioning of cells, as well as proteins and other substances. They have the advantage that, because of the high water content of the phases, biological materials do not become denatured. However, they have the disadvantages of (1) long phase-separation times, as a result of little difference in density between the phases and low interfacial tensions; (2) relatively high viscosities; and (3) high cost of scale-up because the partial miscibility of the components means that a significant percentage of them cannot be recycled.

two-phase reactions \'tü-ˌfāz rē-'ak-shənz\ Enzyme reactions carried out in the presence of a second water-miscible phase, such as an organic solvent. This procedure allows high concentrations of water-insoluble substrate to be maintained. However, enzyme denaturation by the organic solvent can be a problem with this method. Alternatively, the second phase can be a water-immiscible solvent. The reaction is carried out in a stirred reactor that maintains a fine emulsion, with the enzyme present and stable in the aqueous phase but in close association with the substrate in the organic phase.

tylosin \'tī-lō-ˌsin\ An antibiotic produced commercially by strains of *Streptomyces fradiae*. It is used exclusively for animal nutrition (for improved feed efficiency and growth promotion) and in veterinary medicine. Tylosin contains a 16-membered lactam ring with three sugars (mycarose, mycaminose, and mycinose) attached.

U

ultracentrifuge \ˈəl-trə-ˈsen-trə-ˌfyüj\ A high-speed centrifuge used for the separation and analysis of macromolecules (e.g., proteins, nucleic acids). The high g forces (up to 500,000 g) generated by the centrifuge cause sedimentation of the molecules according to their molecular weight, density, and shape. When a density gradient is generated within the supporting fluid, molecules may be separated into bands as their density matches that of the supporting fluid. *See also* density-gradient centrifugation.

ultrafiltration \ˈəl-trə-fil-ˈtrā-shən\ A membrane filtration process using an asymmetric microporous polymeric membrane with pore sizes of 1–50 nm, driven by a hydrostatic pressure of 200–1000 kPa. Separation of components from the mixture takes place by a sieving mechanism, with the open structure of the membrane on the feed side and the fine porous "skin" on the downstream side of the process. This arrangement allows the macromolecules to enter and be retained by the membrane. Thus they are separated, concentrated, and purified from smaller molecules and salts with molecular weights less than about 3000, which pass through the porous skin.

ultrafiltration reactor \ˈəl-trə-fil-ˈtrā-shən rē-ˈak-tər\ A biochemical reactor, used mostly for carrying out enzyme-catalyzed depolymerization reactions, that incorporates an ultrafiltration membrane to separate low- and high-molecular-weight components.

ultrasonic \ˈəl-trə-ˈsän-ik\ Having a frequency of 20 kHz and above.

ultrasonic cleaning \ˈəl-trə-ˈsän-ik ˈklēn-iŋ\ The cleaning of equipment by immersion in a liquid, which may contain detergent, and through which ultrasonic waves are passed. This form of cleaning is efficient and is particularly useful for cleaning devices with small orifices.

ultraviolet (UV) disinfection \ˈəl-trə-ˈvī-ə-lət (ˈyü ˈvē) dis-ᵊn-ˈfek-shən\ The destruction of microorganisms by close proximity to a source of ultraviolet radiation. This technique is used to maintain sterility in contained environments and clean rooms. It has increasing use in water purification.

ultraviolet (UV) radiation \ əl-trə-'vī-ə-lət ('yü 'vē) rād-ē-'ā-shən\ Radiation that occupies a wavelength range between visible radiation and X-radiation, with the normal useful range between 200 and 400 nm. Because many organic molecules absorb in this range, UV spectroscopy can be used for qualitative and quantitative analysis (e.g., scanning gels for the detection of DNA). Because of the increased energy involved in UV radiation, it can also induce bond fission in photochemical reactions and cause the destruction or modification of microorganisms, as in mutation programs and UV sterilization.

undefined medium \ ən-di-'fīnd 'mēd-ē-əm\ A microbial growth medium in which not all the components have been identified. Also known as a complex medium.

upflow \ 'əp-ͺflō\ The passage of a mobile phase against gravity. Upflow is the converse of downflow as a regime for the operation of columns in, for example, chromatography, ion exchange, or extraction processes.

upstream processing \ 'əp-ͺstrēm 'präs-ͺes-iŋ\ The unit operations that come before the main reactor or fermentor. *Contrast with* downstream processing.

urokinase (E.C. 3.4.21.31) \ yùr-ō-'kī-ͺnās\ A serine protease synthesized in the kidney and found in urine. It activates plasminogen to plasmin. *See also* plasmin, tissue plasminogen activator.

V

Vaccinia virus \vak-'sin-ē-a 'vī-rəs\ A large double-stranded DNA virus that has been used extensively to eradicate smallpox. Inoculation with *Vaccinia*, which is harmless in humans, confers resistance to smallpox because of production of antibodies that cross-react to the smallpox virus. *Vaccinia* virus has also been developed as a cloning vector. Live recombinant viruses express foreign antigens on their surfaces, and can therefore be used as immunogens. For example, a *Vaccinia* recombinant containing the coding sequence for the rabies surface glycoprotein has been constructed. When injected into animals, this live hybrid virus protects the animals against later exposure to rabies. This protection stems from the production of antibodies directed against the rabies surface glycoprotein expressed on the surface of the *Vaccinia* virus. Other genes that have been expressed in *Vaccinia* recombinants include hepatitis B virus surface antigen, *Herpes simplex* virus glycoprotein D, Epstein–Barr virus glycoprotein, and influenza virus hemagglutinin.

vacuum centrifuge \'vak-yu̇-əm 'sen-trə-ˌfyüj\ *See* centrifugal concentrator.

vacuum fermentation \'vak-yu̇-əm ˌfər-men-'tā-shən\ The removal of volatile fermentation products during fermentation under vacuum so that the products distill at the usual fermentor operating temperature. A low concentration of products in the fermentor is maintained so that inhibition of the reaction is minimized. The process can also be operated with a bleed stream to remove nonvolatile byproducts.

Van Deemter equation \'van 'dēm-tər i-'kwā-zhən\ In chromatography, an equation that relates the plate height (H) of a gas chromatographic column to the linear gas velocity (u) and has the form

$$H = A + \frac{B}{u} + Cu$$

where A, B, and C are constants. The eddy diffusion term, A, describes the band broadening caused by variation of the gas velocity as a result of the porous structure of the column packing. The term B/u represents band broadening as a result of longitudinal diffusion of the solute molecules in the gas phase, and C_u is related to the mass-transfer resistance that inhibits the equilibration of solute molecules between the gas and stationary phases. The equation represents a hyperbola that produces an optimum gas velocity at which the height of the plate is a minimum.

vapor head temperature \'vā-pər 'hed 'tem-pə(r)-chù(ə)r\ The temperature of vapor above a heated liquid.

variation \ver-ē-'ā-shən\ *See* somatic variants.

VecBase \'vek 'bās\ A database of cloning vector sequences.

vector (cloning vehicle) \'vek-tər ('klōn-iŋ 'vē-‚(h)ik-əl)\ A DNA molecule, usually a small plasmid or bacteriophage DNA capable of self-replication in a host organism, used to introduce a fragment of foreign DNA into a host cell. Commonly used vectors include plasmids, λ-phage, SV40, Epstein Barr virus, and Ti plasmids from *Agrobacterium tumefaciens.*

vegetative propagation (cloning) \'vej-ə-‚tāt-iv ‚präp-ə-'gā-shən ('klōn-iŋ)\ The asexual propagation of plants, either by detaching some part of the plant (e.g., a leaf cutting, shoot cutting) and growing the cutting in an appropriate medium such that a complete plant subsequently develops, or by the generation of whole plants from callus tissue culture or protoplasts.

velocity gradient \ve-'läs-ət-ē 'grād-ē-ənt\ The rate of change of velocity with distance in a fluid. Also known as shear rate.

venturi meter \ven-'tù(ə)r-ē 'mēt-ər\ A device for measuring flow rate of a fluid based on differential pressure measurement upstream and downstream of a constriction in the pipeline. *See also* orifice meter, pitot tube.

Verticillium lecanii \vərt-ə-'sil-ē-əm ‚le-'ken-ē-ī\ A fungus, commercial preparations of which are now marketed for control of aphids in greenhouses and on crops grown under humid conditions.

vesicular–arbuscular (VA) mycorrhiza \və-'sik-yə-lər-är-'bəsk-yə-lər ('vē 'ā) ‚mī-kə-'rī-zə\ A form of endomycorrhiza arising when phycomycete fungi of the family *Endogonaceae* invade root cortical tissues and form a symbiotic association beneficial to the host. Most plants of agricultural importance have VA mycorrhiza. The agricultural importance of VA mycorrhiza lies in its ability to assist phosphate absorption by the plant from the surrounding soil. Because phosphate ions are not very mobile in soil, phosphate depletion zones often form around roots when phosphate is in low supply. Mycorrhizal hyphae extend from the mycelium within the root to beyond the depletion zone and transfer phosphate directly to the host.

vibromixer \'vī-brō-mik-sər\ A culture vessel for plant or animal cells that are sensitive to the mechanical effects of stirring and cannot therefore be grown in spinner bottles, for example. The vessel contains a plate that vibrates in the vertical plane a distance of 0.1–3 mm. Conical perforations in the plate effect the mixing.

vincristine \vin-'kris-‚tēn\ A plant alkaloid isolated from *Catharanthus roseus* and used as an antitumor agent.

vinegar production   \'vin-i-gər prə-'dək-shən\ Oxidative fermentation of dilute ethanol solutions by *Acetobacter*, yielding acetic (ethanoic) acid. Early production systems used trickling filters, but submerged systems are more commonplace in modern processing.

virulence plasmids   \'vir-(y)ə-lən(t)s 'plaz-mədz\ Plasmids that are pathogenic to the host cell (e.g., Ti plasmids of *Agrobacterium tumefaciens* induce crown gall disease on dicotyledonous plants).

virus resistance in plants   \'vī-rəs ri-'zis-tən(t)s 'in 'plants\   *See* transgenic plant.

viscoelasticity   \,vis-kō-ə-ˌlas-'tis-ət-ē\ The partial elastic recovery of a fluid upon the removal of a deforming shear stress.

viscosity   \vis-'käs-ət-ē\ One of the fundamental parameters of rheology. Viscosity relates the shear stress (force applied) to the shear rate of a fluid, which may be time-dependent or time-independent. Many equations relate the two parameters, depending on the nature of the fluid (e.g., dilatant, Newtonian, plastic, pseudoplastic, rheodestructive, rheopectic, thixotropic).

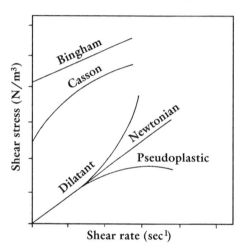

viscous flow   \'vis-kəs 'flō\   *See* laminar flow.

vitamin C   \'vīt-ə-mən 'sē\   *See* L-ascorbic acid.

Vitis vinifera   \'vi-təs vi-'nif(-ə)rə\ Species of grape from which almost all the world's wine is made.

voidage   \'vȯid-ej\ The proportion of a packed column or bed that is not occupied by solid material (e.g., that volume occupied by the fluid surrounding the solid particles).

void volume   \'vȯid 'väl-yəm\ In chromatography, the elution volume of a substance that is not retarded during passage through the column. Void volume may be determined by injecting a substance that is not retained on the column under the conditions used.

volumetric distribution coefficient (*K_d*) \ˌväl-yù-'me-trik ˌdis-trə-'byü-shən kō-ə-'fish-ənt ('kā 'dē)\ A parameter used in exclusion chromatography and gel chromatography to specify the fraction of the volume within the gel that is accessible to the solute. It is determined from elution data by the equation

$$K_d = \frac{V_e - V_o}{V_i}$$

where V_e is elution volume; V_o is void volume; and V_i is internal volume. K_d can take values from zero, for solutes excluded completely from the gel, to unity, for solutes whose permeation equals that of the solvent. When values greater than unity are obtained, binding between the solute and the gel matrix has occurred.

volumetric mass-transfer coefficient \ˌväl-yù-'me-trik 'mas-'tran(t)s-fər kō-ə-'fish-ənt\ *See* $K_L a$.

volumetric oxygen-transfer coefficient \ˌväl-yù-'me-trik 'äk-si-jən-ˌtran(t)s-fər kō-ə-'fish-ənt\ *See* $K_L a$.

Volvariella volvacea \'vȯl-ˌva(ə)r-ē-ˌel-ə ˌvȯl-'vā-sē-ā\ An edible fungus (mushroom) also known as the rice straw fungus. It has been traditionally cultivated, mainly in the Far East, on composted rice straw. More than 50,000 tons of fungus is produced annually.

wall-coated, open-tubular (WCOT) column \'wȯl-ˌkōt-ᵊd ˌōpən-'t(y)ü-byə-lər 'käl-əm\ A capillary gas chromatographic column in which the column wall is coated with the stationary phase. Generally this column has a lower capacity than a support-coated, open-tubular column with the same film thickness. *See also* porous-layer, open-tubular column; support-coated, open-tubular column.

Warburg manometer \'wȯ(ə)r-bərg mə-'näm-ət-ər\ A device used to follow changes in gas uptake or evolution in biological systems.

washout \'wȯsh-ˌaȯt\ The removal of cells from a continuous bioreactor. It occurs when the dilution rate (D) equals the maximum specific growth rate (m_{max}) and is then termed the critical dilution rate (D_{crit}). At this point a steady-state microbial population cannot be maintained because the cell growth cannot match the rate of cell removal and the cell concentration becomes zero as shown:

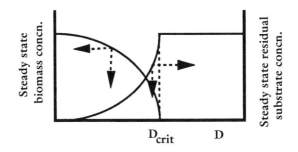

See also critical dilution rate, dilution rate.

water activity \'wȯt-ər ak-'tiv-ə-tē\ A parameter that describes the fraction of free water in a solution. By definition, the water activity of pure water is unity. A low water activity of a solution corresponds to a high osmotic pressure of the solution. Microorganisms can grow only in media with a water-activity level between 0.999 and 0.650. In general, a reduction in the water activity strongly inhibits growth. Microorganisms that can grow at low water-activity levels are called *osmotolerant.*

water-transport number \\'wȯt-ər 'tran(t)s-pō(ə)rt 'nəm-bər\\ In dialysis, a measure of the ratio of the weight of water transported per unit weight of dissolved solute.

WCOT \\'dəb-əl-(ˌ)yü 'sē 'ō 'tē\\ *See* wall-coated, open-tubular column.

weight-average molecular weight \\'wāt-av-(ə-)rij mə-'lek-yə-lər 'wāt\\ *See* molecular weight of polymers.

western blotting \\'wes-tern 'blät-tiŋ\\ *See* protein blotting.

wheat-germ lysate \\'hwēt-jərm 'lī-sāt\\ *See* cell-free translation system.

whey \\'hwā\\ The watery product remaining when the curd is separated from the remaining milk in cheese production. It contains high concentrations of lactose (~50 g/L), as well as minerals, vitamins, and lactic acid. Disposing of waste whey, which is produced daily in considerable quantity, has been a problem for some time in the cheese industry. The recent introduction of the use of lactase (isolated from *Aspergillus niger*) to convert whey into a sweeter solution of glucose and galactose, which is of use to the confectionery and baking industry, has had some success in overcoming this problem. Whey is also used as a substrate for the growth of *Penicillium cyclopium* to produce single-cell protein in the French Heurty process.

whirlpool separator \\'hwər(-ə)l-ˌpül 'sep-(ə-)ˌrāt-ər\\ An alternative name for a hydrocyclone, which is used commonly for the separation of spent grain from brewing mashes. *See also* hydrocyclone.

wild type \\'wī(ə)ld 'tīp\\ The naturally occurring (nonmutated) form of an organism. It is also used to refer to the nonmutated form of a specific gene in an organism.

Winsor-type microemulsions \\'win-sər 'tīp ˌmī-krō-i-məl-shənz\\ A means of classifying microemulsions devised by P. A. Winsor.

- Type I and Type II microemulsions are two-phase systems in which a surfactant-rich phase (microemulsion) is in equilibrium with an excess organic phase (I) or excess aqueous phase (II).
- Type III is a three-phase system in which the microemulsion is in equilibrium with both aqueous and organic phases.
- Type IV is a single isotropic phase system in which the surfactant has completely solubilized both aqueous and organic phases.

Types I, II, and III are interconvertible and depend on the hydrophilic–lipophilic balance (HLB) of the surfactant, the ionic strength of the aqueous phase, the composition of the oil phase, and the temperature. *See also* hydrophilic–lipophilic balance, micellar extraction, microemulsions.

Winsor-type microemulsions

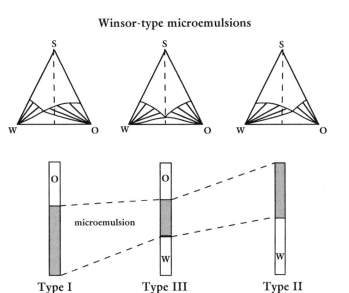

Type I   Type III   Type II

wiped-film evaporator \'wīpt-film i-'vap-ə-ˌrāt-ər\ A modification of a rising- or falling-film evaporator, in which the walls of the heat exchanger are wiped by a rotating device (such as a metal strip or coil) to improve rates of heat and mass transfer and to ensure that the concentrated material does not foul the heat exchanger and to minimize heat degradation of the substance.

wood sugar \'wu̇d 'shu̇g-ər\ *See* xylose.

wort \'wərt\ The aqueous extract obtained after adding warm water to crushed malt (malted barley). In beer and lager manufacture, after separation of the wort and addition of hops, wort is boiled, cooled, inoculated with brewer's yeast, and allowed to ferment.

X

X-gal \\'eks-ˌgal\\ A lactose analogue (5-bromo-4-chloro-3-indolyl-β-D-galactopyrano-side) that is broken down by β-galactosidase to give a blue color. It is used, for example, to differentiate transformants containing the plasmid pUC8 (which synthesize β-galactosidase) and transformants containing plasmids (which have undergone insertional inactivation of the β-galactosidase gene).

X-press \\'eks-ˌpres\\ An apparatus used for rupturing microbial cells (e.g., yeast cells). A frozen paste of cells is disrupted by passage through a perforated disk, with an outlet temperature of ~22 °C. Rupture of cells is caused by the shear forces exerted by the passage of the extruded paste through the small orifice; the shear is aided by ice-crystal formation in the frozen paste. *See also* Hughes press.

xanthan gum \\'zan-thən 'gəm\\ A five-sugar repeat-unit polysaccharide with a molecular weight in the range $0.2-1.5 \times 10^7$, produced extracellularly by *Xanthomonas campestris* strains. The gum produces high viscosity at low concentrations and therefore has a wide range of industrial applications as an agent for viscosity control, in gelling and suspension, as a stabilizer, and in oil-recovery processes. In the food industry it is used to modify the texture of a wide range of foods.

xenobiotic \\ˌzen-ō-bī-'ät-ik\\ A chemical compound not usually produced by living organisms; a synthetic compound. Organic compounds can be classified as *biogenic* (of natural origin) or *anthropogenic* (synthetic). Biogenic compounds are usually degradable by microorganisms, as are natural products produced on a large scale by synthetic routes. However, xenobiotic compounds, because of their unnatural structures, are degraded poorly (*recalcitrant* compounds) or not at all (*persistent* compounds) by microorganisms.

xenograft \\'zen-ə-ˌgraft\\ A graft between individuals of different species.

xylan \\'zī-lan\\ A major component of plant hemicellulose. After cellulose, it is the most abundant renewable polysaccharide in nature. The main component of xylan is D-xylose, a pentose, but the structure of xylan is variable. It ranges from linear

1,4-β-linked polyxylose chains to highly branched heteropolysaccharides, which contain sugars other than xylose. Appreciable quantities of xylan are present in materials released from wood during pulping or pulp processing, but this xylan is regarded as waste and is discarded. The development of enzyme processes that would convert xylan waste to fermentable sugars would be economically beneficial.

xylitol \\'zī-lə-ˌtȯl\\ A C_5 sugar alcohol produced from xylose, which has potential as a natural sweetener. It is manufactured chemically by the catalytic reduction of xylose present in hemicellulose hydrolysates, but it can also be produced with natural xylose by using yeasts. Xylitol can be metabolized by humans and is used as a sugar substitute in diabetic diets.

$$CH_2OH - (CHOH)_3 - CH_2OH$$

xylose (wood sugar) \\'zī-ˌlōs ('wu̇d 'shu̇g-ər)\\ A monosaccharide pentose produced by the hydrolysis of xylans. It can be reduced to give xylitol.

Y

YACs \'yaks\ *See* yeast artificial chromosomes.

yeast(s) \'yēs(t)s\ Fungi that are usually unicellular for part of their life history but may also form pseudomycelium or short lengths of true mycelium. The common method of vegetative reproduction of many yeasts is by budding. There are approximately 500 yeasts that belong to the *Ascomycotina, Deuteromycotina,* and *Basidiomycotina.* Yeasts are important for baking, brewing, wine- and spirit-making, single-cell protein enzymes, and as a vitamin source.

yeast artificial chromosomes (YACs) \'yēst ärt-ə-'fish-əl 'krō-mə-ˌsōmz ('yaks)\ Cloning systems used to clone large contiguous segments of DNA from any organism into suitable vectors in such a way that the recombinants can be transformed into yeast cells where they are stably propagated. The use of such vectors allows the cloning of large DNA fragments up to 500 kilobase pairs. This size is technically important because many genes span such distances. Before the introduction of YACs, such large genes had to be cloned in multiple segments of less than 40 kilobase pairs. The basic vector is YAC2, a plasmid that is readily propagated in *Escherichia coli.* It is cleaved into two fragments that constitute chromosome arms. The left arm has a telomere, a centromere, an autonomously replicating sequence (ARS), and a selectable marker. The right arm has a telomere and a different selectable marker. The chromosome arms are then mixed and ligated with large (>100 kilobase pairs) exogenous DNA. The reconstituted chromosomes are introduced into yeast by standard methods. By screening for two selectable markers, one on each arm, only clones containing chromosomes that have both arms are obtained. Such artificial chromosomes may now be stably propagated. YACs may be conveniently analyzed by pulse-field gel electrophoresis.

yeast episomal plasmids (YEps) \'yēst 'ep-ə-'sō-məl 'plaz-midz ('yeps)\ Any vector derived from the 2-μm circle plasmid of yeast (e.g., plasmid pJDB 219). *See also* shuttle vector.

yeast integrative plasmids (YIps) \\'yēst 'int-ə-ˌgrāt-iv 'plaz-midz ('yips)\\ Bacterial plasmids that carry a yeast gene. They are unable to replicate within a yeast cell unless they are integrated into a chromosome. Integration occurs as a result of crossing-over between the yeast segment on the plasmid and the homologous gene on the chromosome.

yeast replicative plasmids (YRps) \\'yēst 'rep-li-ˌkāt-iv 'plaz-midz ('yerps)\\ Yeast plasmids that can multiply as independent plasmids in yeast because of the presence of a chromosomal DNA sequence containing an origin of replication. *See also* autonomously replicating sequence.

YEps \\'yeps\\ *See* yeast episomal plasmids.

yield factor (Y) \\'yē(ə)ld 'fak-tər\\ A dimensionless constant relating the mass of substrate consumed to the mass of microbes produced.

YIps \\'yips\\ *See* yeast integrative plasmids.

YRps \\'yerps\\ *See* yeast replicative plasmids.

Z

Z-isomers \\'zē-ī-sō-ˌmərz\\ A configuration of geometric isomers in which two atoms or groups are on the same side of the molecule or atom. This term is preferred by IUPAC; it is derived from *zusammen*, German for together. Also known as cis isomers. *Contrast with* E-isomers.

zanflo \\'zan-flō\\ A high-viscosity polysaccharide, comprising fucose, galactose, glucose, and uronic acid residues, produced from *Erwinia tahitica*. It is used for carpet printing.

zeatin \\'zē-ə-tən\\ A naturally occurring cytokinin obtained from corn (maize) grains.

zeolites \\'zē-ə-ˌlīts\\ A group of crystalline aluminosilicates with an open framework structure. The $(Si,Al)_nO_{2n}$ structure has a negative charge that is balanced by cations contained in the cavities. The cations can be easily exchanged for other cations of different charge, water, and gases which can be selectively adsorbed into the cavities. The selection is based on the relative size of the cavity and species to be adsorbed. To provide a range of applications, different zeolites can be obtained with different sizes of cavity. Zeolites are commonly used to remove water from liquids and gases. Also known as molecular sieves.

zonal centrifuge \\'zōn-ᵊl 'sen-trə-fyüj\\ A type of centrifuge intended for the purification of viruses or the isolation of RNA or DNA. The bowl is filled with a liquid that can establish a density gradient, then the feed is introduced into the bowl. The particles sediment into a band in which the density equals that of the surrounding liquid. The bowl rotates in a vacuum at high speed, generating a maximum g number of 100,000 or more.

Zygomycotina \\ˌzī-gə-mī-kō-'tin-ə\\ The division containing the fungi that have hyphae without cross walls and that reproduce sexually by the production of zygospores that result from the fusion of gametangia, which may differ in size. These fungi usually reproduce asexually with the formation of sporangia containing sporangiaspores. Many genera in this group cause spoilage of a range of materials. *Mucor* and *Rhizopus* species are used for enzyme production.

zymogen \\'zī-mə-jən\\ The inactive precursor form of an enzyme. It is usually converted into the active form by limited proteolysis. For example, in the blood clotting process, prothrombin is converted into the active enzyme thrombin by a single cleavage at a specific polypeptide bond. Similarly, in the clot dissolution process, plasminogen is converted into the active form plasmin by a single cleavage step. *See also* plasmin.

Copy editing: Paula M. Bérard
Production: Catherine Buzzell
Acquisition: Barbara E. Pralle
Interior and cover design: Amy Hayes

Typeset by Beacon Graphics Corporation, Ashland, OH
Books printed and bound by Maple Press, York, PA

Bestsellers from ACS Books

The ACS Style Guide: A Manual for Authors and Editors
Edited by Janet S. Dodd
264 pp; clothbound ISBN 0–8412–0917–0; paperback ISBN 0–8412–0943–X

The Basics of Technical Communicating
By B. Edward Cain
ACS Professional Reference Book; 198 pp;
clothbound ISBN 0–8412–1451–4; paperback ISBN 0–8412–1452–2

Chemical Activities (student and teacher editions)
By Christie L. Borgford and Lee R. Summerlin
330 pp; spiralbound ISBN 0–8412–1417–4; teacher ed. ISBN 0–8412–1416–6

Chemical Demonstrations: A Sourcebook for Teachers,
Volumes 1 and 2, Second Edition
Volume 1 by Lee R. Summerlin and James L. Ealy, Jr.;
Vol. 1, 198 pp; spiralbound ISBN 0–8412–1481–6;
Volume 2 by Lee R. Summerlin, Christie L. Borgford, and Julie B. Ealy
Vol. 2, 234 pp; spiralbound ISBN 0–8412–1535–9

Chemistry and Crime: From Sherlock Holmes to Today's Courtroom
Edited by Samuel M. Gerber
135 pp; clothbound ISBN 0–8412–0784–4; paperback ISBN 0–8412–0785–2

Writing the Laboratory Notebook
By Howard M. Kanare
145 pp; clothbound ISBN 0–8412–0906–5; paperback ISBN 0–8412–0933–2

Developing a Chemical Hygiene Plan
By Jay A. Young, Warren K. Kingsley, and George H. Wahl, Jr.
paperback ISBN 0–8412–1876–5

Introduction to Microwave Sample Preparation: Theory and Practice
Edited by H. M. Kingston and Lois B. Jassie
263 pp; clothbound ISBN 0–8412–1450–6

Principles of Environmental Sampling
Edited by Lawrence H. Keith
ACS Professional Reference Book; 458 pp;
clothbound ISBN 0–8412–1173–6; paperback ISBN 0–8412–1437–9

Biotechnology and Materials Science: Chemistry for the Future
Edited by Mary L. Good (Jacqueline K. Barton, Associate Editor)
135 pp; clothbound ISBN 0–8412–1472–7; paperback ISBN 0–8412–1473–5

For further information and a free catalog of ACS books, contact:
American Chemical Society
Product Services Office
1155 16th Street, NW, Washington, DC 20036
Telephone 800–227–5558